IMPEDANCE SOURCE POWER ELECTRONIC CONVERTERS

IMPEDANCE SOURCE POWER ELECTRONIC CONVERTERS

Yushan Liu
Texas A&M University at Qatar, Qatar Foundation, Doha, Qatar

Haitham Abu-Rub
Texas A&M University at Qatar, Qatar Foundation, Doha, Qatar

Baoming Ge
Texas A&M University, College Station, TX, USA

Frede Blaabjerg
Aalborg University, Aalborg East, Denmark

Omar Ellabban
Texas A&M University at Qatar, Qatar Foundation, Doha, Qatar
Helwan University, Cairo, Egypt

Poh Chiang Loh
Aalborg University, Aalborg East, Denmark

IEEE PRESS

WILEY

This edition first published 2016
© 2016 John Wiley & Sons, Ltd

First Edition published in 2016

Registered Office
John Wiley & Sons, Ltd, The Atrium, Southern Gate, Chichester, West Sussex, PO19 8SQ, United Kingdom

For details of our global editorial offices, for customer services and for information about how to apply for permission to reuse the copyright material in this book please see our website at www.wiley.com.

Library of Congress Cataloging-in-Publication Data

Names: Liu, Yushan, 1986– author.
Title: Impedance source power electronic converters / authored by Yushan Liu, Texas A&M University at Qatar, Qatar Foundation, Doha, Qatar, Haitham Abu-Rub, Texas A&M University at Qatar, Qatar Foundation, Doha, Qatar, Baoming Ge, Texas A&M University, College Station, USA, Frede Blaabjerg, Aalborg University, Aalborg East, Denmark, Omar Ellabban, Texas A&M University at Qatar, Qatar Foundation, Doha, Qatar, Helwan University, Cairo, Egypt, Poh Chiang Loh, Aalborg University, Aalborg East, Denmark.
Description: First edition. I Chichester, West Sussex, United Kingdom : John Wiley and Sons, Inc., 2016. I Includes bibliographical references and index.
Identifiers: LCCN 2016014284 (print) I LCCN 2016021902 (ebook) I ISBN 9781119037071 (cloth) I ISBN 9781119037118 (pdf) I ISBN 9781119037101 (epub)
Subjects: LCSH: Electric current converters. I Energy conservation–Equipment and supplies. I Transfer impedance. I Electric power production–Equipment and supplies.
Classification: LCC TK7872.C8 L58 2016 (print) I LCC TK7872.C8 (ebook) I DDC 621.3815/322–dc23
LC record available at https://lccn.loc.gov/2016014284

A catalogue record for this book is available from the British Library.

Front Cover image: Guillermo Perales Gonzale/Getty, TheAYS/Getty, R-J-Seymour/Getty, Cris Haigh/Getty and Stockbyte/Getty.

Set in 10/12pt Times by SPi Global, Pondicherry, India
Printed and bound in Malaysia by Vivar Printing Sdn Bhd

1 2016

Contents

Preface

There are significant research efforts worldwide related to renewable energy conversion, electric transportation, and many other industrial applications that require power converters/inverters. Thus there is a huge scope for developing commercially viable and technically feasible efficient and reliable power converters. The traditional voltage source inverter has to maintain a dead-time between the upper and lower switches in one bridge leg to avoid short circuit, which introduces distortions to ac output waveforms. Also, an extra dc-dc boost converter is usually required when the dc source voltage is insufficient to supply the output voltage, resulting in a two-stage system with high cost and complicated control. The traditional current source inverter has analogous limits. All this has motivated many researchers to work on a single-stage converter. This book brings together state-of-the-art knowledge and cutting-edge techniques in various stages of research related to the most popular and appealing single-stage converter, which is the impedance source converter/inverter and its modifications.

The impedance source network, consisting of inductors, capacitors, and switches/diode, overcomes the above mentioned limitations by offering buck or boost capabilities in a single stage and short-circuit immunity of inverter legs. All this makes it possible to get rid of the dead-time between phase-leg switches and to enhance the reliability of the system. The solution has found widespread investigations for dc-dc, dc-ac, ac-dc, and ac-ac low-/high-power conversion since it was first suggested.

This book presents a systemic view of impedance source converters/inverters, offering comprehensive analysis, control, and comparison of both typical and various derived impedance source topologies reported in literatures and researched by the authors. The impedance source converters/inverters distribute the shoot-through behavior into the inverter/converter phase legs. All the traditional pulse width modulation (PWM) schemes can be used to drive the phase legs. The book addresses and compares different kinds of modified PWM schemes for impedance source converters/inverters, including simple boost control, maximum boost control, maximum constant boost control, space vector modulation, and pulse-width-amplitude modulation. The book also discusses the hardware design of passive components for optimizing the converters/inverters. The impedance values are significant to the system performance.

The approach is to maintain lower size, volume, and weight while ensuring high performance and high quality responses.

Modeling of converters/inverters is essential for understanding the circuit operation and developing control methods. Thus, this book includes the models of impedance source converters/inverters used to design control parameters for corresponding linear control methods, and also to develop model predictive control. The book presents also various existing topics such as applications of impedance source converters/inverters to renewable energy generation and electric drives, multi-leg (four-leg and five-phase) converters/inverters. It includes the configuration, operation, and simulation/experiment results of the discussed topologies and control. Future trends of research and development in this area are also discussed.

The book provides a thorough understanding of the concepts, design, control, and applications of the impedance source converters/inverters. Researchers, senior undergraduate and graduate students, as well as professional engineers investigating vital topics related to power electronic converters will find great value in the content of the book. They will be able to apply the presented design approaches in this book to building and researching the future generation of efficient and reliable power electronics converters/inverters.

The contribution of Dr Poh Chiang Loh in this book is Chapter 9, Section 7.1, Section 12.1, and Subsection 1.1.2. The rest of materials are the contribution of the other authors (except Chapter 17 has the contribution from Dr Sertac Bayhan, Chapter 18 has the contribution from Dr Sertac Bayhan and Mr Mostafa Mosa, and Chapter 19 has the contribution from Dr Mohamed Trabelsi).

Acknowledgment

We would like to take this opportunity to express our sincere appreciation to all the people who directly or indirectly helped in making this book a reality. Our special thanks go to Texas A&M University at Qatar for the support provided to realize this effort. Most of the book chapters (except Chapter 9, Section 7.1, Section 12.1, and Subsection 1.1.2) were made possible by NPRP-EP grant # [X-033-2-007] from the Qatar National Research Fund (a member of Qatar Foundation). The statements made herein are solely the responsibility of the authors.

We are indebted to our family members for their continuous support, patience, and encouragement, without which this book would not have been completed.

Above all we're grateful to the almighty, the most beneficent and merciful, who provided us the continuous confidence and determination in accomplishing this work.

Sincerely,
Authors

Bios

Dr Yushan Liu received her BSc degree in automation from Beijing Institute of Technology, China, in 2008, and her PhD degree in electrical engineering from the School of Electrical Engineering, Beijing Jiaotong University, China, in 2014. She is currently a postdoctoral research associate in the Electrical and Computer Engineering Program, Texas A&M University at Qatar, Qatar Foundation, Doha, Qatar, where she was a research assistant from 2011 to 2014.

Dr Liu is the recipient of *"Research Fellow Excellence Award"* from Texas A&M University at Qatar, one of *"Ten Excellent Doctoral Dissertations"* from Beijing Jiaotong University, and many other prestigious research awards. Her research interests include impedance source inverters, cascade multilevel converters, photovoltaic power integration, renewable energy systems, and pulse-width modulation techniques. She has published more than 50 journal and conference papers and one book chapter in the area of expertise.

Dr Haitham Abu-Rub holds two PhDs, one in electrical engineering and the other in humanities.

Since 2006, Dr Abu-Rub has been associated with Texas A&M University at Qatar, Qatar Foundation, Doha, Qatar, where he was promoted to professor. Currently he is the chair of the electrical and computer engineering program at the same university as well as the managing director of Smart Grid Centre – Extension in Qatar. His main research interests are energy conversion systems, including electric drives, power electronic converters, renewable energy and smart grid.

Dr Abu-Rub is the recipient of many prestigious international awards, such as the American Fulbright Scholarship, the German Alexander von Humboldt Fellowship, the German DAAD Scholarship, and the British Royal Society Scholarship.

Dr Abu-Rub has published more than 250 journal and conference papers, and has supervised many research projects. Currently he is leading many potential projects on photovoltaic and hybrid renewable power generation systems with different types of converters and on electric drives. He is co-author of five books including this, four of which are with Wiley. Dr Abu-Rub is an active IEEE senior member and is an editor in many IEEE journals.

Dr Baoming Ge received his PhD degree in electrical engineering from Zhejiang University, Hangzhou, China, in 2000.

He worked in the Department of Electrical Engineering, Tsinghua University, Beijing, China, from 2000 to 2002. In 2002, he joined the school of electrical engineering in Beijing Jiaotong University, Beijing, China, was promoted to professor in 2006. He worked in the University of Coimbra, Coimbra, Portugal, from 2004 to 2005, and in Michigan State University, East Lansing, MI, USA, from 2007 to 2008 and 2010 to 2014. He is with the Renewable Energy and Advanced Power Electronics Researach Laboratory in the Department of Electrical and Computer Engineering, Texas A&M University, College Station, TX, USA. He has authored more than 200 journal and conference papers, two books, two book chapters, holds seven patents. His main research interests are the renewable energy generation, electrical machine drives, and power electronics.

Dr Frede Blaabjerg received his PhD degree from Aalborg University, Denmark, in 1992. He was with ABB-Scandia, Randers, Denmark, from 1987 to 1988. He became an assistant professor in 1992, an associate professor in 1996, and a full professor of power electronics and drives in 1998. His current research interests include power electronics and its applications such as in wind turbines, PV systems, reliability, harmonics, and adjustable speed drives.

He has received 15 IEEE Prize Paper Awards, the IEEE Power Electronics Society Distinguished Service Award in 2009, the EPE-PEMC Council Award in 2010, the IEEE William E. Newell Power Electronics Award 2014, and the Villum Kann Rasmussen Research Award 2014. He was an editor-in-chief of the *IEEE Transactions on Power Electronics* from 2006 to 2012. He has been a Distinguished Lecturer for the IEEE Power Electronics Society from 2005 to 2007 and for the IEEE Industry Applications Society from 2010 to 2011.

Dr Omar Ellabban was born in Egypt in 1975. He received his BSc degree in electrical machines and power engineering from Helwan University, Egypt, his MSc degree in electrical machines and power engineering from Cairo University, Egypt, and his PhD degree in electrical engineering from Vrije Universiteit Brussels, Brussels, Belgium, in 1998, 2005, and 2011, respectively.

He joined the R&D Department, Punch Powertrain, Sint-Truiden, Belgium, in 2011, where he and his team developed a next-generation, high-performance hybrid powertrain. In 2012, he joined Texas A&M University at Qatar, first as a post-doctoral research associate and then an assistant research scientist in 2013, where he is involved in different renewable energy projects. He has authored more than 50 journal and conference papers, one book chapter, and two conference tutorials. His current research interests include renewable energies, smart grid, automatic control, motor drives, power electronics, and electric vehicles.

Dr Poh Chiang Loh received the B Eng and M Eng degrees in electrical engineering from the National University of Singapore, Singapore, in 1998 and 2000, respectively, and the PhD degree in electrical engineering from Monash University, Clayton, Australia, in 2002.

He is currently with the Department of Energy Technology, Aalborg University, Aalborg, Denmark. His research interests are in power converters and their grid applications.

1

Background and Current Status

Yushan Liu[1], Haitham Abu-Rub[1], Baoming Ge[2], Frede Blaabjerg[3],
Poh Chiang Loh[3] and Omar Ellabban[1,4]

[1] *Electrical and Computer Engineering Program, Texas A&M University at Qatar,*
Qatar Foundation, Doha, Qatar
[2] *Department of Electrical and Computer Engineering, Texas A&M University,*
College Station, TX, USA
[3] *Department of Energy Technology, Aalborg University, Aalborg East, Denmark*
[4] *Department of Electrical Machines and Power Engineering, Helwan University, Cairo, Egypt*

Significant research efforts are underway to develop commercially viable, technically feasible, highly efficient, and highly reliable power converters for renewable energy, electric transportation, and various industrial applications. This chapter presents state-of-the-art knowledge and cutting-edge techniques in various stages of research related to impedance source converters/inverters, including the concepts, advantages compared to existing technology, classification, current status, and future trends.

1.1 General Introduction to Electrical Power Generation

1.1.1 Energy Systems

Electric power generation comprises traditional power generation, such as hydroelectric, thermal and nuclear power production, and renewable energy sources, which already has a large penetration joined by photovoltaic (PV) and wind energy [1]. Climatic constraints and large amounts of pollution require us to limit our development and utilization of traditional energy. Renewable energy and energy savings are receiving a greater attention as a sustainable and environmentally friendly alternative. Figure 1.1 shows the the levels of annual global renewable energy in gigawatts (GW), including solar PV, concentrating solar power (CSP), wind, bioenergy, geothermal, ocean, and hydropower [2]. It can be seen that globally installed

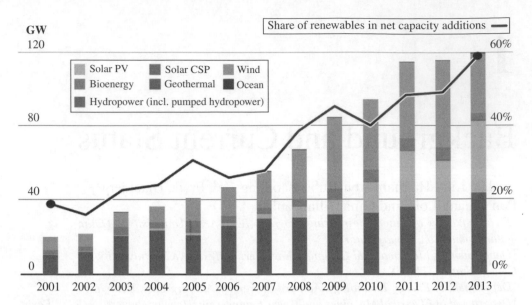

Figure 1.1   Global renewable energy annual changes in gigawatts (2001–2013) [2] (*Source:* Reproduced with permission of REN21).

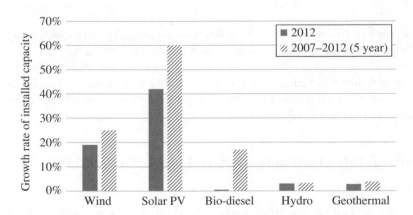

Figure 1.2   Growth rates of installed capacity of different renewable energies [2] (*Source*: Reproduced with permission of REN21).

renewable energy shows a rapid increase since 2007. To 2013, the share of renewables in net capacity additions has reached 60%, doubling the share in 2007.

Among global renewable energy sources, wind and solar energy are the leading potential sources of electricity for the 21st century for several reasons: they utilize an abundant energy source (the sun or wind) and have no emissions. Furthermore, solar power can be easily integrated into buildings, and so on. Figure 1.2 shows the growth rates of installed capacity of different renewable energies in 2012 and in five years from 2007 to 2012 [2]; Figure 1.3(a) and (b) show globally installed wind and PV power capacity to 2014 [3, 4]. The cumulative capacity of wind reached 369.6 GW in 2014, and that of PV in 2014 is 177 GW. It can be seen

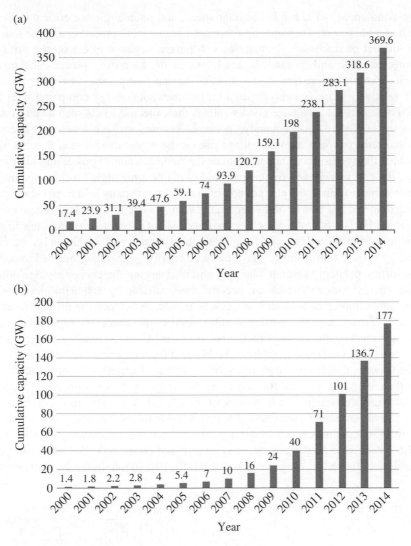

Figure 1.3 Globally installed (a) wind power capacity [3] (*Source*: Delphi234, https://commons.wikimedia.org/wiki/File:Global_Wind_Power_Cumulative_Capacity.svg. Used under CC0 1.0 Universal Public Domain Dedication https://creativecommons.org/publicdomain/zero/1.0/deed.en) and (b) PV power capacity (to 2014) [4] (*Source*: Reproduced with permission of IEA Photovoltaic Power System Programme).

that they have had a fast growth rate since 2007. In addition, fuel cells (FCs) have achieved global attention as an alternative power source for hybrid electric vehicles (HEVs). Fuel cell vehicles (FCVs) have generated interests among industrialists, environmentalists, and consumers. An FCV ensures the air quality, with the wide driving range and convenience of a conventional internal combustion engine vehicle.

Nevertheless, power generated by renewable energy sources is intermittent and heavily depends on the environmental conditions. For instance, the power incident on a solar panel, the panel temperature, and the solar panel voltage affect the utilization of solar power generation;

similarly, wind speed, wind turbine angular speed, and pitch angle are critical to the amount of harvested wind power. They are unpredictable because of the weather and the seasons. The resultant impact of stochastic fluctuations will have a negative effect on the utility grid in grid-connected mode and on loads in standalone mode. Moreover, power consumption also presents its own characteristics of seasonal and human living habits. In spring and autumn, there are relatively more fine days with a lot of renewable energy compared with the other seasons. These seasons also have good weather; thus, electric loads such as air conditioners may be used less often. Consequently, increased generation from renewable energy power systems and reduced loads cause a voltage rise on the power distribution line. At weekends, during which the systems continue to produce the same amount of power and industrial loads are lower, the grid voltage and frequency could easily become high. Overvoltage may exceed the upper tolerance limit at the point of common coupling; usually grid overvoltage protection will regulate the output power of the renewable energy system if the AC voltage exceeds the control range. Fuel cells prefer to be operated at constant power to prolong their lifetime and it is also more efficient in this way. However, the traction power of a vehicle is ever-changing.

An energy storage unit installed in a renewable energy system may be used to compensate for the insufficient energy through charging and discharging the energy storage unit, so that renewable energy power systems can become more reliable by acquiring the possibility to cope with some important auxiliary services. Similarly, to balance the difference and also to handle regenerative energy, a battery is often used as an energy storage device in FCVs. Basically, the main source of the vehicle's power is the FC; the secondary power source is the battery, which stores excess energy from the FC, and from regenerative braking [5].

Efficient energy transfer and high reliability of power electronics, involved in the interface between the energy sources and the grid or loads, are essential for converting the fluctuating powers into suitable voltage and frequency AC power [6]. According to the configurations of PV panels between power converters, PV power systems are categorized into AC-module, string, multi-string, and central inverter-based topologies, as shown in Figure 1.4 [6–8]. Figure 1.5 shows wind power generation systems based on induction/synchronous generators (I/SG) or doubly fed induction generators (DFIG) [9, 10]. To maximize the energy production

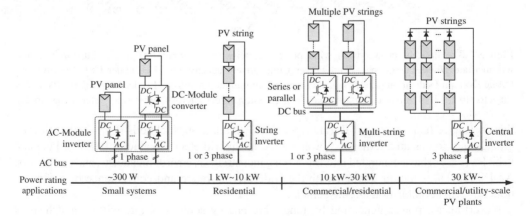

Figure 1.4 PV power systems categorized by configurations of PV panels between power converters (*Source*: Kjaer 2005 [8]. Reproduced with permission of IEEE).

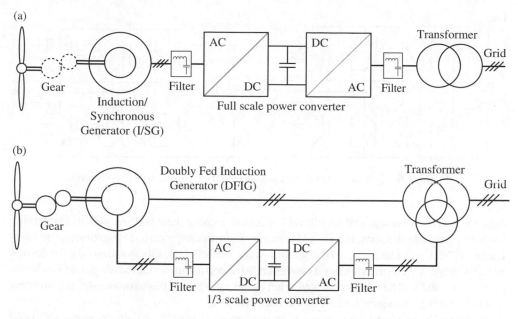

Figure 1.5 Wind power generation systems based on (a) induction/synchronous generator (I/SG), and (b) doubly fed induction generator (DFIG) (*Source*: Blaabjerg 2013 [9]. Reproduced with permission of IEEE).

in all operating conditions, various maximum power point tracking (MPPT) methods, such as hill-climbing, perturb and observe, incremental conductance, fuzzy theory, and genetic algorithms, have been developed for solar panel, wind turbine, and FC [11, 12], which are fulfilled by back-end power conversion devices. Appropriate energy management for energy sources, energy storage batteries, and grids has been explored for smoothing the power integrated into utility grids, which is also achieved by the power electronics-based conditioning units [13, 14]. Therefore, the efficiency of the whole power generation system finally depends on the inverters/converters used.

1.1.2 Existing Power Converter Topologies

Converters are widely used in industry for performing energy conversion from one form to another [15–19]. Typical examples are adjustable speed motor drives, electric power interfaces, uninterruptible power supplies, rectifiers, and power factor correctors. Traditionally, these applications use a two-level voltage-source converter, but as the power and switching ranges increase, the two-level topology is increasingly viewed as inappropriate. The main restriction is related to the semiconductor manufacturing technology: current power devices have limited voltage rating, current rating, and switching frequency. In addition, there is currently no immediately available low-cost solution for mass-producing devices made from silicon carbide (SiC) or gallium nitride (GaN), even though there are some promising developments. Therefore, instead of waiting for a breakthrough in semiconductor technology, a more effective and immediate approach for resolving present restrictions is to use multilevel

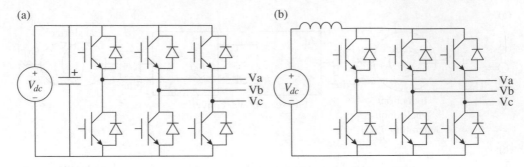

Figure 1.6 Typical dc-ac (a) voltage-source and (b) current-source converters.

converters. The advantages of multilevel converters include their well-recognized suitability for high power applications, improved harmonic performance, reduced electromagnetic inter-ference (EMI), and a larger pool of discrete voltage levels for flexibly synthesizing the desired output voltage waveform. Multilevel converters are therefore important to the power electronic community, and will hence be reviewed briefly after introducing the basic two-level converters that have been in existence for several decades [18, 19].

Traditionally, two-level converters can be implemented as either a voltage-source (VS) or a current-source (CS) converter. Their representative topologies are given in Figure 1.6(a) and (b), configured for dc-ac inversion. The same topologies can be used for ac-dc rectification, but for convenience of the description, the dc-ac inversion mode is usually assumed. Beginning with the VS converter shown in Figure 1.6(a), it is fed from a dc voltage source usually filtered by a relatively large capacitor connected in parallel. For dc-to-ac conversion, a dc voltage source feeds the main converter circuit – a three-phase inverter bridge that converts the dc power to ac for an ac load. The dc voltage source is usually a dc capacitor fed by a battery, fuel cell stack, or diode rectifier. For ac-to-dc conversion, an ac source feeds the converter bridge which converts ac to dc for a dc load. Six switches are used in the main circuit, each traditionally composed of a power transistor and an anti-parallel (or free-wheeling) diode to provide bidirec-tional current flow and unidirectional voltage blocking. Its maximum linear ac output voltage is known to be limited to 1.15 times half the dc source voltage if triplen offsets are included for modulation [14, 15]. The VS converter is therefore a buck or step-down dc-ac converter (or boost or step-up ac-dc converter) if no additional dc-dc converter is used for boosting its dc-link voltage.

By contrast, a CS converter is fed from a dc current source, which in most cases is imple-mented by connecting a dc voltage source in series with a comparably large inductor, as shown in Figure 1.6(b). The dc inductor to some extent behaves as a boosting component, whose presence allows the ac output voltage of the CS converter to be always greater than the dc source voltage. For dc-to-ac conversion, a dc current source feeds the main converter circuit – a three-phase inverter bridge that converts dc to ac, to power an ac load. The dc current source is usually a relatively large dc inductor fed by a voltage source such as a battery, fuel cell stack, diode rectifier, or thyristor converter. For ac-to-dc conversion, an ac source feeds the main converter bridge which converts ac to dc to power a dc load. Six switches are used in the main circuit, each traditionally composed of a semiconductor switching device with reverse block capability such as gate-turn-off thyristor (GTO), silicon controlled rectifier (SCR), or a power transistor with a series diode to provide unidirectional current flow and bidirectional voltage blocking. The CS converter is therefore only suitable for boost dc-ac power conversion (or buck ac-dc power rectification).

The VS and CS converters are thus capable of performing different types of power conversions, and should therefore be chosen based on the requirements under consideration. However, for certain well-accepted reasons such as the absence of large dc inductors and fewer semiconductors conducting in series, the VS converter has dominated most applications. This happens even though the desired system requirements might theoretically favor the CS converter more than the VS converter. This is true, for example, in photovoltaic generation, which usually demands a dc-ac boost, and the direct application of the CS converter is still not popular even though the dc-ac boost ability is inherently found with the CS converter. Instead, the VS converter is almost always adopted, even though an additional dc-dc boost converter might at times be needed. The VS converter is therefore the clear "winner." Many converters developed subsequently have adopted the same VS characteristics, including the multilevel converters to be described next.

Among the multilevel converters proposed, the three gaining most attention are the diode-clamped, cascaded, and flying-capacitor converters. The diode-clamped converter's implementation is realized by connecting multiple switches in series, and then adding clamping diodes for distributing the dc voltage stress evenly among the switches per phase-leg. For illustration, a three-level and a five-level diode-clamped converter are shown in Figure 1.7(a) and (b), respectively. The former is also referred to as the neutral-point-clamped (NPC) converter, whose dc-link voltage has been divided by capacitors C_1 and C_2 to form three distinct voltage levels, to which each phase output can be tied. For example, by turning on only switches SA1 and SA2 of the leftmost phase-leg in Figure 1.7-(a), the phase output will be tied to $+V_{dc}/2$. Similarly, by turning on SA2 and SA'1, the output will be zero, while turning on SA'1 and SA'2 causes the output to be $-V_{dc}/2$. A three-level phase voltage waveform is thus produced by the four switches per phase-leg operating as two complementary pairs. The first pair consists of SA1 and SA'1, while the second pair consists of SA2 and SA'2.

The same switching principles apply to the five-level diode-clamped converter, as shown in Figure 1.7-(b), which will therefore not be explicitly described. Instead, to demonstrate its operation, Table 1.1 has been included to show how the five discrete voltage levels per phase-leg can be obtained by switching its eight switches. The diode-clamped converter is thus an ideally feasible topology, but when implemented physically, it will face a tough dc capacitor voltage balancing problem. This problem, if not resolved satisfactorily, will give rise to low-order harmonic distortion and increased voltage stress across the capacitors and semiconductor switches. Although a number of voltage balancing techniques have since been proposed with most relying on extra hardware or redundant state swapping, the balancing concern has limited a practical diode-clamped converter to only three discrete voltage levels which, as mentioned, is called the NPC converter.

For higher level converters, the cascaded multilevel inverter is preferred, and is simply assembled by connecting multiple single-phase full-bridges or H-bridges in series. A typical cascaded five-level inverter is shown in Figure 1.8, where each H-bridge is powered by a separate dc source. Since each H-bridge generates three distinct voltage levels, $+V_{dc}$, 0, and $-V_{dc}$, the cascading of two H-bridges in series per phase-leg produces five distinct voltage levels, $+2V_{dc}$, $+V_{dc}$, 0, $-V_{dc}$, and $-2V_{dc}$. Although the cascaded converter offers many advantages, its requirement for multiple isolated dc sources makes it an expensive topology used only for

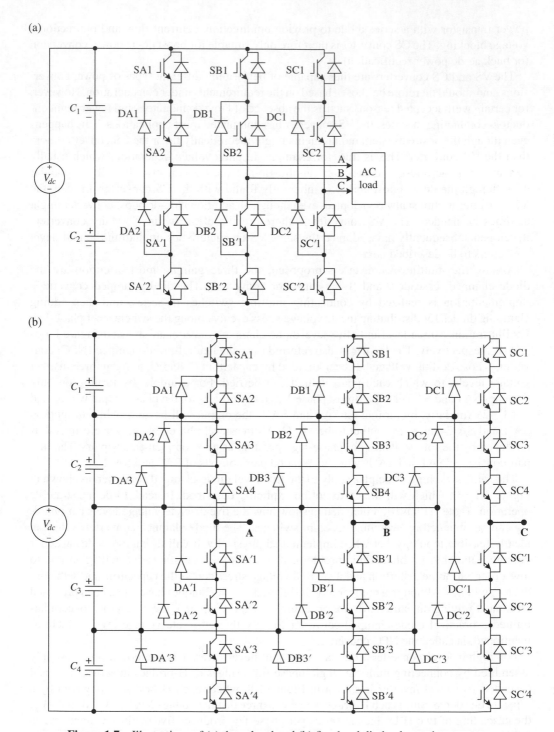

Figure 1.7 Illustrations of (a) three-level and (b) five-level diode-clamped converters.

Table 1.1 Switching states of five-level diode-clamped converter per phase-leg (X = A, B or C)

ON Switches	Voltage
SX1, SX2, SX3, SX4	$+V_{dc}/2$
SX2, SX3, SX4, SX'1	$+V_{dc}/4$
SX3, SX4, SX'1, SX'2	0
SX4, SX'1, SX'2, SX'3	$-V_{dc}/4$
SX'1, SX'2, SX'3, SX'4	$-V_{dc}/2$

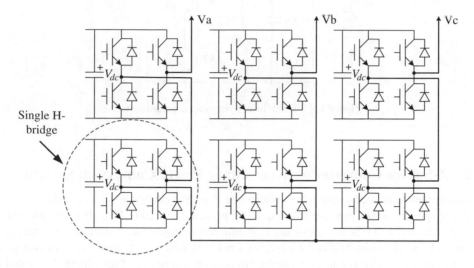

Figure 1.8 Five-level cascaded multilevel inverter [20].

converters with more than five distinct voltage levels. However, this limitation is slowly being diluted by the introduction of renewable sources, where multiple sources are readily available. Upon solving the source availability, the modular structure of the cascaded converter will again appear as an attractive feature since it allows easier converter construction and a more streamlined manufacturing process.

The third popular multilevel converter is the flying-capacitor converter, whose representative three-level phase-leg is shown in Figure 1.9. With its isolated capacitor C_1 regulated at half the dc-link voltage, the converter can produce three distinct output voltage levels: $V_{dc}/2$, 0, and $-V_{dc}/2$. Supposing that the voltage rating of the capacitors used is the same as that of the active semiconductor switches, an n-level flying-capacitor converter can be realized with $(n-1) \times (n-2)/2$ clamping capacitors per phase-leg, in addition to $(n-1)$ dc-link capacitors. That is obviously a drawback since the system size will increase tremendously with the number of output voltage levels. The increase will, however, create fewer technical problems than the diode-clamped topology because with an appropriate phase-shifted modulation scheme implemented, the capacitor voltages of the flying-capacitor converter are naturally self-balanced. A higher level flying-capacitor converter can therefore be implemented more easily, but it is unlikely to surpass the cascaded multilevel converter.

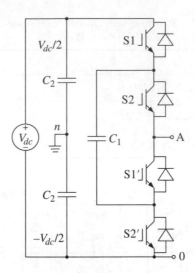

Figure 1.9 Three-level flying-capacitor converter.

1.2 Z-Source Converter as Single-Stage Power Conversion System

In applications that need to boost the voltage, the Z-source converter is a possibility.

The Z-source power converter was initially presented in 2002 [20]. It is an impedance source converter and it is different from conventional voltage source and current source converters. The main feature is that the impedance network consists of inductors, capacitors, and switches/diodes, and it is employed in the circuit to boost or buck the voltage. It is a circuit to combine the dc-dc boost converter and inverter, using a single-stage power conversion system. Figure 1.10 shows the general topology of the Z-source converter. A two-port network that consists of inductors L_1 and L_2 and capacitors C_1 and C_2 connected in an X shape is employed to provide an impedance source (Z-source) coupling the converter (or inverter) to the dc source or load.

A three-phase voltage-fed Z-source inverter, as shown in Figure 1.11. is used as an example to briefly illustrate the operating principle. The traditional three-phase voltage-source inverter has six active states and two zero states. For the Z-source inverter, several extra zero states are possible by gating on both the upper and lower devices of any one phase leg, any two phase legs, or all three phase legs [20]. These shoot-through zero states are forbidden in the traditional voltage-source inverter, because they would cause a short-circuit across the source. The Z-source network and shoot-through zero states provide a unique buck–boost feature of the inverter. All the traditional pulse width modulation (PWM) schemes can be used to control the Z-source inverter and their theoretical input–output relationships still hold true. In addition, with the unique feature of the shoot-through zero states, several new PWM methods: simple boost control, maximum boost control, maximum constant boost control, and space vector modulation, have been developed [20–29].

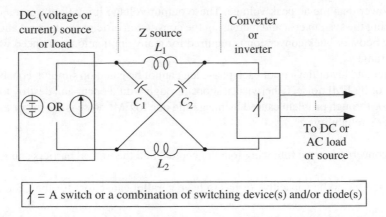

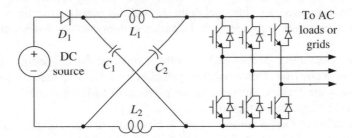

Figure 1.10 A general topology of the Z-source converter (*Source*: Peng 2003 [20]. Reproduced with permission of IEEE).

Figure 1.11 Three-phase voltage-fed Z-source inverter—an example of Z-source inverter.

1.3 Background and Advantages Compared to Existing Technology

As mentioned before, power electronics is the enabling technology for the energy processing necessary for electronic and electrical equipment of all types, from renewable energy to power systems, from house appliances to aerospace systems. Both the VS converter and CS converter have their conceptual and theoretical limitations and barriers that prevent economical and efficient solutions to many applications.

The VS converter has the following three conceptual and theoretical limitations:

1. The ac load has to be inductive or the ac source has to have series inductors.
2. The ac output voltage has an upper limit because it cannot exceed the dc rail voltage or the dc rail voltage has to be greater than the ac input voltage. Therefore, the VS inverter is a buck (i.e. step-down) inverter for dc-to-ac power conversion and the VS converter is a boost (i.e. step-up) rectifier (or boost converter) for ac-to-dc power conversion. For dc-to-ac power conversion applications where the dc voltage is given, the obtainable maximum ac output can not be greater than the given dc voltage. For ac-to-dc power conversion applications, where the ac source voltage is fixed, the minimum dc output is limited because it can

not be lower than the ac peak voltage. These output voltage limitations are major barriers to reducing the system cost and increasing the conversion efficiency, because another stage of dc-dc boost or buck conversion is required for many applications that need wide output voltage ranges.

3. The upper and lower devices of each phase leg cannot be gated on simultaneously either on purpose or by EMI noise. Otherwise, a shoot-through would occur and destroy the devices. The shoot-through problem caused by misgating due to EMI noise is the major threat to the converter's reliability.

The CS converter has the following four conceptual and theoretical barriers and limitations:

1. The ac side has to be capacitive or has to have parallel capacitors for the CS converter to operate.
2. The ac output voltage has to be greater than the original dc voltage that feeds the dc inductor or the dc voltage produced is always smaller than the ac input voltage. Therefore, the CS inverter is a boost inverter for dc-to-ac power conversion and the CS converter is a buck rectifier (or buck converter) for ac-to-dc power conversion. For applications where a wide voltage range is required, an additional dc-to-dc buck (or boost) converter is needed. This additional power conversion stage again increases system cost and lowers the conversion efficiency.
3. At least one of the upper devices and one of the lower devices have to be gated on and maintained on at any time. Otherwise, an open circuit of the dc inductor would occur and this will destroy the devices. The open-circuit problem of EMI noise causing misgating off is a threat to the converter's reliability.
4. The main switches of the CS converter have to block any reverse voltage which requires a series diode to be used in combination with high-speed and high-performance transistors such as insulated gate bipolar transistors (IGBTs). This prevents the direct use of low-cost and high-performance IGBT modules and integrated power modules (IPMs).

In summary, both the VS converter and the CS converter have the following three common problems:

1. They are either a boost or a buck converter and cannot be a buck–boost converter. Hence, their obtainable output voltage range is limited to either lower or higher than the input voltage.
2. Their main circuits cannot be interchangeable. In other words, the main circuit of the VS converter can not be used as a CS converter and vice versa.
3. They are vulnerable to EMI noise and thus are not as reliable as desired.

The Z-source power converter provides a new single-stage converter topology and theory with the intention of achieving the functions of two-stage power converters. The Z-source converter compromises an impedance network to couple the main converter circuit to the power source or load. As a result, the Z-source converter is neither a VS converter nor a CS converter and has none of their inherent problems.

The major unique feature of the Z-source network is that, unlike the traditional VS or CS, it can be open- and short-circuited, which provides a mechanism for the main converter circuit to step up or step down the voltage as desired. The Z-source network provides the following three aspects of flexibility for the source, main circuit, and load:

1. The source of the Z-source converter can be either a voltage source or a current source. Therefore, unlike the traditional VS or CS converters, the Z-source converter's dc source can be anything, such as a battery, a diode rectifier, a thyristor converter, a fuel cell stack, an inductor, a capacitor, or a combination of these.
2. The main circuit of the Z-source converter can be either a traditional VS configuration or a traditional CS configuration. In addition, switches used in the Z-source converter can be a combination of switching devices and diodes such as the anti-parallel combination shown in Figure 1.6-(a) or the series combination shown in Figure 1.6 (b).
3. The load of the Z-source converter can be inductive or capacitive.

1.4 Classification and Current Status

As shown in Figure 1.10, a two-port network that consists of inductors L_1 and L_2 and capacitors C_1 and C_2 connected in a unique X shape is employed to provide an impedance source (Z-source) coupling the converter (or inverter) to the dc source or load. The unique feature of the Z-source network is that, unlike the traditional VS or CS, it can be open- or short-circuited, which provides a mechanism for the main converter circuit to step up and step down voltage as desired in a single-stage power conversion. These features give the Z-source converter not only a high reliability against EMI, but also fewer components and lower cost to achieve the same function as a conventional two-stage converter/inverter (dc-dc converter plus inverter). The configuration allows the elimination of dead time, which improves the output current/ voltage waveform quality (no distortion). The Z-source concept has been extended to all dc-to-ac, ac-to-dc, ac-to-ac, and dc-to-dc converter power conversions [30, 31].

The involved application fields include PV power generation [32–38], wind power generation [39–46], electric vehicles [47–51], etc. Without doubt, the Z-source converter has been a significant branch of power electronic converters, following on from VS and CS converters.

Z-source related research has been a hot topic since its birth. The number of modifications and new Z-source topologies has exploded. Figure 1.12-(a) and (b) shows a summary of the Z-source converter categories and Z-source network topologies that can be found in recent literature, with the chapter numbers for this book. They can be sorted into four main categories according to conversion functionality: ac-dc rectifiers, dc-ac inverters, ac-ac converters, and dc-dc converters. A further breakdown leads to two-level and multilevel [20–29, 32–38], ac-ac regulators [52–56], matrix converters [57–63], and non-isolated and isolated dc-dc converters [64–68]. From the Z-source network topology standpoint, it can be voltage-fed or current-fed. The Z-source networks can be divided into four types: original Z-source [20], quasi Z-source [69], trans-Z-source [70], and other Z-source, such as embedded Z-source, semi-Z-source, distributed Z-network, switched inductor Z-source, tapped-inductor Z-source, and diode-assisted and capacitor-assisted quasi-Z-source, LCCT-Z-source [71–75]. All are derived from the original Z-source and quasi-Z-source topologies.

All the diverse Z-source network topologies in the literature have been derived by modifying the original Z-source network, or by rearranging the connections of inductors and capacitors. Each Z-source network topology has its unique features for different or particular application needs.

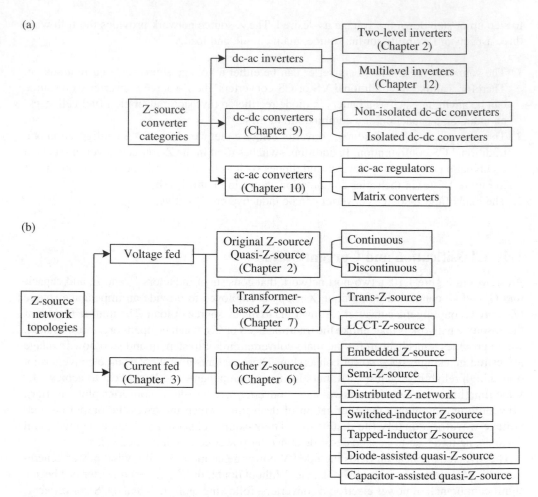

Figure 1.12 Summary of different Z-source converters in the literature, plus chapter numbers for this book: (a) Z-source converter categories; (b) Z-source network topologies.

New Z-source topologies are still being developed for three main reasons: (1) reduction of the Z-source network component count and rating, (2) extension of voltage gain range, and (3) application-oriented optimization and improvement.

For example, four quasi-Z-source (two voltage-fed and two current-fed) inverters were proposed to reduce the passive component ratings, to achieve continuous input current, and to provide a common negative dc rail between the source and inverter bridge. These new benefits spread this topology to applications of renewable energy generation and motor drives. The embedded Z-source was proposed to achieve continuous input current and lower capacitor voltage rating, and its multi-source feature is suitable for photovoltaic power generation. Theoretically, the original Z-source, quasi-Z-source, and embedded Z-source, all have unlimited voltage gain. However, a high voltage gain (>2–3), will result in a high voltage stress imposed on the switches. Trans-Z-source (two voltage-fed and two current-fed) inverters were proposed to have higher voltage gains while keeping voltage stress low and reducing component count to one transformer (or one coupled inductor) and one capacitor. There are many other ways to increase voltage gain by integrating traditional switched-inductor, switched-capacitor, tapped-inductor, diode-assisted, and capacitor-assisted extensions to the Z-source/quasi-Z-source network, but they require more components. Semi-Z-source inverters were proposed to achieve low cost and high efficiency in applications such as single-phase grid-tie PV power systems. A semi-Z-source inverter with only two active switches has a voltage boost function and a double-ground feature that eliminates the need to float/isolate PV panels without leakage current, and improves safety. Distributed impedance networks such as transmission lines and hybrid LC components can be used for a Z-source network. These distributed Z-source networks are difficult to implement, but, a distributed Z-source inverter does not need any extra diode or switch to achieve the voltage boost function, thus having minimum component count.

1.5   Future Trends

Obviously, the Z-source concept has opened up a new research area in the power electronics field. The above description only provides a brief summary of the major Z-source network topologies. There are many modifications and twists of the Z-source topologies mentioned. Each topology has its own unique features and suited applications. There is no one-size-fits-all solution. It is feasible that new Z-source topologies will continue to appear to meet needs and improve performance in different applications. Motor drives and renewable energy generation, such as PV and wind power, will be perspective applications for Z-source converters, because of the unique voltage buck–boost ability with minimum component count and potential low cost. New power electronic devices, such as the SiC and GaN, will definitely improve the Z-source converters' performance. Their high switching frequency, low loss, and high temperature capacity will contribute to smaller size Z-source passive components and high converter efficiency. Z-source converters are still driving forward in terms of topologies and applications. Chapter 20 will provide detailed prospects.

1.6   Contents Overview

A brief introduction of the concept, advantages, classification, current status, and future trends have been reviewed in this chapter, and detailed system contents are presented throughout the rest of this book. As illustrated in Figure 1.12, Chapters 2 and 3 provide comprehensive analysis of the operating principle and modeling of voltage-fed and current-fed Z-source/quasi-Z-source inverters (ZSI/qZSI); different modulation methods and shoot-through duty cycle control methods are discussed in Chapters 4 and 5; an overview of various improved Z-source/quasi-Z-source (ZS/qZS) networks is addressed in Chapter 6; typical extended converter topologies, such as transformer-based Trans-ZSI/qZSI and LCCT-ZSI/qZSI, ZS/qZS ac-dc rectifiers, Z-source dc-dc converters, ZS/qZS matrix converters, energy stored ZSI/qZSI, and ZS/qZS multilevel inverters are detailed in Chapters 7–12, including their principle, modeling, control, and examples; hardware design, especially the impedance value design, is demonstrated in Chapter 13; their applications to PV and wind generation systems, adjustable speed drives, and multi-leg (four-leg and five-phase) converters/inverters are described in Chapters 14–17; the model predictive control (MPC) of ZSI/qZSI is illustrated in Chapter 18;

grid integration of the qZS-based PV multilevel inverter is discussed for the traditional method and MPC in Chapter 19; finally, Chapter 20 presents future prospects of this promising topology.

Acknowledgment

Most of this chapter (except Subsection 1.1.2) was made possible by NPRP-EP grant # [X-033-2-007] from the Qatar National Research Fund (a member of Qatar Foundation). The statements made herein are solely the responsibility of the authors.

References

[1] D. M. Tagare, *Electricity Power Generation: The Changing Dimensions*, Wiley & Sons Ltd, 2011.

[2] REN21 (2014, June). The First Decade (2004-2014), 10 Years of Renewable Energy Progress. [Online]. Available: http://www.ren21.netPortals/0/documents/activities/Topical%20Reports/REN21_10yr.pdf.

[3] Wikipedia, Wind power. [Online]. Available: https://en.wikipedia.org/wiki/Wind_power#cite_note-GWEC_Market-1.

[4] IEA PVPS International Energy Agency Photovoltaic Power Systems Programme (2015, March). IEA PVPS Report: 2014 Snapshot of Global PV Markets. [Online]. Available: http://www.iea-pvps.org/fileadmin/dam/public/report/technical/PVPS_report_-_A_Snapshot_of_Global_PV_-_1992-2014.pdf.

[5] W. Vielstich, *Handbook of Fuel Cells*, John Wiley & Sons Ltd, 2009.

[6] H. Abu-Rub, M. Malinowski, K. Al-Haddad, *Power Electronics for Renewable Energy Systems, Transportation and Industrial Applications*. John Wiley & Sons, 2014.

[7] F. Blaabjerg, Z. Chen, S. B. Kjaer, "Power electronics as efficient interface in dispersed power generation systems," *IEEE Trans. Power Electron.*, vol.19, no.4, pp.1184–1194, 2004.

[8] S. B. Kjaer, J. K. Pedersen, F. Blaabjerg, "A review of single-phase grid-connected inverters for photovoltaic modules," *IEEE Trans. Ind. Appl.*, vol.41, no.5, pp.1292–1306, Sep. 2005.

[9] F. Blaabjerg, M. Ke, "Future on power electronics for wind turbine systems," *IEEE Journal of Emerging and Selected Topics in Power Electronics*, vol.1, no.3, pp.139–152, Sept. 2013.

[10] M. Liserre, R. Cardenas, M. Molinas, J. Rodriguez, "Overview of Multi-MW wind turbines and wind parks," *IEEE Trans. Ind. Electron.*, vol.58, no.4, pp.1081–1095, April 2011.

[11] M. A. Khattak, M. Asif Khattak, *A Review & Analysis of Solar MPPT Algorithms & Hardware Architectures*. LAP LAMBERT Academic Publishing, 2014.

[12] M. A. G. de Brito, L. Galotto, L. P. Sampaio, G. de Azevedo e Melo, C. A. Canesin, "Evaluation of the main MPPT techniques for photovoltaic applications," *IEEE Trans. Ind. Electron.*, vol.60, no.3, pp.1156–1167, March 2013.

[13] D. A. Howey, S. M. Mahdi Alavi, *Rechargeable Battery Energy Storage System Design*. John Wiley & Sons, 2015.

[14] Y. Brunet, *Energy Storage*, John Wiley & Sons, 2010.

[15] K. Thorborg, *Power Electronics*, Prentice Hall International (UK) Ltd., London, 1988.

[16] M. H. Rashid, *Power Electronics*, 2nd Edition, Prentice Hall, 1993.

[17] N. Mohan, W. P. Robbin, T. Undeland, *Power Electronics: Converters, Applications, and Design*, 2nd Edition, John Wiley and Sons, 1995.

[18] A. M. Trzynadlowski, *Introduction to Modern Power Electronics*, John Wiley and Sons, 1998.

[19] B. K. Bose, *Modern Power Electronics and AC Drives*, Prentice Hall PTR, 2002.

[20] F. Z. Peng, "Z-Source Inverter," *IEEE Trans. Ind. Appl.*, vol.39, no.2, pp.504–510, March/April 2003. The paper was originally presented at the 2002 IEEE/IAS Annual Meeting.

[21] F. Z. Peng, M. Shen, Z. Qian, "Maximum boost control of the Z-source inverter," *IEEE Trans. Power Electron.*, vol.20, no.4, pp.833–838, July 2005.

[22] M. Shen, J. Wang, A. Joseph, F. Z. Peng, Tolbert L. M., Adams D. J., "Constant boost control of the Z-source inverter to minimize current ripple and voltage stress," *IEEE Trans. Ind. Appl.*, vol.42, no.3, pp.770–778, May–Jun 2006.

[23] Y. Liu, B. Ge, F. J. T. E. Ferreira, A. T. de Almeida, H. Abu-Rub, "Modeling and SVM control of quasi-Z-source inverter," in *Proc. 11th International Conference on Electrical Power Quality and Utilisation (EPQU)*, 2011, pp.1–7.

[24] Y. Tang, S. Xie, J. Ding, "Pulse-width Modulation of Z-Source Inverters With Minimum Inductor Current Ripple," *IEEE Trans. Ind. Electron.*, vol.61, pp.98–106, 2014.

[25] J. Jung, A. Keyhani, "Control of a fuel cell based Z-source converter," *IEEE Trans. Energy Convers.*, vol.22, no.2, pp.467–476, June 2007.

[26] U. S. Ali, V. Kamaraj, "A novel space vector PWM for Z-source inverter," in *Proc. 2011 1st International Conference on Electrical Energy Systems (ICEES)*, 2011, pp.82–85.

[27] H. Rostami, D. A. Khaburi, "Voltage gain comparison of different control methods of the Z-source inverter," in *Proc. International Conference on Electrical and Electronics Engineering (ELECO)*, 2009, pp.I-268–I-272.

[28] O. Ellabban, J. Van Mierlo, P. Lataire, "Experimental study of the shoot-through boost control methods for the Z-source inverter," *European Power Electronics and Drives Association Journal*, vol.21, no.2, pp.18–29, June 2011.

[29] Y. Liu, B. Ge, H. Abu-Rub, F. Z. Peng, "Overview of Space Vector Modulations for Three-phase Z-Source/Quasi-Z-Source Inverters," *IEEE Trans. Power Electron.*, vol.29, no.4, pp.2098–2108, April 2014.

[30] Y. Liu, H. Abu-Rub, B. Ge, "Z-Source/Quasi-Z-Source Inverters – Derived Networks, Modulations, Controls, and Emerging Applications to Photovoltaic Conversion," *IEEE Ind. Electron. Mag.*, vol.8, no.4, pp.32–44, Dec. 2014.

[31] Y. P. Siwakoti, F. Z. Peng, F. Blaabjerg, P. C. Loh, G. E. Town, "Impedance-Source Networks for Electric Power Conversion Part I: A Topological Review," *IEEE Trans. Power Electron.*, vol.30, no.2, pp.699–716, Feb. 2015.

[32] L. Liu, H. Li, Y. Zhao, X. He, Z. J. Shen, "1 MHz cascaded Z-source inverters for scalable grid-interactive photovoltaic (PV) applications using GaN device," in *Proc. 2011 IEEE Energy Conversion Congress and Exposition (ECCE)*, pp.2738–2745, 2011.

[33] Y. Zhou, L. Liu, H. Li, "A High-Performance Photovoltaic Module-Integrated Converter (MIC) Based on Cascaded Quasi-Z-Source Inverters (qZSI) Using eGaN FETs," *IEEE Trans. Power Electron.*, vol.28, no.6, pp.2727–2738, June 2013.

[34] D. Sun, B. Ge, F. Z. Peng, H. Abu-Rub, D. Bi, Y. Liu, "A New Grid-Connected PV System Based on Cascaded H-bridge Quasi-Z Source Inverter," in *Proc. 2012 IEEE International Symposium on Industrial Electronics (ISIE)*, pp.951–956, 2012.

[35] Y. Fayyad, L. Ben-Brahim, "Multilevel cascaded Z source inverter for PV power generation system," in *Proc. 2012 International Conference on Renewable Energy Research and Applications (ICRERA)*, pp.1–6, 2012.

[36] Y. Liu, B. Ge, H. Abu-Rub, F. Z. Peng, "A Modular Multilevel Space Vector Modulation for Photovoltaic Quasi-Z-Source Cascade Multilevel Inverters", in *Proc. 2013 Twenty-Eighth Annual IEEE Applied Power Electronics Conference and Exposition (APEC)*, pp.714–718, 2013.

[37] Y. Liu, B. Ge, H. Abu-Rub, F. Z. Peng, "An Effective Control Method for Quasi-Z-Source Cascade Multilevel Inverter-based Grid-tie Single-Phase Photovoltaic Power System", *IEEE Trans. Ind. Informat.*, vol.10, no.1, pp.399–407, Feb. 2014.

[38] Y. Xue, B. Ge, F. Z. Peng, "Reliability, Efficiency, and Cost Comparisons of MW Scale Photovoltaic Inverters," in *Proc. IEEE Energy Conversion Congress and Exposition (ECCE)*, pp.1627–1634, 2012.

[39] S. Qu, W. Yongyu, "On control strategy of Z-source inverter for grid integration of direct-driven wind power generator," in *31st Chinese Control Conference (CCC)*, pp.6720–6723, 25–27 July 2012.

[40] X. Wang, D. M. Vilathgamuwa, K. J. Tseng, C. J. Gajanayake, "Controller design for variable-speed permanent magnet wind turbine generators interfaced with Z-source inverter," in *Proc. International Conference on Power Electronics and Drive Systems (PEDS)*, pp.752–757, 2009.

[41] S. M. Dehghan, M. Mohamadian, A. Y. Varjani, "A New Variable-Speed Wind Energy Conversion System Using Permanent-Magnet Synchronous Generator and Z-Source Inverter," *IEEE Trans. Energy Convers.*, vol.24, no.3, pp.714–724, Sept. 2009.

[42] U. Supatti, F. Z. Peng, "Z-source inverter with grid connected for wind power system," in *Proc. 2009 IEEE Energy Conversion Congress and Exposition (ECCE)*, pp.398–403, 2009.

[43] T. Maity, H. Prasad, V. R. Babu, "Study of the suitability of recently proposed quasi Z-source inverter for wind power conversion," in *Proc. 2014 International Conference on Renewable Energy Research and Application (ICRERA)*, pp.837–841, 2014.

[44] W.-T. Franke, M. Mohr, F. W. Fuchs, "Comparison of a Z-source inverter and a voltage-source inverter linked with a DC/DC-boost-converter for wind turbines concerning their efficiency and installed semiconductor power," in *Proc. 2008 IEEE Power Electronics Specialists Conference (PESC)*, pp.1814–1820, 2008.

[45] Y. Liu, B. Ge, F. Z. Peng, H. Abu-Rub, A. T. De Almeida, F. J. T. E. Ferreira, "Quasi-Z-Source inverter based PMSG wind power generation system," in *Proc. 2011 IEEE Energy Conversion Congress and Exposition (ECCE)*, pp.291–297, 2011.

[46] B. K. Ramasamy, A. Palaniappan, S. M. Yakoh, "Direct-drive low-speed wind energy conversion system incorporating axial-type permanent magnet generator and Z-source inverter with sensorless maximum power point tracking controller," *IET Renewable Power Generation*, vol.7, no.3, pp.284–295, May 2013.

[47] F. Z. Peng, M. Shen, K. Holland, "Application of Z-Source Inverter for Traction Drive of Fuel Cell—Battery Hybrid Electric Vehicles," *IEEE Trans. Power Electron.*, vol.22, no.3, pp.1054–1061, May 2007.

[48] S. M. Dehghan, M. Mohamadian, A. Yazdian, "Hybrid Electric Vehicle Based on Bidirectional Z-Source Nine-Switch Inverter," *IEEE Trans. Veh. Commun.*, vol.59, no.6, pp.2641–2653, July 2010.

[49] F. Guo, L. Fu, C. Lin, C. Li, W. Choi, J. Wang, "Development of an 85-kW Bidirectional Quasi-Z-Source Inverter With DC-Link Feed-Forward Compensation for Electric Vehicle Applications," *IEEE Trans. Power Electron.*, vol.28, no.12, pp.5477–5488, Dec. 2013.

[50] P. Liu, H. P. Liu, "Permanent-magnet synchronous motor drive system for electric vehicles using bidirectional Z-source inverter," *IET Electrical Systems in Transportation*, vol.2, no.4, pp.178–185, December 2012.

[51] Q. Lei, D. Cao, F. Z. Peng, "Novel Loss and Harmonic Minimized Vector Modulation for a Current-Fed Quasi-Z-Source Inverter in HEV Motor Drive Application," *IEEE Trans. Power Electron.*, vol.29, no.3, pp.1344–1357, March 2014.

[52] X. Fang, Z. Qian, F. Z. Peng, "Single-phase Z-source PWM AC-AC converters," *IEEE Power Electron Lett.*, vol.3, no.4, pp.121–124, Dec. 2005.

[53] Y. Tang, S. Xie, C. Zhang, "Z-Source AC-AC Converters Solving Commutation Problem," *IEEE Trans. Power Electron.*, vol.22, no.6, pp.2146–2154, Nov. 2007.

[54] M. Nguyen, Y. Jung, Y. Lim, "Single-Phase AC-AC Converter Based on Quasi-Z-Source Topology," *IEEE Trans. Power Electron.*, vol.25, no.8, pp.2200–2210, Aug. 2010.

[55] L. He, S. Duan, F. Z. Peng, "Safe-Commutation Strategy for the Novel Family of Quasi-Z-Source AC-AC Converter," *IEEE Trans. Ind. Informat.*, vol.9, no.3, pp.1538–1547, Aug. 2013.

[56] M. Nguyen, Y. Lim, Y. Kim, "A Modified Single-Phase Quasi-Z-Source AC-AC Converter," *IEEE Trans. Power Electron.*, vol.27, no.1, pp.201–210, Jan. 2012.

[57] W. Song, Y. Zhong, H. Zhang, X. Sun, Q. Zhang, W. Wang, "A study of Z-source dual-bridge matrix converter immune to abnormal input voltage disturbance and with high voltage transfer ratio," *IEEE Trans. Ind. Informat.*, vol.9, no.2, pp.828–838, May 2013.

[58] X. Liu, P. C. Loh, P. Wang, X. Han, "Improved modulation schemes for indirect Z-source matrix converter with sinusoidal input and output waveforms," *IEEE Trans. Power Electron.*, vol.27, no.9, pp.4039–4050, Sept. 2012.

[59] B. Ge, Q. Lei, W. Qian, F. Z. Peng, "A family of Z-source matrix converters," *IEEE Trans. Ind. Electron.*, vol.59, no.1, pp.35–46, Jan. 2012.

[60] S. Liu, B. Ge, H. Abu-Rub, F. Z. Peng, and Y. Liu, "Quasi-Z-source matrix converter based induction motor drives," in *Proc. 38th Annual Conference on IEEE Industrial Electronics Society (IECON)*, pp.5303–5307, 2012.

[61] O. Ellabban, H. Abu-Rub, B. Ge, "A Quasi-Z-Source Direct Matrix Converter Feeding a Vector Controlled Induction Motor Drive," *IEEE Journal of Emerging and Selected Topics in Power Electronics*, vol.3, no.2, pp.339–348, June 2015.

[62] L. Huber, D. Borojevic, "Space vector modulated three-phase to three-phase matrix converter with input power factor correction," *IEEE Trans. Ind. Appl.*, vol.31, no.6, pp.1234–1246, Nov/Dec 1995.

[63] M. Shen, A. Joseph, J. Wang, F. Z. Peng, D. J. Adams, "Comparison of traditional inverters and Z-source inverter for fuel cell vehicles," *IEEE Trans. Power Electron.*, vol.22, no.4, pp.1453–1463, July 2007.

[64] D. Vinnikov, I. Roasto, "Quasi-Z-Source-based isolated DC/DC converters for distributed power generation," *IEEE Trans. Ind. Electron.*, vol.58, pp.192–201, Jan 2011.

[65] I. Roasto, D. Vinnikov, "New voltage mode control method for the quasi-Z-Source-based isolated DC/DC converters," in *Proc. 2012 IEEE International Conference on Industrial Technology (ICIT)*, pp.644–649, 2012.

[66] A. Chub, O. Husev, D. Vinnikov, F. Blaabjerg, "Novel family of quasi-Z-source DC/DC converters derived from current-fed push-pull converters," in *Proc. 2014 16th European Conference on Power Electronics and Applications (EPE'14-ECCE Europe)*, pp.1–10, 2014.

[67] Y. P. Siwakoti, F. Blaabjerg, P. C. Loh, G. E. Town, "High-voltage boost quasi-Z-source isolated DC/DC converter," *IET Power Electronics*, vol.7, no.9, pp.2387–2395, September 2014.

[68] V. Fernão Pires, E. Romero-Cadaval, D. Vinnikov, I. Roasto, J. F. Martins, "Power converter interfaces for electrochemical energy storage systems – A review," *Energy Conversion and Management*, vol.86, pp.453–475, 2014.

[69] J. Anderson and F. Z. Peng, "Four quasi-Z-Source inverters," in *Proc. PESC '08 – 39th IEEE Annual Power Electronics Specialists Conference*, 2008, pp.2743–2749.

[70] W. Qian, F. Z. Peng, H. Cha, "Trans-Z-Source Inverters," *IEEE Trans. Power Electron.*, vol.26, no.12, pp.3453–3463, Dec. 2011.

[71] C. J. Gajanayake, F. Luo, H. B. Gooi, P. L. So, L. K. Siow, "Extended-BoostZ-Source Inverters," *IEEE Trans. Power Electron.*, vol.25, no.10, pp.2642–2652, Oct. 2010.

[72] M. Zhu, K. Yu, F. Luo, "Switched Inductor Z-Source Inverter," *IEEE Trans. Power Electron.*, vol.25, no.8, pp.2150–2158, Aug. 2010.

[73] N. Minh-Khai, Y. Lim, G. Cho, "Switched-Inductor Quasi-Z-Source Inverter," *IEEE Trans. Power Electron.*, vol.26, no.11, pp.3183–3191, Nov. 2011.

[74] M. Adamowicz, J. Guzinski, R. Strzelecki, F. Z. Peng, H. Abu-Rub, "High step-up continuous input current LCCT-Z-source inverters for fuel cells," in *Proc. 2011 IEEE Energy Conversion Congress and Exposition (ECCE)*, pp.2276–2282, 2011.

[75] M. Adamowicz, R. Strzelecki, F. Z. Peng, J. Guzinski, H. Abu-Rub, "New type LCCT-Z-source inverters," in *Proc. of 14th European Conference on Power Electronics and Applications (EPE)*, pp.1–10, 2011.

2

Voltage-Fed Z-Source/ Quasi-Z-Source Inverters

Voltage-fed Z-source/quasi-Z-source inverters are discussed in this chapter, including their configurations, working principles, voltage and current stresses, and a comparison with their Z-source counterpart.

There are several types of voltage-fed ZSI/qZSI, but they have the same principle of operation, in that the shoot-through behavior between power devices of one phase leg is utilized to achieve a voltage boost in single-stage power conversion. Thus, this chapter investigates the voltage-fed qZSI with continuous input current as an example to illustrate the principle of ZSI/qZSI in terms of the steady-state model, small-signal dynamic model, and also some simulation results.

2.1 Topologies of Voltage-Fed Z-Source/Quasi-Z-Source Inverters

Figure 2.1(b)–(d) shows the kind of voltage-fed qZSI in comparison with that of ZSI in Figure 2.1(a). The voltage-fed qZSIs with discontinuous input current is shown in Figure 2.1(b), with continuous input current is shown in Figure 2.1(c), and the bidirectional one replacing the diode with a switch is shown in Figure 2.1(d). Because of the input inductor L_1, the qZSI in Figure 2.1(c) draws a continuous constant DC current from the DC source. Compared with the ZSI, which draws a discontinuous current, the constant current will significantly reduce input stress; the qZSI with continuous input current is especially well suited to PV system applications [1–4].

The qZSI has two general types of operational state on the DC side: the non-shoot-through state (i.e. the six active states and two conventional zero states), and the shoot-through state (i.e. both switches in at least one phase conduct simultaneously). In the non-shoot-through state, the inverter bridge, viewed from the DC side, is equivalent to a current source, whereas in the shoot-through state, the inverter bridge is a short-circuit. The equivalent circuits of the

Impedance Source Power Electronic Converters, First Edition. Yushan Liu, Haitham Abu-Rub, Baoming Ge, Frede Blaabjerg, Omar Ellabban, and Poh Chiang Loh.
© 2016 John Wiley & Sons, Ltd. Published 2016 by John Wiley & Sons, Ltd.

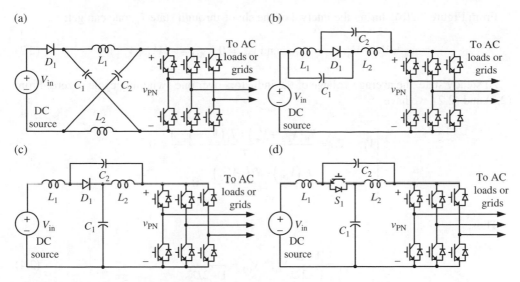

Figure 2.1 Topologies of voltage-fed ZSI/qZSI: (a) ZSI, (b) qZSI with discontinuous input current, (c) qZSI with continuous input current, and (d) bidirectional qZSI (*Source*: Anderson 2008 [1]. Reproduced with permission of IEEE).

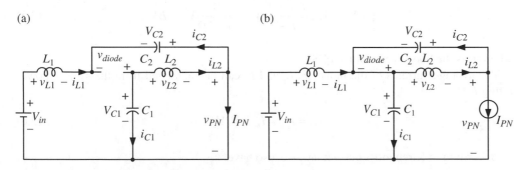

Figure 2.2 Equivalent circuit of the voltage-fed qZSI with continuous input current: (a) non-shoot-through state, (b) shoot-through state.

two states are shown in Figure 2.2(a) and (b), respectively. It is well known that the shoot-through state is strictly forbidden in the traditional VSI because it will cause a short-circuit of the voltage source and damage the devices. However, in the qZSI and ZSI, the unique LC and diode network connected to the inverter bridge modifies the operation of the circuit, allowing the shoot-through states. Furthermore, by using the shoot-though state, the (quasi-)Z-source network boosts the dc-link voltage. This feature will effectively protect the circuit from damage; thus, it improves system reliability significantly.

Assuming that during one switching cycle T, the interval of the shoot-through state is T_0, then the interval of the non-shoot-through state is T_1; thus, $T=T_0+T_1$ and the shoot-through duty ratio $D=T_0/T$. From Figure 2.2(a), during the interval of the non-shoot-through state T_1, we have [2, 3]:

$$v_{L1} = V_{in} - V_{C1}, v_{L2} = -V_{C2}, v_{PN} = V_{C1} - v_{L2} = V_{C1} + V_{C2}, v_{diode} = 0. \tag{2.1}$$

From Figure 2.2(b), during the interval of the shoot-through state T_0, one can get:

$$v_{L1} = V_{C2} + V_{in}, v_{L2} = V_{C1}, \quad \text{and } v_{PN} = 0, v_{diode} = -\left(V_{C1} + V_{C2}\right). \tag{2.2}$$

At steady state, the average voltage of the inductors over one switching cycle is zero. From (2.1) and (2.2), we have

$$\begin{cases} V_{L1} = \overline{v}_{L1} = \dfrac{T_0\left(V_{C2} + V_{in}\right) + T_1\left(V_{in} - V_{C1}\right)}{T} = 0 \\ V_{L2} = \overline{v}_{L2} = \dfrac{T_0\left(V_{C1}\right) + T_1\left(-V_{C2}\right)}{T} = 0 \end{cases}. \tag{2.3}$$

Thus,

$$V_{C1} = \frac{1-D}{1-2D} V_{in}, \quad V_{C2} = \frac{D}{1-2D} V_{in}. \tag{2.4}$$

From (2.1) and (2.2), the peak dc-link voltage across the inverter bridge is

$$V_{PN} = V_{C1} + V_{C2} = \frac{T}{T_1 - T_0} V_{in} = \frac{1}{1-2D} V_{in} = BV_{in}, \tag{2.5}$$

where B is the boost factor of the qZSI.

The average currents of the inductors L_1 and L_2 can be calculated from the system power rating P

$$I_{L1} = I_{L2} = I_{in} = P/V_{in}. \tag{2.6}$$

According to Kirchhoff's current law, we also have

$$I_{C1} = I_{C2} = I_{PN} - I_{L1}, \quad I_D = 2I_{L1} - I_{PN}. \tag{2.7}$$

In summary, the voltage and current stress of the qZSI are shown in Table 2.1 [2, 3], where

1. M is the modulation index.
2. $m = \left(1-D\right)/\left(1-2D\right); n = D/\left(1-2D\right); B = 1/\left(1-2D\right).$

The stress on the ZSI is shown as well for comparison.

From Table 2.1 we can establish that the qZSI inherits all the advantages of the ZSI. It can buck or boost a voltage, cope with a wide range of input voltage, and produce the desired voltage for the load or connection to the grid in a single stage. This feature results in the reduced number of switches involved in the PV system, and therefore reduced cost and improved system efficiency. When the voltage of the PV panel is low, it boosts the dc-link

Table 2.1 Voltage and current characteristics of the voltage-fed qZSI and ZSI (*Source*: Li 2009 [2]. Reproduced with permission of IEEE.)

$v_{L1}=v_{L2}$		v_{PN}		v_{diode}		V_{C1}	V_{C2}	v_{In}	$I_{in}=I_{L1}=I_{L2}$	$I_{C1}=I_{C2}$	I_D
In T_0	In T_1	In T_0	In T_1	In T_0	In T_1						
ZSI mV_{in}	$-nV_{in}$	0	BV_{in}	BV_{in}	0	mV_{in}	mV_{in}	$MBV_{in}/2$	P/V_{in}	$I_{PN}-I_{L1}$	$2I_{L1}-I_{PN}$
qZSI mV_{in}	$-nV_{in}$	0	BV_{in}	BV_{in}	0	mV_{in}	nV_{in}	$MBV_{in}/2$	P/V_{in}	$I_{PN}-I_{L1}$	$2I_{L1}-I_{PN}$

voltage, which helps avoid redundant PV panels for higher DC voltage or unessential inverter overrating. As mentioned, it is able to handle the shoot-through state, so it is more reliable than conventional VSI. For the same reason, there is no need to add any dead time into the control schemes, which reduces the output distortion.

In addition, there are some unique merits of the qZSI when compared with conventional ZSI [3]. The ZSI has discontinuous input current in the boost mode, whereas the input current of the qZSI is continuous, owing to the input inductor L_1, which reduces the input stress significantly; thus, it can reduce the capacitance for the output of the PV panels. The two capacitors in the ZSI sustain the same high voltage, whereas the voltage on capacitor C_2 in the qZSI is lower, which allows a lower capacitor voltage rating. For the qZSI, there is a common DC rail between the source and inverter, which is easier to assemble and causes less problem with EMI.

2.2 Modeling of Voltage-Fed qZSI

2.2.1 Steady-State Model

Figure 2.3 shows the equivalent circuit of the qZSI [5–10]. In Figure 2.3(a), the inverter bridge is equivalent to a short-circuit when shoot-through zero vectors are working; in Figure 2.3(b), the inverter bridge is replaced by a constant current source in non-shoot-through state as demonstrated before. The qZS inductor currents and capacitor voltages and their reference directions are as shown in Figure 2.3, where R denotes the series resistance of the capacitors and r denotes the parasitic resistance of the inductors.

From Figure 2.3(a), in the shoot-through state, the circuit equations represented by the state-space equations are [9]

$$\begin{bmatrix} L_1 & 0 & 0 & 0 \\ 0 & L_2 & 0 & 0 \\ 0 & 0 & C_1 & 0 \\ 0 & 0 & 0 & C_2 \end{bmatrix} \cdot \begin{bmatrix} \dot{i}_{L1}(t) \\ \dot{i}_{L2}(t) \\ \dot{v}_{C1}(t) \\ \dot{v}_{C2}(t) \end{bmatrix} = \begin{bmatrix} -(R+r) & 0 & 0 & 1 \\ 0 & -(R+r) & 1 & 0 \\ 0 & -1 & 0 & 0 \\ -1 & 0 & 0 & 0 \end{bmatrix} \cdot \begin{bmatrix} i_{L1}(t) \\ i_{L2}(t) \\ v_{C1}(t) \\ v_{C2}(t) \end{bmatrix} + \begin{bmatrix} 1 & 0 \\ 0 & 0 \\ 0 & 0 \\ 0 & 0 \end{bmatrix} \begin{bmatrix} V_{in}(t) \\ I_{PN}(t) \end{bmatrix}$$

(2.8)

which is written as $F\dot{x} = A_1 x + B_1 u$.

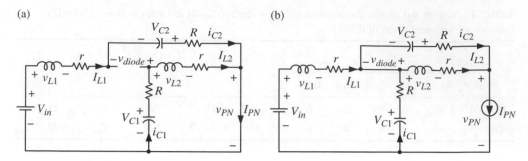

Figure 2.3 Equivalent circuit of qZSI shown in Figure 2.1(c): (a) shoot-through states, (b) non-shoot-through states (*Source*: Liu 2011 [9]. Reproduced with permission of IEEE).

From Figure 2.3(b), in non-shoot-through state, the circuit equations represented by state space equations are

$$
\begin{bmatrix} L_1 & 0 & 0 & 0 \\ 0 & L_2 & 0 & 0 \\ 0 & 0 & C_1 & 0 \\ 0 & 0 & 0 & C_2 \end{bmatrix} \cdot \begin{bmatrix} i_{L1}(t) \\ i_{L2}(t) \\ \dot{v}_{C1}(t) \\ \dot{v}_{C2}(t) \end{bmatrix} = \begin{bmatrix} -(R+r) & 0 & -1 & 0 \\ 0 & -(R+r) & 0 & -1 \\ 1 & 0 & 0 & 0 \\ 0 & 1 & 0 & 0 \end{bmatrix} \cdot \begin{bmatrix} i_{L1}(t) \\ i_{L2}(t) \\ v_{C1}(t) \\ v_{C2}(t) \end{bmatrix} + \begin{bmatrix} 1 & R \\ 0 & R \\ 0 & -1 \\ 0 & -1 \end{bmatrix} \cdot \begin{bmatrix} V_{in}(t) \\ I_{PN}(t) \end{bmatrix}
$$

$$(2.9)$$

which is written as $F\dot{x} = A_2 x + B_2 u$.

Using the state space average method, we can obtain $A = D \cdot A_1 + (1-D) \cdot A_2$, $B = D \cdot B_1 + (1-D) \cdot B_2$, where D represents the shoot-through duty cycle while $1-D$ represents the non-shoot-through duty cycle. Then we get

$$
F\dot{x} = Ax + Bu = \begin{bmatrix} -(R+r) & 0 & D-1 & D \\ 0 & -(R+r) & D & D-1 \\ 1-D & -D & 0 & 0 \\ -D & 1-D & 0 & 0 \end{bmatrix} \cdot \begin{bmatrix} i_{L1}(t) \\ i_{L2}(t) \\ v_{C1}(t) \\ v_{C2}(t) \end{bmatrix} + \begin{bmatrix} 1 & (1-D)R \\ 0 & (1-D)R \\ 0 & D-1 \\ 0 & D-1 \end{bmatrix} \cdot \begin{bmatrix} V_{in}(t) \\ I_{PN}(t) \end{bmatrix}
$$

$$(2.10)$$

When the system is in steady state, then $AX + BU = 0$, where $X = \begin{bmatrix} I_{L1} & I_{L2} & V_{C1} & V_{C2} \end{bmatrix}^T$, $U = \begin{bmatrix} V_{in} & I_{PN} \end{bmatrix}^T$. That is

$$
\begin{cases} -(R+r)I_{L1} + (D-1)V_{C1} + DV_{C2} + V_{in} + (1-D)RI_{PN} = 0 \\ -(R+r)I_{L2} + DV_{C1} + (D-1)V_{C2} + (1-D)RI_{PN} = 0 \\ (1-D)I_{L1} - DI_{L2} + (D-1)I_{PN} = 0 \\ -DI_{L1} + (1-D)I_{L2} + (D-1)I_{PN} = 0 \end{cases}
$$

This can be solved to give

$$
\begin{cases}
V_{C1} = \dfrac{1-D}{1-2D}V_{in} - V_{22} \\[3mm]
V_{C2} = \dfrac{D}{1-2D}V_{in} - V_{22}, \quad \text{where} \quad V_{22} = \dfrac{(1-D)(r+2DR)}{(1-2D)^2}I_{PN} \\[3mm]
I_{L1} = I_{L2} = \dfrac{1-D}{1-2D}I_{PN}
\end{cases}
\tag{2.11}
$$

It can be seen that the currents of the two inductors in the quasi-Z source network are the same in steady state. Moreover, if we ignore the series resistances of capacitors and the parasitic resistances of inductors, that is $V_{22} = 0$ in (2.11), this gives

$$
V_{C1} = \frac{1-D}{1-2D}V_{in}, \quad V_{C2} = \frac{D}{1-2D}V_{in}
\tag{2.12}
$$

Equation (2.12) is the same as the conclusion in [2], and verifies the correctness of the state-space average model.

2.2.2 Dynamic Model

2.2.2.1 Small-Signal Model

Mathematically, the small perturbation of state variables is denoted by $\hat{x} = \begin{bmatrix} \hat{i}_{L1}(t) & \hat{i}_{L2}(t) & \hat{v}_{C1}(t) & \hat{v}_{C2}(t) \end{bmatrix}^T$, input signals are denoted by $\hat{u} = \begin{bmatrix} \hat{V}_{in}(t) & \hat{I}_{PN}(t) \end{bmatrix}^T$, and shoot-through duty ratio is denoted by $\hat{d}(t)$. Substituting these perturbed variables into (2.10), the small-signal state equations are [9]

$$
F\hat{\dot{x}} = A\hat{x} + B\hat{u} + \left[(A_1 - A_2) \cdot X + (B_1 - B_2) \cdot U \right] \cdot \hat{d}
$$

$$
=
\begin{bmatrix}
-(R+r) & 0 & D-1 & D \\
0 & -(R+r) & D & D-1 \\
1-D & -D & 0 & 0 \\
-D & 1-D & 0 & 0
\end{bmatrix}
\begin{bmatrix}
\hat{i}_{L1}(t) \\
\hat{i}_{L2}(t) \\
\hat{v}_{C1}(t) \\
\hat{v}_{C2}(t)
\end{bmatrix}
+
\begin{bmatrix}
1 & (1-D)R \\
0 & (1-D)R \\
0 & D-1 \\
0 & D-1
\end{bmatrix}
\cdot
\begin{bmatrix}
V_{in}(t) \\
I_{PN}(t)
\end{bmatrix}
+
\begin{bmatrix}
V_{C1} + V_{C2} - I_{PN}R \\
V_{C1} + V_{C2} - I_{PN}R \\
-I_{L1} - I_{L2} + I_{PN} \\
-I_{L1} - I_{L2} + I_{PN}
\end{bmatrix}
\cdot \hat{d}(t)
$$

$$
\tag{2.13}
$$

Let $I_{11} = I_{PN} - 2I_L$, $V_{11} = V_{C1} + V_{C2} - I_{PN}R$. By Laplace transforms, (2.13) is converted into

$$
sL_1\hat{i}_{L1}(s) = -(R+r)\hat{i}_{L1}(s) + (D-1)\hat{v}_{C1}(s) + D\hat{v}_{C2}(s) + \hat{V}_{in}(s) + (1-D)R\hat{I}_{PN}(s) + V_{11}\hat{d}(s)
\tag{2.14}
$$

$$
sL_2\hat{i}_{L2}(s) = -(R+r)\hat{i}_{L2}(s) + D\hat{v}_{C1}(s) + (D-1)\hat{v}_{C2}(s) + (1-D)R\hat{I}_{PN}(s) + V_{11}\hat{d}(s)
\tag{2.15}
$$

$$sC_1 \hat{v}_{C1}(s) = (1-D)\hat{i}_{L1}(s) - D\hat{i}_{L2}(s) + (D-1)\hat{I}_{PN}(s) + I_{11}\hat{d}(s) \tag{2.16}$$

$$sC_2 \hat{v}_{C2}(s) = -D\hat{i}_{L1}(s) + (1-D)\hat{i}_{L2}(s) + (D-1)\hat{I}_{PN}(s) + I_{11}\hat{d}(s) \tag{2.17}$$

Using $L_1 = L_2 = L, C_1 = C_2 = C$, subtracting (2.15) from (2.14) and (2.17) from (2.16), we get

$$\hat{i}_{L1}(s) - \hat{i}_{L2}(s) = \frac{sC}{LCs^2 + (R+r)Cs + 1} \cdot \hat{V}_{in}(s) \tag{2.18}$$

$$\hat{v}_{C1}(s) - \hat{v}_{C2}(s) = \frac{1}{LCs^2 + (R+r)Cs + 1} \cdot \hat{V}_{in}(s) \tag{2.19}$$

Comparing (2.11) and (2.18), we can see that inductor currents, i_{L1} and i_{L2}, of qZSI are different in small perturbation models, indicating the differences in their dynamics. Note that in ZSI, the small perturbation models are the same [5, 6]. Thus, (2.18) shows the differences between qZSI and ZSI. Also, the deduced small-signal model of qZSI can reflect the dynamic process very well. From (2.14) to (2.17), the signal flow graph of qZSI can be obtained, as shown in Figure 2.4.

Therefore, the small-signal transfer functions of V_{C1} and i_{L2} can be solved as [9, 10]

$$\hat{v}_{C1}(s) = \frac{K_1(D-1) + (1-2D)(1-D)}{(K_1+1)\left[K_1 + (1-2D)^2\right]} \cdot \hat{V}_{in}(s) + \frac{(1-2D)(1-D)R + K_2(1-D)}{K_1 + (1-2D)^2} \cdot \hat{I}_{PN}(s)$$

$$+ \frac{(1-2D)V_{11} + K_2 I_{11}}{K_1 + (1-2D)^2} \cdot \hat{d}(s) \tag{2.20}$$

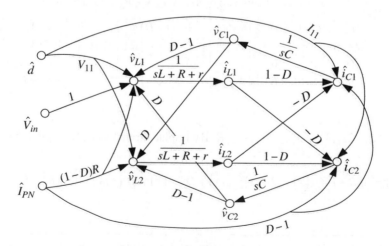

Figure 2.4 Signal flow graph of qZSI (*Source*: Liu 2011 [9]. Reproduced with permission of IEEE).

$$
\hat{i}_{L2}(s) = \frac{2K_1 D(1-D)}{K_2\left(K_1+1\right)\left[K_1+\left(1-2D\right)^2\right]} \cdot \hat{V}_{in}(s) + \frac{K_2\left(1-D\right)\left(1-2D\right)+K_1 R\left(1-D\right)}{K_2\left[K_1+\left(1-2D\right)^2\right]} \cdot \hat{i}_{PN}(s)
$$
$$
+ \frac{K_2\left(1-2D\right)I_{11}+K_1 V_{11}}{K_2\left[K_1+\left(1-2D\right)^2\right]} \cdot \hat{d}(s)
$$

(2.21)

where $K_1 = LCs^2 + \left(R+r\right)Cs$, $\quad K_2 = Ls + R + r$.

2.2.2.2 Analysis in Frequency Domain

Parameter scanning of the Bode diagram can be used to optimize the parameter design of a given system in the frequency domain. In this section, pole-zero maps and Bode diagrams are analyzed, in order to provide an important method for the parameter design of qZSI. For designing closed-loop control of qZSI, the most important transfer function is that from the shoot-through duty cycle to the output capacitor voltage. Therefore, using this function as an example, in this section we obtain its pole-zero maps and Bode diagrams under changes of parameters L and C, where L denotes the inductance of L_1 and L_2, and C denotes the capacitance of C_1 and C_2. The transfer function from the shoot-through duty cycle to the output capacitor voltage is [9]

$$
G_{Vcd}(s) = \left.\frac{\hat{v}_{C1}(s)}{\hat{d}(s)}\right|_{\substack{\hat{v}_{in}(s)=0 \\ \hat{i}_{PN}(s)=0}} = \frac{LI_{11}s+\left(R+r\right)I_{11}+\left(1-2D\right)V_{11}}{LCs^2+\left(R+r\right)Cs+\left(1-2D\right)^2}
$$

(2.22)

By varying the inductance L from $100\,\mu\text{H}$ to $500\,\mu\text{H}$, Figure 2.5(a) shows the shifting of poles and zeros towards the imaginary axis [10]. The shifting of zeros increases the non-minimum-phase undershoots, while the shifting of poles increases the system settling time and response. Then by changing capacitance C from $100\,\mu\text{F}$ to $500\,\mu\text{F}$, Figure 2.5(b) shows the shifting of poles vertically towards the real axis, while the zeros stay constant. It is observed that to increase system damping reduces the amplitude of overshoot and undershoot, but will increase the rise time. For sources that have wider operating ranges, it is important to study the movements of zero and pole in order to maintain acceptable performance and stability.

With the changes of L and C, the Bode diagrams of the transfer function G_{Vcd} are shown in Figure 2.6. It can be seen that when inductance increases, the amplitude–frequency characteristic changes gently and the quality factor decreases. However, when the capacitance increases, the amplitude–frequency characteristic changes steeply and the quality factor increases. Similarly, the resonant frequency reduces with the two parameters increasing.

The established transfer function model (2.20) of the capacitor voltage V_{C1} is simulated in the time domain. Take the step responses of V_{C1} when L changes from $100\,\mu\text{H}$ to $500\,\mu\text{H}$ at $C=300\,\mu\text{F}$ and C changes from $100\,\mu\text{F}$ to $500\,\mu\text{F}$ at $L=100\,\mu\text{H}$. Simulation results are shown

(a)

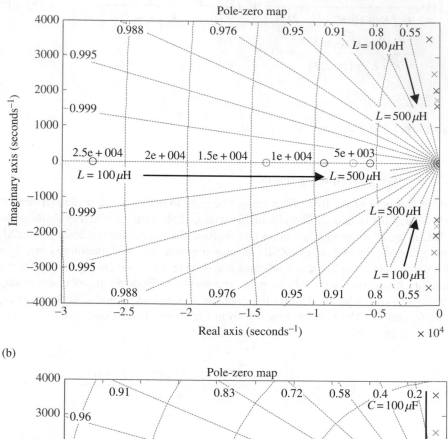

(b)

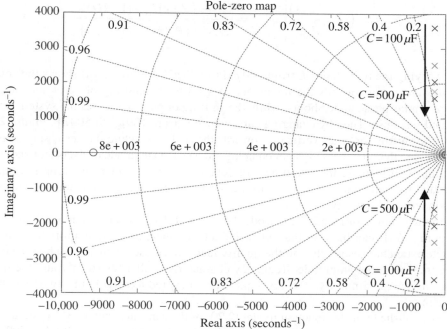

Figure 2.5 Zero-pole trajectory: (a) inductance changing; (b) capacitance changing (*Source*: Liu 2014 [10]. Reproduced with permission of Liu.).

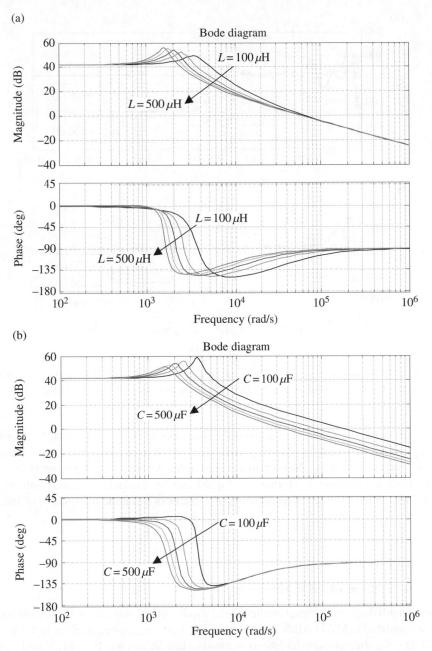

Figure 2.6 Bode diagram. (a) Inductance changing when $C=300\ \mu F$, (b) capacitance changing when $L=100\mu H$ (*Source*: Liu 2014 [10]. Reproduced with permission of Liu.).

in Figure 2.7. From Figure 2.7(a), it can be seen that the value of L is smaller, the overshoot of the step response is less, and the settling time is shorter. From Figure 2.7(b), it can be seen that when C increases, the overshoot decreases but the rise time becomes longer. The simulation results in the time domain are consistent with the frequency domain analysis.

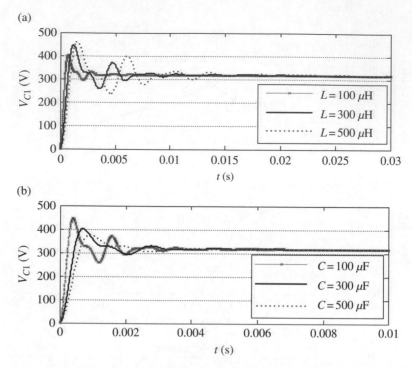

Figure 2.7   Step responses of capacitor voltage: (a) inductance changing when $C=300\,\mu\text{F}$, (b) capacitance changing when $L=100\,\mu\text{H}$ (*Source*: Liu 2011 [9]. Reproduced with permission of IEEE).

Thus, the values of L and C in a quasi-Z-source network can be selected according to the compromise between the damping response and settling time from pole–zero maps and Bode diagrams of the deduced transfer functions. As required, the impacts from other parameters on the dynamic performance of qZSI can be similarly obtained, such as shoot-through duty cycle D, the series resistance R of capacitors, parasitic resistance r of inductors, and so on.

2.3   Simulation Results

2.3.1   Simulation of qZSI Modeling

First, the established small-signal mathematical model and actual circuit simulation model of qZSI are simulated in MATLAB/Simulink, using the system parameters given in Table 2.2 [9]. From (2.11), the theoretical calculations of steady-state values are $V_{C1}=317\text{V}$ and $i_{L2}=25\text{A}$. From equations (2.20) and (2.21) and Figure 2.4, simulation results of capacitor voltage V_{C1} and inductor current i_{L2} are respectively compared with a small-signal mathematical model and actual circuit simulation model, as shown in Figure 2.8. It can be seen that the steady-state values are basically coincident with each other and also with theoretical calculations, while the dynamic response of these two models is different. In summary, the simulation and calculation results indicate that the established small-signal mathematical model of qZSI is correct and credible.

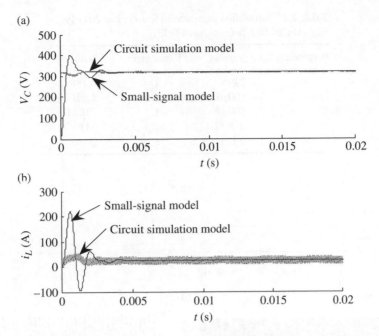

Figure 2.8 Comparison of the simulation and small-signal models: (a) capacitor voltage, (b) inductor current (*Source*: Liu 2011 [9]. Reproduced with permission of IEEE).

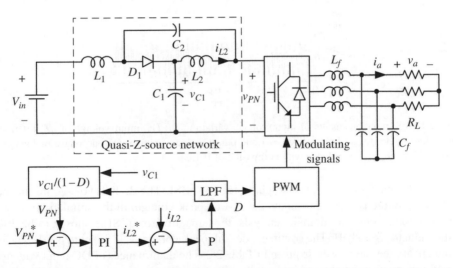

Figure 2.9 Block diagram of simulated qZSI control system (*Source*: Liu 2011 [9]. Reproduced with permission of IEEE).

2.3.2 Circuit Simulation Results of Control System

Based on small-signal modeling, the qZSI system using a dc-link voltage control is presented in Figure 2.9 [9]. After the voltage–current closed-loop control and then filtered by a low-pass filter (LPF), the shoot-through duty cycle D is delivered to the PWM scheme (this will be

Table 2.2 Simulation parameters (*Source*: Liu 2011 [9].
Reproduced with permission of IEEE)

Parameters	Values	Parameters	Values
V_{in}	249 V	D	0.18875
L	100 μH	L_f	1 mH
C	300 μF	C_f	110 μF
R	0.08 Ω	R_L	5 Ω
r	0.15 Ω	I_{PN}	20 A

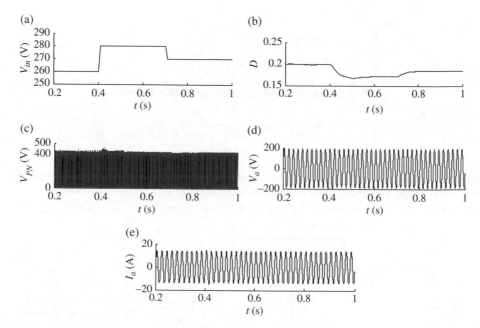

Figure 2.10 Simulation results of input voltage variation: (a) DC input voltage of qZSI, (b) shoot-through duty cycle, (c) dc-link voltage, (d) one phase of load voltage, (e) one phase of load current (*Source*: Liu 2011 [9]. Reproduced with permission of IEEE).

detailed in Chapter 4) to operate the phase legs of qZSI. The dc-link voltage instantaneous calculation module is used to compute the dc-link peak voltage. In the simulations, referring to the root locus and time-domain analysis, the capacitor for qZSI is chosen to be 300 μF, and the inductor is 500 μH. The command dc-link peak voltage V_{PN}^* is set to 400 V. The other parameters are the same as those given in Table 2.2. The disturbances of DC input voltage and AC load variations are discussed in the following.

The operating conditions as input voltage changes are: $0 < t \leq 0.4$ s, the input DC voltage V_{in} is 260 V; 0.4 s $< t \leq 0.7$ s, V_{in} is 280 V; 0.7 s $< t \leq 1$ s, V_{in} is 270 V. The simulation results are shown in Figure 2.10, from which it can be seen that when the input voltage increases or decreases, the shoot-through duty cycle D will respectively reduce or increase in order to obtain the desired dc-link voltage, with a transient regulation. The ac output voltage and current are kept constant for the entire simulation time.

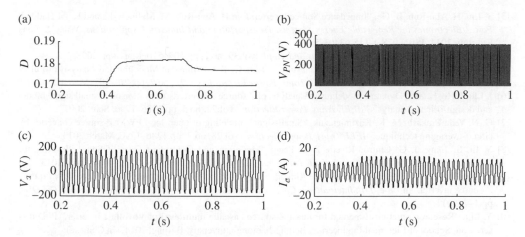

Figure 2.11 Simulation results of load variation: (a) Shoot-through duty cycle, (b) dc-link voltage, (c) one phase of load voltage, (d) one phase of load current (*Source*: Liu 2011 [9]. Reproduced with permission of IEEE).

The simulation results of the AC load variation are shown in Figure 2.11 with the operating conditions: $0 < t \leq 0.4$ s, the output resistive load R_L is $20\,\Omega$; 0.4 s $< t \leq 0.7$ s, R_L is $10\,\Omega$; 0.7 s $< t \leq 1$ s, R_L is $15\,\Omega$. It can be seen that when the load changes, the dc-link voltage and ac output voltage remain unchanged, resulting in the load current at the constant loads.

In brief, it can be concluded that no matter how the DC input voltage varies or the AC load varies, due to the proposed control schemes, the dc-link voltage of qZSI can be kept constant, demonstrating a reliable performance to external disturbances.

2.4 Conclusion

The three-phase two-level voltage-fed ZSI/qZSI were presented in this chapter. Their topologies, operation, steady-state and dynamic model were introduced. The shoot-through behavior was inserted into the phase leg bridges, bringing in the voltage boosting in single-stage inverter topology to the ZSI/qZSI. There are four types of voltage-fed ZSI/qZSI, categorized by input current continuity or discontinuity. The qZSI with continuous input current and lower voltage on one of the capacitors has attracted more attention in later research. Its modeling and operation have been investigated in detail with simulation results, providing the basics for other derived impedance topologies and applications.

References

[1] J. Anderson, and F. Z. Peng, "A class of quasi-Z-source inverters," in *Proc. IEEE Industry Applications Society Annual Meeting*, pp.1–7, 2008.

[2] Y. Li, J. Anderson, F. Z. Peng, D. Liu, "Quasi-Z-source inverter for photovoltaic power generation systems," in *Proc. Twenty-Fourth Annual IEEE Applied Power Electronics Conference and Exposition, (APEC)*, pp.918–924, 2009.

[3] Y. Liu, H. Abu-Rub, B. Ge, "Impedance Source Inverters," *in* H. Abu-Rub, M. Malinowski, and K. Al-Haddad, *Power Electronics for Renewable Energy Systems, Transportation and Industrial Applications*, Wiley & Sons Ltd, July 2014, ISBN 978-1-118-63403-5.

[4] F. Z. Peng, "Z-source inverter," *IEEE Trans. Ind. Appl.*, vol.39, no.2, pp.504–510, Mar./Apr. 2003.

[5] P. C. Loh, D. M. Vilathgamuwa, C. J. Gajanayake, Y. Lim, C. Teo, "Transient Modeling and Analysis of Pulse-Width Modulated Z-Source Inverter," *IEEE Trans. Power Electron.,* vol.22, no.2, pp.498–507, March 2007.

[6] J. Liu, J. Hu, L. Xu, "Dynamic modeling and analysis of Z source converter – derivation of ac small signal model and design-oriented analysis," *IEEE Trans. Power Electron.*, vol.22, no.5, pp.1786–1796, Sept. 2007.

[7] G. N. Veda Prakash, M. K. Kazimierczuk, "Small-signal modeling of open-loop PWM Z-source converter by circuit-averaging technique," *IEEE Trans. Power Electron.*, vol.28, no.3, pp.1286–1296, March 2013.

[8] Y. Li, S. Jiang, J. G. Cintron-Rivera, F. Z. Peng, "Modeling and Control of Quasi-Z-Source Inverter for Distributed Generation Applications," *IEEE Trans. Ind. Electron.,* vol.60, no.4, pp.1532–1541, April 2013.

[9] Y. Liu, B. Ge, F. J. T. E. Ferreira, A. T. de Almeida, H. Abu-Rub, "Modeling and SVPWM control of quasi-Z-source inverter," in *Proc. 11th International Conference on Electrical Power Quality and Utilisation (EPQU)*, pp.1–7, 2011.

[10] Y. Liu, "Research on control method for quasi-Z-source cascade multilevel photovoltaic inverter," PhD dissertation, School of Electrical Engineering, Beijing Jiaotong University, Beijing, 2014. (In Chinese)

3

Current-Fed Z-Source Inverter

The current-fed Z-source inverter/quasi-Z-source inverter (CF-ZSI/qZSI) – unlike the traditional current source inverter (CSI) which can only boost voltage and allow unidirectional power flow, and unlike the voltage-fed Z-source inverter/quasi-Z-source inverter that needs to replace the diode with a bidirectional conducting, unidirectional blocking switch to achieve bidirectional power flow – can buck–boost voltage, and achieve bidirectional power flow with a single-stage structure. This chapter presents recent topology modifications to the CF-ZSI/qZSI, its operation principles, modulation, modeling and control, passive components design guidelines, and applications.

3.1 Introduction

There are two types of three-phase inverters: voltage source inverter (VSI) as in Figure 3.1(a), and a current source inverter (CSI), Figure 3.1(b). The VSI uses six switches with anti-parallel diodes and an electrolytic capacitor at its dc-link. It produces an ac output that is always lower or equal to half of its input voltage. Being its duality, the CSI uses six reverse-blocking switches and a large inductor at its dc-link to produce the demanded input current source. However, the CSI has the following limitations [1]:

- The ac output voltage has to be greater than the original dc-bus voltage that feeds the dc inductor. Therefore, the CSI is a boost inverter for dc-ac power conversion. For applications where a wide voltage range is desirable, an additional dc-dc buck converter is needed. The additional power conversion stage increases the system cost and lowers the efficiency.

Impedance Source Power Electronic Converters, First Edition. Yushan Liu, Haitham Abu-Rub, Baoming Ge, Frede Blaabjerg, Omar Ellabban, and Poh Chiang Loh.
© 2016 John Wiley & Sons, Ltd. Published 2016 by John Wiley & Sons, Ltd.

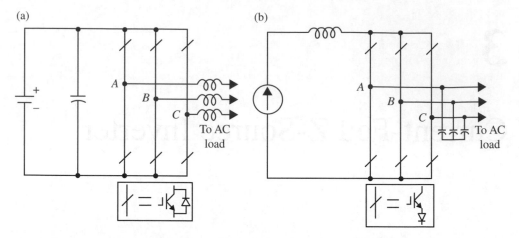

Figure 3.1   Conventional inverters: (a) voltage source inverter, (b) current source inverter.

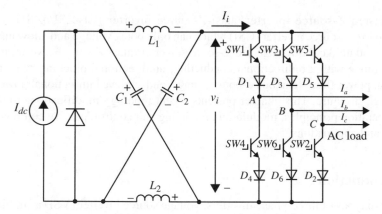

Figure 3.2   Current-fed Z-source inverter (CF-ZSI) configuration (*Source*: Fang 2004 [2]. Reproduced with permission of IEEE).

- At least one of the upper devices and one of the lower devices have to be gated ON and maintained ON at any time. Otherwise, an open circuit of the dc inductor would occur and destroy the devices. The overlap time for safe current commutation is needed in the CSI, which also causes waveform distortion and low frequency harmonic problems.

An alternative approach for getting the buck–boost capability for CSI is adding the Z-source network shown in Figure 3.2 [2]. Unlike a traditional CSI where an upper and a lower power switch are always turned ON to conduct the dc-link current, a current-fed Z-source inverter (CF-ZSI) can assume an additional open-circuit state by turning OFF all switches without breaking any inductive current. The open-circuit state of the CF-ZSI can be used to boost the output current (voltage-buck operation) without the unwanted interruption of source current. Functionally, the CF-ZSI is a robust single-stage buck–boost converter derived from the voltage-fed Z-source inverter (VF-ZSI) topology and can easily be controlled by adding unconventional open-circuit states to the inverter pulse-width-modulated state sequence.

3.2 Topology Modification

In recent years, various VF-ZSI topologies have been presented in a wide range of studies. There are many topology modifications either to overcome the drawbacks of the basic VF-ZSI topology or to increase its voltage gain. Most of these topology modifications can be adjusted for the CF-ZSI.

Inductors in the Z-source network introduce significant power losses due to their parasitic components, which make the CF-ZSI inefficient. To mitigate this negative effect, the input current source can alternatively be embedded in the Z-source network, as shown in Figure 3.3, where two dc sources are embedded in the Z-source network, Figure 3.3(a), resulting in the symmetrical operation of LC elements. Alternatively, one dc source can further be removed, as illustrated in Figure 3.3(b), which can still possess the unique benefits of embedded topology. Another possibility is inserting an input source into the right-hand side of the Z-source network as

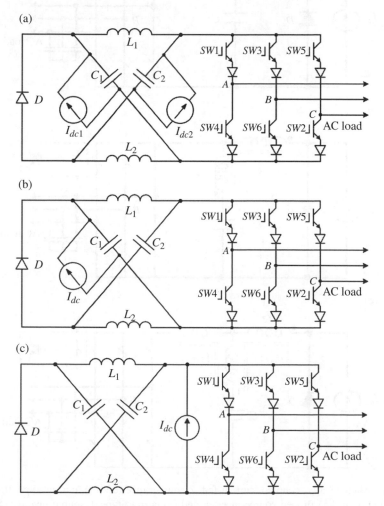

Figure 3.3 Embedded current-fed Z-source inverter with (a) two input current source, (b) single input current source and dc-link embedded CF-ZSI (*Source*: Li 2008 [7]. Reproduced with permission of IEEE).

shown in Figure 3.3(c). Furthermore, by comparison, the dc-link embedded Z-source CSI shows its unique advantage of having the smallest Z-source inductor current flow, which can help to reduce system size and increase efficiency [3].

Magnetically coupled techniques are more attractive since they can produce a high voltage/current gain with the support of coupled transformers or inductors with minimum component counts. CF-ZSI, Figure 3.4(a), can boost the current gain by increasing the turns ratio of a coupled

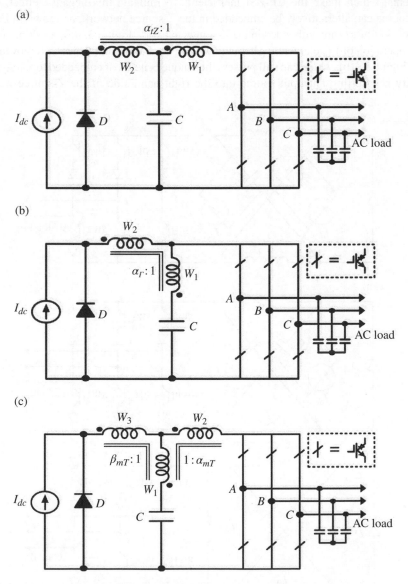

Figure 3.4 Magnetically coupled current-fed Z-source inverter: (a) current-type trans-Z-source inverter, (b) current-type flipped-source inverter, and (c) current-type T-source inverter (*Source*: Tran 2014 [5]. Reproduced with permission of IEEE).

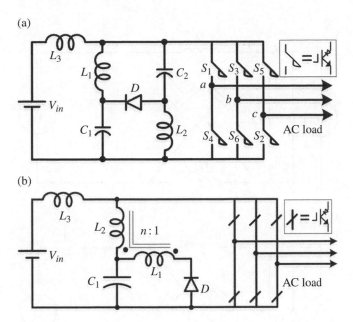

Figure 3.5 (a) Current-fed quasi-Z-source inverter and (b) current-fed trans-quasi-Z-source inverter (*Source*: Qian 2011 [6]. Reproduced with permission of IEEE).

transformer [4], whereas CF-flipped Γ-SI, Figure 3.4(b), can step up its gain by reducing the transformer turns ratio [4]. Consequently, for demanding high current gain applications, the trans-ZSI becomes bulky and costly, whereas the flipped Γ-source inverter is difficult to design since its gain is sensitive at small transformer turn ratios. The CF-T-source inverter shown in Figure 3.4(c) is proposed in [5] to overcome these constraints of the CF-flipped Γ-source and the CF-trans-Z-source inverter. Unlike the CF-trans-Z-source and CF-flipped Γ-source inverters, the current gain of the CF-T-source inverter can be raised not only by increasing the first turn ratio α_{mT} but also by decreasing the second turn ratio β_{mT}. This is a unique feature, not available in other impedance source inverters.

The qZSIs have been proposed to further improve on the traditional ZSIs. The qZSIs have several of their own merits, such as reduced passive component ratings, continuous input current configuration, and a common dc rail between the source and inverter. Extending this modification to the CF-ZSI, the CF-qZSI has been proposed, Figure 3.5(a). Compared to VF-qZSI, the CF-qZSI can regenerate power without changing the front switch into an active switch, thus reducing the switch count. Also, the CF-qZSI can be further modified to a current-fed trans-quasi-Z-source inverter, Figure 3.5(b) [6].

3.3 Operational Principles

3.3.1 Current-Fed Z-Source Inverter

Being a counterpart of the VF-ZSI, the CF-ZSI assumes an open-circuit state, which is forbidden in the traditional CSI, to perform its unique current boost operation. The Z-source network with two inductors (L_1 and L_2) and two capacitors (C_1 and C_2) can also be used with

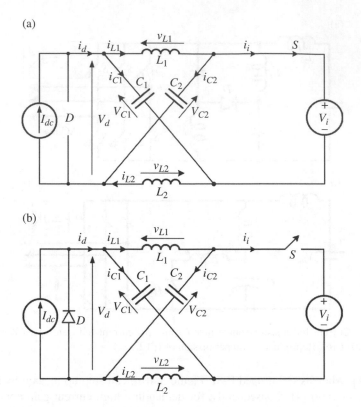

Figure 3.6 Equivalent circuit of the CF-ZSI CSI: (a) non-open-circuit states and (b) open-circuit states (*Source*: Li 2008 [7]. Reproduced with permission of IEEE).

the traditional CSI, as shown in Figure 3.2. Since the right inverter bridge is a traditional CSI, it can assume all active (finite output current) and null (zero output current) switching states of the traditional CSI. However, unlike the traditional CSI, the CF-ZSI can now be open-circuited by turning on either only one or no switch without breaking any dc inductive current. The impact of this open-circuit state on the inverter can be analyzed by considering its open-circuit and non-open-circuit equivalent circuits, Figure 3.6 [7].

To carefully analyze its operational principle, first assume that the inverter operates in its non-open-circuit states for an interval T_1 during the switching cycle T, an upper and a lower switch of the inverter conducting to connect the line voltage to the inverter side of the Z-source network. This voltage is represented by a voltage source V_i, Figure 3.6(a), whose value is finite if the upper and lower conducting switches are from different phase legs (active state, e.g. *SW*1 and *SW*2). However, its value will be zero if both switches are from the same phase leg (null state, e.g. *SW*1 and *SW*4). During the same time, diode D at the source side reverse-biases, hence permitting all source current to flow to the impedance network for charging L_1 and L_2, while C_1 and C_2 release their stored energy to boost the inverter dc-link current i_i. The circuit equations for the Z-source network can thus be rewritten as (assuming that $L_1 = L_2 = L$ and $C_1 = C_2 = C$) [7]:

$$i_d = I_{dc}; \quad i_c = I_{dc} - I_L; \quad i_i = I_L - i_c = 2I_L - I_{dc} \tag{3.1}$$

More specifically, for the equivalent circuit shown in Figure 3.6(b) during the time interval T_0, the inverter side of the Z-source network is opened by turning off all switches. At the same time, diode D turns on to conduct the excess current of i_d ($i_d > I_{dc}$), and to allow magnetic energy stored in L_1 and L_2 to be transferred to electrostatic energy stored in C_1 and C_2; the circuit equations can be written as:

$$I_{L1} = I_{L2} = I_L = i_{C1} = i_{C2} = i_C; i_d = I_L + i_C = 2I_L; i_i = 0 \tag{3.2}$$

Averaging the current i_C through any of the two Z-source capacitors over a switching period $(T = T_0 + T_1)$ then results in the following expressions for the peak dc current $\hat{i}_i$, peak ac current $\hat{i}_x$ ($x = a$, b, or c), and output-to-input voltage ratio:

$$\hat{i}_i = 2I_L - I_{dc} = \frac{1}{1 - 2T_0/T} I_{dc} = BI_{dc} \tag{3.3}$$

$$\hat{i}_x = M\hat{i}_i = B\{MI_{dc}\} \tag{3.4}$$

$$V_d I_{dc} = \frac{3}{2} \hat{v}_x \hat{i}_x \cos\delta \Rightarrow \frac{\hat{v}_x}{V_d} = \frac{2}{3BM\cos\delta} \tag{3.5}$$

where the term in { } gives the ac output of a traditional CSI, M is its modulation ratio, B is the boost factor resulting from the open circuit zero state that can be controlled by the duty cycle of the open-circuit zero state over the non-open circuit states of the inverter PWM, and δ is the angle between the ac voltage and current. Obviously, (3.3)–(3.5) show that the ac output current and voltage of the CF-ZSI can be boosted and bucked respectively by increasing B (always ≥ 1). This capability is not provided by the traditional CSI.

3.3.2 Current-Fed Quasi-Z-Source Inverter

Figure 3.7(a) and (b) show the CF-qZSI topologies with discontinuous and continuous input current, respectively. The CF-qZSI shown in Figure 3.7(a), when compared to the CF-ZSI, features reduced current in inductor L_2, as well as reduced passive component count, while the CF-qZSI shown in Figure 3.7(b) features lower current in inductors L_1 and L_2. Due to the input inductor, L_3, the CF-qZSI does not require input capacitance. All voltage and current-fed qZSI topologies feature a common dc rail between the source and inverter, unlike the traditional ZSI circuits. Furthermore, these qZSI circuits have no disadvantages when compared to the traditional ZSI topologies. These qZSI topologies therefore can be used in any application in which the ZSI would traditionally be used [8–13]. Unlike the traditional CSI that can only boost the voltage and allow unidirectional power flow, also differently from the VF-ZSI/qZSI that needs to replace the diode with a bidirectional conducting, unidirectional blocking switch to achieve bidirectional power flow, the CF-qZSI can buck–boost voltage, and achieve bidirectional power flow with a single-stage structure [10].

Figure 3.8 shows the equivalent circuits of the CF-qZSI when viewed from the dc-link. The inverter has the following different operation states: In active states, as shown in Figure 3.8(a), the inverter bridge becomes an equivalent dc voltage source (denoted V_{out}) when in one of the

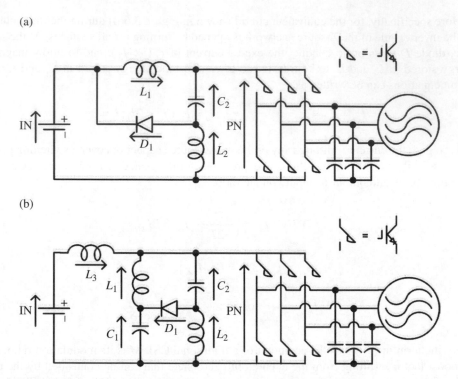

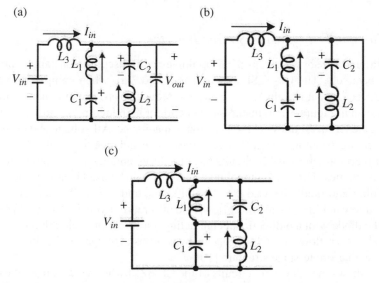

Figure 3.7   Topology of current-fed qZSI with: (a) discontinuous input current, (b) continuous input current (*Source*: Yang 2009 [9]. Reproduced with permission of IEEE).

Figure 3.8   Equivalent circuits of current-fed qZSI: (a) active state, (b) short zero state, and (c) open zero state (*Source*: Yang 2009 [9]. Reproduced with permission of IEEE).

six active states, and the diode D is off. In shoot-through states, the inverter bridge is equivalent to a short-circuit, the dc-link voltage is zero, the diode D is off, and the output ac voltage is blocked by the switches, as shown in Figure 3.8(b). In traditional zero states, the inverter bridge is equivalent to an open circuit, and the currents of the inductors flow through the diode D to charge the capacitors, as shown in Figure 3.8(c).

Considering that the average voltage of the inductors over one switching period should be zero in steady state, we have [10]:

$$V_{C1} = V_{C2} = V_C = V_{in}, v_{L1} = v_{L2} = v_{L3} = v_L, V_{out} = \frac{D_A + D_{sh} + D_{op}}{D_A} V_{in} = \frac{1 - 2D_{op}}{D_A} V_{in}, \quad (3.6)$$

where D_A, D_{sh}, and D_{op} are the duty cycles of active states, shoot-through states, and traditional zero states, respectively. In the same way, considering that the average current of the capacitors over one switching period should be zero in steady state:

$$I_{L1} = I_{L2} = \frac{D_{op}}{D_A + D_{sh} - D_Z} I_{in} = \frac{D_{op}}{1 - 2D_Z} I_{in}. \quad (3.7)$$

The dc voltage gain V_{out}/V_{in} expressed in (3.6) can be varied by adjusting two independent degrees of control freedom, given as D_A and D_{op}. Figure 3.9(a) graphically illustrates the V_{out}/V_{in} versus D_A, where in mode 1 there are no traditional zero states, and in mode 2, entire short-through states are turned to zero states. In mode 1, $D_Z=0$, the CF-qZSI operates like the conventional CSI. The gain of mode 1 is the potential maximum gain that the CF-qZSI can obtain with a given active duty ratio D_A. While in mode 2, $D_Z=1-D_A$, the voltage gain is the potential minimum gain that the CF-qZSI can obtain with a given active duty ratio D_A. When parts of the short-through states turn into the zero states, the voltage gain will be located between the gain of Mode 1 and the gain of Mode 2, as given in Figure 3.9(a). Therefore, the desired voltage gain can be obtained by tuning the two degrees of control freedom, D_Z and D_A. Figure 3.9(b) shows the relationship between I_{L1}/I_{in} and D_Z as given in (3.7). Figure 3.9(a)

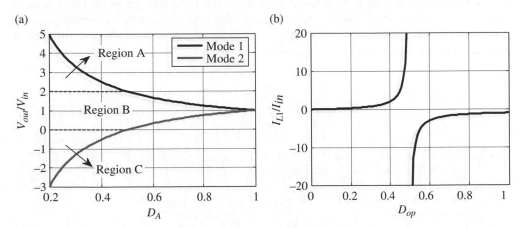

Figure 3.9 Characteristics of CF-qZSI: (a) dc voltage gain V_{out}/V_{in} versus D_A, (b) I_{L1}/I_{in} versus D_{op} (*Source*: Yang 2009 [9]. Reproduced with permission of IEEE).

reveals that a desired dc voltage gain V_{out}/V_{in} corresponds to different combinations of D_Z and D_A. However, from (3.6), it would better to operate the CF-qZSI in Mode 1 or Mode 2 to keep minimum current stress on the switches.

3.4 Modulation

Many PWM control methods have been developed and used for the traditional three-phase CSI. The traditional CSI has six active vectors (or switching states) when the dc current is impressed across the load, and three zero vectors when the input terminals are shorted through both switches conducting simultaneously in the same phase leg. These nine switching states and their combinations have spawned many PWM control schemes [14–16].

The CF-ZSI/CF-qZSI has an additional zero vector: an open-circuit switching state, which is forbidden in traditional CSI. For traditional CSI, both switches of any phase leg can never be gated off at the same time because an open circuit would occur and destroy the inverter. The CF-ZSI advantageously utilizes the open circuit states to boost the dc bus current by gating off both upper and lower switches of a phase leg. Therefore the CF-ZSI/CF-qZSI can boost the current and produce the desired output current which is greater than the available dc bus current. In addition, the reliability of the inverter is greatly improved because the open circuit can no longer destroy the circuit. Thus it provides a low-cost, reliable, and high efficiency single-stage structure for buck and boost power conversion.

Similarly to a ZSI, by using a sinusoidal voltage reference for the carrier-based PWM method, the CSI can use a sinusoidal current reference instead. Therefore, the three carrier-based control methods for the VF-ZSI – simple, constant, and maximum boost – can be used for the CF-ZSI/CF-qZSI, except that the voltage reference signals are replaced by current reference signals, as indicated in Figure 3.10. Similar to the voltage boost control methods of the VF-ZSI, three different current reference signals are used for the current boost control. In order to generate an open zero state, the simple boost control uses two straight lines as reference, the maximum boost control uses the three-phase current reference envelope as reference, the constant boost control utilizes two sinusoidal envelope curves as reference. These three current references for current boost control use the same formula as the three voltage boost controls. Hence, the D_{op} relationship with M of these three current boost control methods is the same as the shoot-through duty cycle relationship with M in the voltage boost control methods, which is summarized in the second column of Table 3.1.

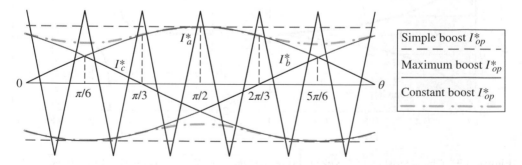

Figure 3.10   Current-fed Z-source/quasi-Z-source inverter current boost control methods.

Table 3.1 Comparison of normalized output voltage and normalized inductor currents with three different current boost control strategies for the CF-ZSI/CF-qZSI

	D_{op} values due to M	$\dfrac{\hat{v}_{ll}}{V_{in}}$ versus M	$\dfrac{\hat{v}_{ll}}{V_{in}}$ versus D_{op}	$\dfrac{I_{in}}{I_{l\_rms}}$ versus D_{op}	$\dfrac{I_{L1}}{I_{l\_rms}}$ versus D_{op}
Simple boost	$D_{op}=1-M$	$\dfrac{4}{3\cos\varphi}\left(\dfrac{2M-1}{M}\right)$	$\dfrac{4}{3\cos\varphi}\left(\dfrac{1-2D_{op}}{1-D_{op}}\right)$	$\dfrac{2\sqrt{6}}{3}\left(\dfrac{1-2D_{op}}{1-D_{op}}\right)$	$\dfrac{2\sqrt{6}}{3}\left(\dfrac{D_{op}}{1-D_{op}}\right)$
Constant boost	$D_{op}=1-\dfrac{\sqrt{3}M}{2}$	$\dfrac{4}{3\cos\varphi}\left(\dfrac{\sqrt{3}M-1}{M}\right)$	$\dfrac{2\sqrt{3}}{3\cos\varphi}\left(\dfrac{1-2D_{op}}{1-D_{op}}\right)$	$\sqrt{2}\left(\dfrac{1-2D_{op}}{1-D_{op}}\right)$	$\sqrt{2}\left(\dfrac{D_{op}}{1-D_{op}}\right)$
Maximum boost	$D_{op}=1-\dfrac{3\sqrt{3}M}{2\pi}$	$\dfrac{4}{3\pi\cos\varphi}\left(\dfrac{3\sqrt{3}M-\pi}{M}\right)$	$\dfrac{2\sqrt{3}}{\pi\cos\varphi}\left(\dfrac{1-2D_{op}}{1-D_{op}}\right)$	$\dfrac{3\sqrt{2}}{\pi}\left(\dfrac{1-2D_{op}}{1-D_{op}}\right)$	$\dfrac{3\sqrt{2}}{\pi}\left(\dfrac{D_{op}}{1-D_{op}}\right)$

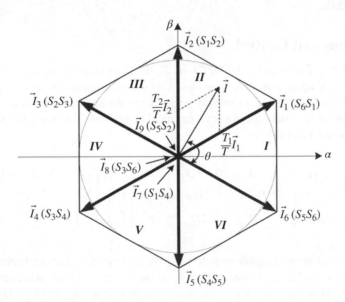

Figure 3.11 Space vectors modulation of the traditional CSI.

The most simple and effective modulation method for the CSIs is space vector PWM (SVPWM), which explicitly selects active and zero space vectors and places them within a carrier period. Figure 3.11 shows the space vectors of the CSI, where the nine current vectors $\vec{I}_1,\ldots,\vec{I}_9$ correspond to the switch states $\vec{I}_1=[S_1,S_6],\ldots,\vec{I}_9=[S_5,S_2]$, respectively. In one sampling interval T, the output current vector $\vec{I}$ is commonly split into the two nearest adjacent current vectors ($\vec{I}_i$ and $\vec{I}_{i+1}$, $i=1, \ldots, 6$) and one of the three short-zero vectors ($\vec{I}_7,\vec{I}_8$, or $\vec{I}_9$). Besides the aforementioned nine possible states in the conventional CSI, the CF-ZSI has two extra open-zero states by turning some of the short-zero states to open-zero states, as shown in Figure 3.12 [13]. These two open-zero states can be assumed by turning off all the upper switches (S_1, S_3, and S_5) or all the lower switches (S_4, S_6, and S_2).

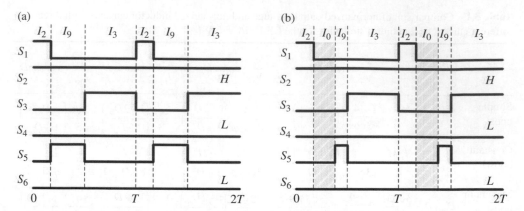

Figure 3.12 SVPWM signals for (a) conventional current source inverter and (b) current-fed Z-source inverter/current-fed quasi Z-source inverter in sextant III (*Source*: Lei 2014 [13]. Reproduced with permission of IEEE).

3.5 Modeling and Control

To use the state-space average model for CF-qZSI, the three states in Figure 3.8 – active state, short-zero state, and open-zero state – are classified into non-open-zero state, representing the former two, and open-zero state. In non-open-zero state, the inverter is considered as a voltage source, and in open-zero state it is an open circuit. The state average model of this equivalent circuit state is written as [18]

$$
\begin{bmatrix} L & 0 & 0 \\ 0 & C & 0 \\ 0 & 0 & L \end{bmatrix} \cdot \begin{bmatrix} \dot{I}_L(t) \\ \dot{V}_C(t) \\ \dot{I}_{dc}(t) \end{bmatrix} = \begin{bmatrix} 0 & 1-2D_{op} & 0 \\ 2D_{op}-1 & 0 & D_{op} \\ 0 & -2D_{op} & -r \end{bmatrix} \cdot \begin{bmatrix} I_L(t) \\ V_C(t) \\ I_{dc}(t) \end{bmatrix} + \begin{bmatrix} -V_{out}\left(1-D_{op}\right) \\ 0 \\ V_{in}-V_{out}\left(1-D_{op}\right) \end{bmatrix} \tag{3.8}
$$

There are three differential equations in terms of the Z-network inductor current I_L, Z-network capacitor voltage V_c, and input current I_{dc}. The D_{op} is considered as a constant in every steady state. The average output voltage in active state can be expressed as: $V_{out}=D_a V_{ca}+D_b V_{cb}+D_c V_{cc}$. In order to calculate V_{out}, assume the output capacitor voltage to be a sinusoidal function with amplitude V_m:

$$
\begin{bmatrix} v_{ca}(t) \\ v_{cb}(t) \\ v_{cc}(t) \end{bmatrix} = \begin{bmatrix} V_m \sin(\omega t) \\ V_m \sin\left(\omega t - \dfrac{2\pi}{3}\right) \\ V_m \sin\left(\omega t + \dfrac{2\pi}{3}\right) \end{bmatrix} \tag{3.9}
$$

The duty cycles can also be considered as sinusoidal with amplitude M (modulation index) and angle β with respect to the voltage:

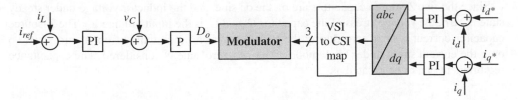

Figure 3.13 Schematic diagram of grid-connected CF-ZSI.

$$
\begin{bmatrix} D_a(t) \\ D_b(t) \\ D_c(t) \end{bmatrix} = \begin{bmatrix} M\sin(\omega t + \beta) \\ M\sin\left(\omega t + \beta - \dfrac{2\pi}{3}\right) \\ M\sin\left(\omega t + \beta + \dfrac{2\pi}{3}\right) \end{bmatrix}
\tag{3.10}
$$

Through calculation it can be found that the average output voltage in active state is a constant value: $V_{out} = 3M'V_m/2$, where $M' = M$ for motoring mode and $M' = -M$ for regenerating mode. So the final simplified equation is

$$
\begin{bmatrix} \dot{I}_L(t) \\ \dot{V}_C(t) \\ \dot{I}_{dc}(t) \end{bmatrix} = \begin{bmatrix} 0 & \dfrac{1-2D_{op}}{L} & 0 \\ \dfrac{2D_{op}-1}{C} & 0 & \dfrac{D_{op}}{C} \\ 0 & \dfrac{-2D_{op}}{L} & \dfrac{-r}{L} \end{bmatrix} \cdot \begin{bmatrix} I_L(t) \\ V_C(t) \\ I_{dc}(t) \end{bmatrix} + \begin{bmatrix} -\dfrac{3M'V_m}{2L}\left(1-D_{op}\right) \\ 0 \\ \dfrac{2V_{in} - 3M'V_m\left(1-D_{op}\right)}{2L} \end{bmatrix}
\tag{3.11}
$$

Thus the state-space average mode of this circuit in motoring and regenerating mode is shown in Equation (3.11), with $M' = M$ in motoring mode and $M' = -M$ in regenerating mode.

Figure 3.13 is an example of a design for a closed-loop controller for a grid-connected CF-ZSI. There is an inherent feature of CSI, where oscillations at the resonant frequency of the output LC filter can occur. In order to attenuate such oscillations, a closed-loop charge controller is implemented in the synchronous reference frame. In grid-connected mode, the inverter is synchronized with the grid frequency using a phased-locked loop (PLL). The inverter charge vector Q is used to force the inverter current to follow its reference, while actively damping oscillations associated with the output LC filter [19].

3.6 Passive Components Design Guidelines

The current stress and voltage stress of the Z-source network passive components and input inductor also need to be analyzed in order to properly design the capacitor and inductor in the circuit [12].

Since the two Z-source capacitors are on the dc side, and the inductor voltage under steady state is zero, the voltage stress of the capacitor C_1 and C_2 is the input voltage V_{in}. The Z-source capacitor current is near zero during the boost operation mode without open-zero state, so only the current stress under buck mode with open-zero state is considered. The capacitance requirement can be calculated as

$$C_1 = C_2 = \frac{(1-D_{op})D_{op}}{M} \frac{2\sqrt{2}I_{lrms}}{\sqrt{3}\Delta V_C f_s} \tag{3.12}$$

where $D_{op}, I_{lrms}, M, \Delta V_C$, and f_s are the open-zero state duty ratio, the output line RMS current, the modulation index, the capacitor voltage ripple, and the switching frequency, respectively.

Also, the inductor values can be calculated based on the desired current ripples as

$$L_1 = L_2 = \frac{V_{in}D_{op}}{\Delta I_L f_s}, L_3 = \frac{V_{in}D_{op}}{\Delta I_{in} f_s} \tag{3.13}$$

where ΔI_L and ΔI_{in} are the desired inductor and input current ripples, respectively.

3.7   Discontinuous Operation Modes

The CF-ZSI/CF-qZSI has three basic operation modes on the assumption that the capacitor voltage is kept constant and equal to the input voltage. However, when the capacitors are small or the load power factor is low, the circuit can enter new modes, which are called discontinuous modes. The condition for this to happen is that the capacitor voltage becomes less than half of the output line-to-line peak voltage [20].

Mode 1–3 (continuous modes): Mode 1: active state, the inverter bridge is in an active state and the dc-link voltage V_{PN} is equal to the equivalent output voltage V_{eq}; the diode is off if the equivalent voltage satisfies $V_{eq} < 2V_{in}$. Mode 2: short-zero state, the inverter bridge is equivalent to a short-circuit by turning on the upper and lower switches in the same phase leg or in the same two phase legs, or three phase legs together. The dc-link voltage is zero, so the diode is off. Mode 3: In open-zero state, the inverter bridge is equivalent to an open circuit by turning off all the upper switches or turning off all the lower switches. The diode is turned on, so the dc-link voltage is equal to the sum of the two capacitor voltages.

Mode 4 (discontinuous mode I): In open-zero state, the diode is on and the inverter bridge is off. So the capacitor is being charged. But in the other two states, the capacitor keeps discharging because of the unchanged inductor current. At the end of mode 2, if the capacitor voltage decreases to be smaller than half of the output line-to-line peak voltage, at the moment that the inverter is switched to active state again, the diode in the quasi-Z network will be turned on because the voltage drop on the diode is positive. In this case, the diode will be reverse-biased; the inverter is equivalent to an open circuit. The capacitor will be charged again in this new open zero state. This state is described in Figure 3.14(d); in mode 4, the capacitor voltage satisfies $V_C < V_{l\text{-}lpeak}/2$.

Mode 5 (discontinuous mode II): At the end of mode 4, when the capacitor voltage increases to be equal to half of the output peak line-to-line voltage and the inverter bridge is still switched in the active state, the diode in the qZ-network will be reverse-biased. So the capacitor voltage

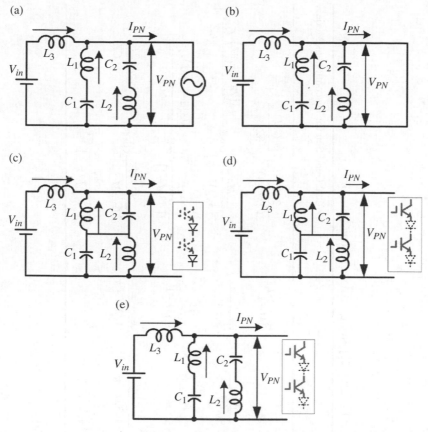

Figure 3.14 Possible operation modes of the CF-qZSI shown in: (a) mode 1, (b) mode 2, (c) mode 3, (d) mode 4, and (e) mode 5 (*Source*: Lei 2011 [18]. Reproduced with permission of IEEE).

will stay at $V_{l\text{-}lpeak}/2$ until the end of the active state. However, the voltage drop on the reverse-blocking IGBT is still smaller than or equal to zero, and the inverter bridge is still in the open circuit state. The equivalent circuit of mode 5 is shown in Figure 3.14(e). This mode will also happen when the capacitor voltage decreases to half of the $V_{l\text{-}lpeak}$ in mode 1. The switch will still be switched in the active state, but the circuit will enter mode 5 until the end of active state. In mode 5, the capacitor voltage and inductor current satisfies

$$V_C = \frac{V_{l\text{-}lpeak}}{2}, \quad I_L = -\frac{I_{in}}{2} \tag{3.14}$$

The discontinuous mode is defined as a new mode when the capacitor voltage satisfies $V_C \leq V_{l\text{-}lpeak}/2$. So modes 4 and 5 are included as discontinuous modes. When the circuit starts to boost the current by changing some of the short-zero state into the open-zero state, the discontinuous mode can happen, depending on the load condition and capacitance value. Different control methods will yield different circuit characteristics due to different sequences and combinations of the operation modes, as shown in Table 3.2.

Table 3.2 Characteristics of different control strategies (*Source*: Lei 2011 [20]. Reproduced with permission of IEEE).

Control Strategy	D_{op}	B	G	I_s	Critical Condition for Discontinuous Mode
Constant Boost	$1-M$	$\dfrac{1}{2M-1}$	$\dfrac{M}{2M-1}$	$\dfrac{1}{2M-1}I_{in}$	$\dfrac{3M}{2(2M-1)}-\dfrac{2(1-M)T_s}{3ZC}\leq 1$
Maximum boost	$1-\dfrac{3\sqrt{3}M}{2\pi}$	$\dfrac{\pi}{3\sqrt{3}M-\pi}$	$\dfrac{\pi M}{3\sqrt{3}M-\pi}$	$\dfrac{\pi}{3\sqrt{3}M-\pi}I_{in}$	$\dfrac{3M\pi}{2(3\sqrt{3}M-\pi)}\cos\phi-\dfrac{\sqrt{3}(2\pi-3\sqrt{M})}{2\pi^2}\dfrac{T_s}{CZ}\leq 1$
Maximum constant boost	$1-\dfrac{\sqrt{3}M}{2}$	$\dfrac{1}{\sqrt{3}M-1}$	$\dfrac{M}{\sqrt{3}M-1}$	$\dfrac{I_{in}}{\sqrt{3}M-1}$	$\dfrac{3M}{2(\sqrt{3}M-1)}\cos\phi-\dfrac{\sqrt{3}(2-\sqrt{3}M)T_s}{6CZ}\leq 1$
SVPWM (Buck mode)	$1-\dfrac{3\sqrt{3}}{2\pi}M$	$\dfrac{\pi}{3\sqrt{3}M-\pi}$	$\dfrac{\pi M}{3\sqrt{3}M-\pi}$	$\dfrac{\pi I_{in}}{3\sqrt{3}M-\pi}$	$\dfrac{3}{2}\dfrac{\sqrt{3}\pi M}{6\sqrt{3}M-2\pi}\cos\phi-\dfrac{2\pi-3\sqrt{3}M}{2\pi^2}\dfrac{2T_s}{CZ}<1$
SVPWM (Boost mode)	0	1	$\dfrac{2\pi}{3\sqrt{3}M}$	I_{in}	$\dfrac{3}{2}\dfrac{\cos\phi}{3\sqrt{3}M/\pi-1}<1$

3.8 Current-Fed Z-Source Inverter/Current-Fed Quasi-Z-Source Inverter Applications

The current-fed trans-ZSIs are suitable for plug-in hybrid electric vehicles (PHEVs). First, they have sinusoidal voltage and current to directly couple the motor/generator and the battery, as illustrated in Figure 3.15. The current-fed trans-ZSIs accomplish bidirectional power flow and voltage buck–boost using the fewest switching devices (only one extra diode and six IGBTs). Also, they can have direct connection to the grid. This makes it possible to save an additional on-board charger for single-phase 120 V and three-phase 220 V power [6].

Also, the CF-qZSI can easily be applied to hybrid electric vehicles (HEVs) and general-purpose variable-speed motor drives. Applying the CF-qZSI in HEVs, both the motor drive and the control of the state of charge of the battery can be implemented in a single-stage configuration, which is less complex, more reliable, and more cost-effective when compared to a conventional two-stage configuration (i.e. a bidirectional dc-dc converter combined with a VSI). Figure 3.16 shows the parallel and series hybrid system with CF-qZSI used [17].

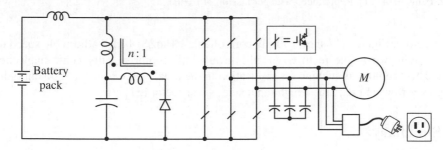

Figure 3.15   The current-fed trans-qZSI for plug-in hybrid electric vehicle application.

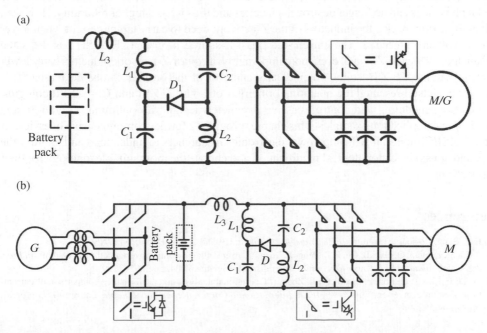

Figure 3.16   Current-fed qZSI for HEV application: (a) parallel HEV and (b) series HEV (*Source*: Lei 2009 [17]. Reproduced with permission of IEEE).

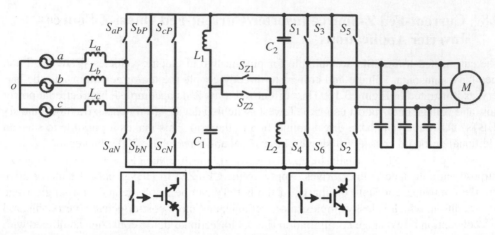

Figure 3.17 A bidirectional fully controlled current-fed quasi-Z-source inverter-based ASD (*Source*: Fang 2014 [21]. Reproduced with permission of IEEE).

Furthermore, Figure 3.17 shows a fully controllable CF-qZSI-based adjustable speed drive (ASD) system, where the rectifier and the inverter all comprise fully controllable power devices. This topology is suitable for application in the area, which needs to transmit bidirectional power flow and requires a high input side power factor [21].

3.9   Summary

The current-fed Z-source inverter (CF-ZSI), which is derived from primary Z-source inverter topology, has many attractive characteristics, such as: buck–boost function, open-circuit state of the phase legs can no longer destroy the inverter and thus it has a higher reliability. However, it has some drawbacks: the inductors of the Z-network need to carry larger current when lower output voltage is needed. The current-fed quasi-Z-source inverter (CF-qZSI) has the same advantages as the CF-ZSI and, even more importantly, its quasi-Z-network inductors carry lower current than the CF-ZSI, so the volume, size, and cost of the inverter can be decreased. This chapter has also presented the operating principles of both CF-ZSI and CF-qZSI topologies, modulation, modeling and control, passive parameter design, discontinuous operation, and finally some application examples. This chapter provides a further comprehensive overview of the CF-ZSI/CF-qZSI. It will help power electronics researchers to understand and identify the pros and cons of each topological modification and choose the most suitable topology for their applications.

References

[1] F. Z. Peng, "Z-source inverter," *IEEE Trans. Ind. Appl.*, vol.39, no.2, pp.504–510, Mar/Apr. 2003.

[2] X. Fang, Z. Qian, Q. Gao, B. Gu, F. Z. Peng, X. Yuan, "Current mode Z-source inverter-fed ASD system," in *Proc. 2004 35th Annual IEEE Power Electronics Specialists Conference*, 2004, pp.2805–2809.

[3] D. Li, F. Gao, P. C. Loh, F. Blaabjerg, L. Zhang, "Operational analysis and performance comparison of current-type Z-source inverters," in *Proc. IEEE International Conference on Sustainable Energy Technologies*, pp.798–803, 2008.

[4] P. C. Loh, D. Li, F. Blaabjerg, "Current-type flipped-Γ-source inverters," in *Proc. 7th International in Power Electronics and Motion Control Conference (IPEMC)*, pp.594–598, 2–5 June 2012.

[5] Q. Tran, K. Low, A. Ho, T. Chun, "A new current-type magnetically coupled T-source inverter," in *Proc. IEEE International Conference on Industrial Technology (ICIT)*, pp.318–323, 26 Feb.–1 March 2014.

[6] W. Qian, F. Z. Peng, H. Cha, "Trans-Z-source inverters," *IEEE Trans. Power Electron.*, vol.26, no.12, pp.3453–3463, Dec. 2011.

[7] D. Li, F. Gao, P. C. Loh, F. Blaabjerg, L. Zhang, "Operational analysis and performance comparison of current-type Z-source inverters," in *Proc. IEEE International Conference on Sustainable Energy Technologies (ICSET)*, pp.798–803, 24–27 Nov. 2008.

[8] X. Fang, Z. Chen, X. Wang, "Current-Fed Quasi-Z-Source Inverter-Based Adjustable Speed Drive System with Bidirectional Power Flow," *IEEE Trans. Appl. Supercond.*, vol.24, no.5, pp.1–6, Oct. 2014.

[9] S. Yang, F. Z. Peng, Q. Lei, R. Inoshita, Z. Qian, "Current-fed quasi-Z-source inverter with voltage buck-boost and regeneration capability," in *Proc. IEEE Energy Conversion Congress and Exposition (ECCE)*, pp.3675–3682, 20–24 Sept. 2009.

[10] S. Yang, Q. Lei, F. Peng, R. Inoshita, Z. Qian, "Current-fed quasi-Z-source inverter with coupled inductors," in *Proc. IEEE Energy Conversion Congress and Exposition (ECCE)*, pp.3683–3689, 20–24 Sept. 2009.

[11] Q, Lei, F. Peng, L. He, S. Yang, "Power loss analysis of current-fed quasi-Z-Source inverter," in *Proc. IEEE Energy Conversion Congress and Exposition (ECCE)*, pp.2883–2887, 12–16 Sept. 2010.

[12] D. Cao, Q. Lei, F. Z. Peng, "Development of high efficiency current-fed quasi-Z-source inverter for HEV motor drive," in *Proc. Twenty-Eighth Annual IEEE Applied Power Electronics Conference and Exposition (APEC)*, pp.157–164, 17–21 March 2013.

[13] Q. Lei, D. Cao, F. Z. Peng, "Novel loss and harmonic minimized vector modulation for a current-fed quasi-Z-source inverter in HEV motor drive application," *IEEE Trans. Power Electron.*, vol.29, no.3, pp.1344–1357, March 2014.

[14] X. Fang, M. Zhu, Z. Chen, J. Liu, X. Zhao, "Current-fed Z-source inverter modulation," in *Proc. International Conference on Electrical Machines and Systems (ICEMS)*, 2011, pp.1–6, 20–23 Aug. 2011.

[15] X. Fang, "Maximum boost control of the current-fed Z-source inverter," in *Proc. IEEE International Conference on Industrial Technology*, pp.1–6, 21–24 April 2008.

[16] S. Rezapoor, A. Abedini, S. M. T. Bathaee, "An improved MPWM for current-fed Z-source inverter with high current gain and low current stress," in *Proc. 3rd Power Electronics and Drive Systems Technology Conference (PEDSTC)*, pp.18–24, 15–16 Feb. 2012.

[17] Q. Lei, S. Yang, F. Z. Peng, R. Inoshita, "Application of current-fed quasi-Z-source inverter for traction drive of hybrid electric vehicles," in *Proc. IEEE Vehicle Power and Propulsion Conference (VPPC)*, pp.754–760, 7–10 Sept. 2009.

[18] Q. Lei, F. Z. Peng, B. Ge, "Transient modeling of current-fed quasi-Z-source inverter," in *Proc. IEEE Energy Conversion Congress and Exposition (ECCE)*, pp.2283–2287, 17–22 Sept. 2011.

[19] D. M. Vilathgamuwa, P. C. Loh, M. N. Uddin, "Transient modeling and control of Z-source current type inverter," in *Proc. IEEE Industry Applications Conference*, pp.1823–1830, 23–27 Sept. 2007.

[20] Q. Lei, F. Z. Peng, S. Yang, "Discontinuous operation modes of current-fed quasi-Z-source inverter," in *Proc. Twenty-Sixth Annual IEEE Applied Power Electronics Conference and Exposition (APEC)*, pp.437–441, 2011.

[21] X. Fang, Z. Chen, X. Wang, "Current-fed quasi-Z-source inverter-based adjustable speed drive system with bidirectional power flow," *IEEE Trans. Appl. Supercond.*, vol.24, no.5, pp.1–6, Oct. 2014.

4

Modulation Methods and Comparison

Pulse-width modulation (PWM) control methods for ZSI/qZSI have been proposed to achieve a wider modulation range, lower voltage stress on the switches, and simpler real-time implementation. They can be summarized as carrier-based sinewave pulse-width modulation (SPWM), space vector modulation (SVM), and pulse-width amplitude modulation (PAM). This chapter details the continuous algorithms of SPWM and SVM and their comparison, and PAM will be addressed in later chapters. The three-phase two-level voltage-fed ZSI/qZSI from Chapter 2 is considered for illustration, and all addressed modulation methods are applicable to most three-phase ZS/qZS derivations.

4.1 Sinewave Pulse-Width Modulations

There are three classic carrier-based PWM methods for the qZSI: simple boost control (SBC) [1], maximum boost control (MBC) [2], and maximum constant boost control (MCBC) [3]. They generate the shoot-through states by applying different shoot-through references to the traditional carrier-based SPWM.

Sketch maps of SBC and MBC are shown in Figure 4.1. When the carrier is greater than the upper shoot-through reference v_p and the upper envelope of three-phase modulating waves (v_a^*, v_b^*, v_c^*), or lower than the lower shoot-through reference v_n and the lower envelope of modulating waves, the three bridge legs conduct together, i.e. a shoot-through state is produced. In between, the switches behave in the same way as traditional carrier-based SPWM, whereas the different shoot-through references result in different boost capability, voltage gain, and stress on the inverter power switches.

Impedance Source Power Electronic Converters, First Edition. Yushan Liu, Haitham Abu-Rub, Baoming Ge, Frede Blaabjerg, Omar Ellabban, and Poh Chiang Loh.
© 2016 John Wiley & Sons, Ltd. Published 2016 by John Wiley & Sons, Ltd.

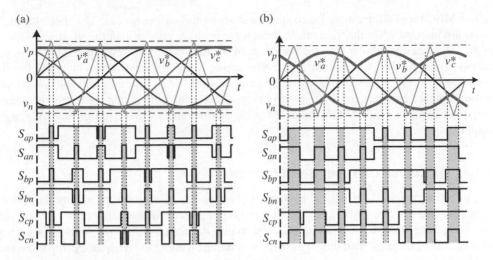

Figure 4.1 Modulation methods of (a) simple boost control and (b) maximum boost control for the three-phase two-level ZSI/qZSI.

4.1.1 Simple Boost Control

The shoot-through reference of SBC is a straight line equal to or greater than the upper envelope of the modulating waves, or equal to or lower than the lower envelope of the modulating waves, as Figure 4.1(a) shows. In this way, the maximum shoot-through duty ratio D_{max} is $(1 - M)$, which will decrease with the increase of modulation index M. Additionally, the shoot-through duty ratio will be zero when M increases to one, that is, the inverter then operates as the traditional VSI.

In the simple boost method [1], two straight lines equal to or greater than the peak value of the three phase references are used to insert the shoot-through duty ratio. In this way, the shoot-through time per switching cycle is constant, meaning that the boost factor is constant. For this method, the obtainable shoot-through duty ratio decreases with the increase of modulation index M. Thus, the maximum shoot-through duty ratio is limited to $(1 - M)$. That is, the shoot-through duty ratio reaches zero when M increases to its maximum value, at which point the inverter operates the same as the traditional VSI. The dc inductor current and capacitor voltage have no ripple that is associated with the output frequency.

4.1.2 Maximum Boost Control

Figure 4.1(b) illustrates the MBC, where the shoot-through reference is just equal to the upper or lower envelope of modulating waves. All traditional zero states are replaced by the shoot-through states; in other words, D_{max} can achieve the total zero state duty ratio of $(1 - 3\sqrt{3}M / 2\pi)$ per switching cycle. However, the shoot-through duty ratio of the MBC varies at six times of the output frequency, introducing a low frequency ripple into the quasi-Z-source capacitor voltage and inductor current. This becomes especially significant and will cause a higher requirement of the Z-network components when the output frequency is very low. Therefore, the MBC is suitable for application fields that have a fixed or comparatively high output frequency.

The MBC turns all traditional zero states into shoot-through states [2]. The shoot-through states are inserted when the triangular carrier wave is either greater than the maximum curve of the references or smaller than the minimum of the references. However, the shoot-through period of this method varies each cycle, introducing a low frequency current ripple that is associated with the output frequency in the inductor current and the capacitor voltage, which causes a higher requirement of the passive components when the output frequency becomes very low. Therefore, the MBC is suitable for applications that have a fixed or relatively high output frequency.

4.1.3 Maximum Constant Boost Control

The implementation of MCBC [3] is shown in Figure 4.2. By slightly modifying the shoot-through references of the MBC, the MCBC can keep the shoot-through duty ratio constant per switching cycle. The MCBC is a popular compromise between the SBC and MBC, as a higher voltage gain of the qZSI than that in the SBC is achieved without low-frequency ripples on the voltage and current in the impedance components.

For the maximum constant boost control with third harmonic injection, the inverter turns to a shoot-through zero state when the carrier triangle wave is greater than the upper shoot-through envelope or smaller than the lower shoot-through envelope [3]. Although it causes a slightly higher voltage stress across the devices than the maximum control method, and the voltage gain ability of this method is slightly smaller than that of maximum boost control, it has a much lower voltage stress than the simple control method. It always keeps the shoot-through duty ratio constant, no low-frequency ripple being associated with the output frequency. Therefore, the inductor and capacitor requirements of the Z-network are greatly reduced.

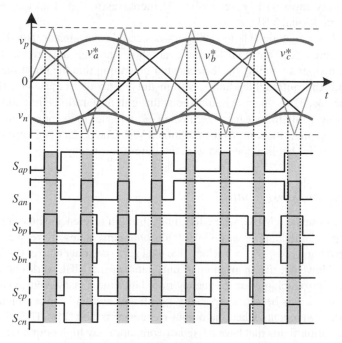

Figure 4.2   Maximum constant boost control of the ZSI/qZSI.

4.2 Space Vector Modulations

4.2.1 Traditional SVM

As shown in Figure 4.3(a), the traditional SVM technique for three-phase VSIs generates eight voltage space vectors, and there are six sectors, I to VI. The well-known algorithm can be defined as [4]

$$
\begin{cases}
T_1 = T_s M \sin\left[\dfrac{\pi}{3} - \theta + \dfrac{\pi}{3}(i-1)\right] \\[2mm]
T_2 = T_s M \sin\left[\theta - \dfrac{\pi}{3}(i-1)\right] \\[2mm]
T_0 = T_s - T_1 - T_2
\end{cases}
\tag{4.1}
$$

$$
U_{ref} = U_1 \frac{T_1}{T_s} + U_2 \frac{T_2}{T_s}
\tag{4.2}
$$

where $i \in \{1, 2, \ldots, 6\}$ denotes the ith sector; T_0 is the time interval of the traditional zero vector U_0; T_1 and T_2 are the time intervals of active vectors U_1 and U_2, respectively; and θ is the inclined angle of voltage reference vector U_{ref} and U_1. The modulation index M is defined as $M = \sqrt{3} U_{ref} / V_{dc}$, and V_{dc} is the dc-link voltage.

Figure 4.3(b) shows the three-phase VSI switching time sequence in sector I, where T_{max}, T_{mid}, T_{min} are the maximum, middle, and minimum switching times for the three bridge legs, $T_{min} = T_0/4$, $T_{mid} = T_0/4 + T_1/2$, and $T_{max} = T_s/2 - T_0/4$. Additionally, the zero state repeats periodically every $\pi/3$, so the average zero state duty ratio over one control cycle is

$$
\frac{\overline{T_0}}{T_s} = \frac{1}{\pi/3} \int_0^{\pi/3} \left[1 - \frac{\sqrt{3}}{2} M \sin(\pi/3 + \theta)\right] d\theta = 1 - \frac{3\sqrt{3}M}{2\pi}
\tag{4.3}
$$

4.2.2 SVMs for ZSI/qZSI

The qZSI has the traditional VSI's six active voltage vectors, two traditional zero vectors, and one additional shoot-through zero vector. Referring to (4.2), in sector I, the qZSI reference voltage vector becomes

$$
U_{ref} = U_1 \frac{T_1}{T_s} + U_2 \frac{T_2}{T_s} + U_0 \frac{T_0}{T_s} + V_{sh} \frac{T_{sh}}{T_s}
\tag{4.4}
$$

where V_{sh} is the shoot-through voltage vector.

For the aforementioned SVMs of the qZSI, the desired total shoot-through time interval is equally divided into several parts per control cycle and each part is separately combined into the switching moment between traditional zero vector and active vector. If only one bridge leg turns on each time, there are three implementation ways $V_{sh\{A,B,C\}}$ of shoot-through zero vector

(a)

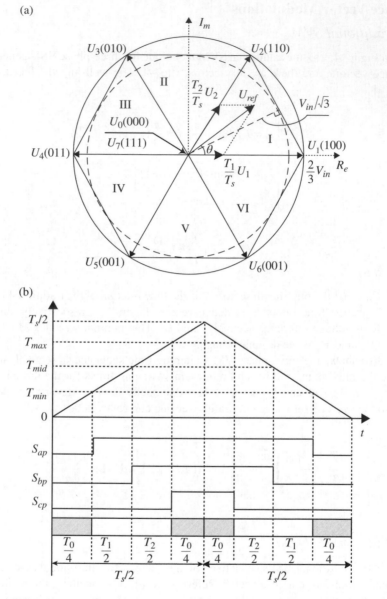

(b)

Figure 4.3 SVM for a traditional voltage source inverter: (a) basic voltage space vectors and (b) switching time sequence.

for three phase legs. Figure 4.4(a) shows the corresponding voltage space vectors of the qZSI. The shoot-through state is produced during the time intervals of the traditional zero vectors, hence [4–6]

$$T_s = T_0 + T_1 + T_2 + T_{sh} \quad \text{and} \quad U_0 \frac{T_0}{T_s} + V_{sh} \frac{T_{sh}}{T_s} = 0 \tag{4.5}$$

(a)

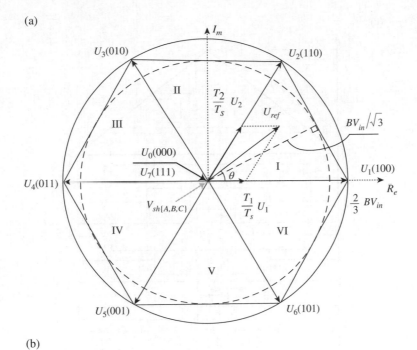

(b)

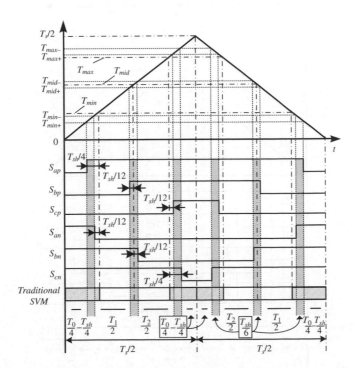

Figure 4.4 SVMs for the qZSI: (a) voltage space vectors; switching time sequences of (b) ZSVM6, (c) ZSVM4, (d) ZSVM2, (e) ZSVM1-I, and (f) ZSVM1-II (*Source*: Liu 2014 [9]. Reproduced with permission of IEEE).

(c)

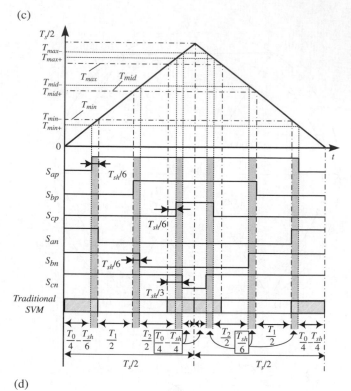

(d)

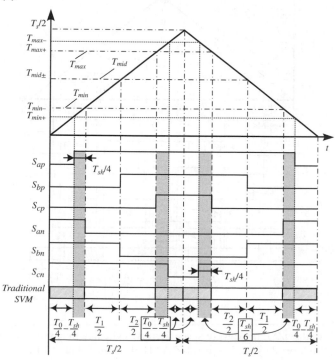

Figure 4.4   (*Continued*)

(e)

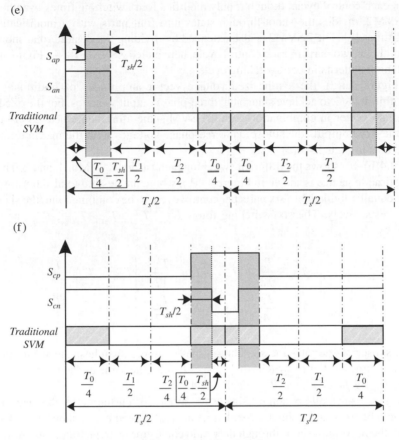

(f)

Figure 4.4 (*Continued*)

From (4.4) and (4.5), it is known that the qZSI will output the same voltage vector when compared to (4.2) in the traditional VSI.

Because of the advantages of low harmonics and high dc-link voltage utilization, the traditional space-vector concept has been applied to the ZSI/qZSI [5–7]. Besides the six active and two zero voltage vectors in the traditional SVM scheme, additional shoot-through zero vectors are introduced into the SVM technique of qZSI by modifying the switching times, which are used to compare with the microprocessor counter and then generate the on–off control signals for power switches. It is especially beneficial for digital realization of the shoot-through of the qZSI.

In the SVM technique of qZSI, the total shoot-through time interval in one switching cycle is divided into several identical parts and each part is distributed to the transition moment of the active vector and traditional zero vector, without any extra switching operations. According to the different distributions, there are three main types of SVM for the qZSI. Reference [5] divides the shoot-through states into six parts and evenly inserts them into six switches by modifying all the six switching control signals. This approach is here called the ZSVM6. Reference [6] also addresses a method, termed ZSVM4, with a six-part shoot-through time interval in each control cycle, though it only modifies four switching times. Another method, called ZSVM2, divides the shoot-through states into four parts with a modification of two switching times [7]. The ZSVM2 is also extended by dividing the desired total shoot-through time interval into two parts in each control cycle, here called ZSVM1 [8, 9]. Figure 4.4 reveals their switching patterns in one switching cycle.

From Figure 4.3(a), the synthesized voltage vector amplitudes of traditional VSIs are limited within $V_{in}/\sqrt{3}$ to achieve sinusoidal three-phase output voltages. For the qZSI, with the shoot-through vector in combination with the SVM – the aforementioned ZSVMx ($x = 6$, 4, and 2) – the qZSI outputs the sinusoidal three-phase voltages with the maximum amplitude of $BV_{in}/\sqrt{3}$, as in Figure 4.4(a).

Figure 4.4(b)–(d) shows the different switching patterns for ZSVM6, 4, and 2. The ZSVM6 switching time sequence is shown in Figure 4.4(b), where the desired total shoot-through time interval is equally divided into six parts per control cycle to be combined into the six switching moments, respectively. The six switching times T_{max+}, T_{mid+}, T_{min+}, T_{max-}, T_{mid-}, and T_{min-} are calculated by

$$\begin{cases} T_{max+} = T_{max} + \dfrac{T_{sh}}{12} \\[2mm] T_{max-} = T_{max} + \dfrac{T_{sh}}{4} \end{cases} \begin{cases} T_{mid+} = T_{mid} - \dfrac{T_{sh}}{12} \\[2mm] T_{mid-} = T_{mid} + \dfrac{T_{sh}}{12} \end{cases} \begin{cases} T_{min+} = T_{min} - \dfrac{T_{sh}}{4} \\[2mm] T_{min-} = T_{min} - \dfrac{T_{sh}}{12} \end{cases} \tag{4.6}$$

where the subscript $+$ denotes the switching times of upper switches; the subscript $-$ denotes the switching times of lower switches; T_{max}, T_{mid}, T_{min} are the same as the traditional SVM, i.e. $T_{min} = T_0/4$, $T_{mid} = T_0/4 + T_1/2$, and $T_{max} = T_s/2 - T_0/4$.

From Figure 4.4(b), all six control signals are modified. Furthermore, the entire traditional zero state can be used as the shoot-through time interval if need be, as soon as $(T_0/4 - T_{sh}/4) \geq 0$. As a result, the maximum shoot-through duty ratio equals the average traditional zero state duty ratio shown in (4.3).

Figure 4.4(c) shows the ZSVM4 switching pattern, and the switching times are calculated by

$$\begin{cases} T_{max+} = T_{max} + \dfrac{T_{sh}}{6} \\[2mm] T_{max-} = T_{max} + 2 \times \dfrac{T_{sh}}{6} \end{cases} \begin{cases} T_{mid+} = T_{mid} \\[2mm] T_{mid-} = T_{mid} + \dfrac{T_{sh}}{6} \end{cases} \begin{cases} T_{min+} = T_{min} - \dfrac{T_{sh}}{6} \\[2mm] T_{min-} = T_{min} \end{cases} \tag{4.7}$$

Even though the desired total shoot-through time interval is also divided into six parts in one control cycle, it only needs to modify four switching times in (4.7). From (4.7) and the switching pattern of ZSVM4 shown in Figure 4.4(c), it can be seen that T_{mid-} performs a $T_{sh}/6$ time delay. Accordingly, there is $T_{sh}/6$ delay for T_{max+} to ensure the unchanged action time of active states, (i.e., $T_2/2$) in Figure 4.4(c). Meanwhile, the $T_{sh}/3$ time delay for T_{max-} is performed in order to achieve each shoot-through time interval of $T_{sh}/6$ (T_{sh} is fulfilled through six times), as shown in Figure 4.4(c). However, this process should not affect the active states, so the available maximum shoot-through time interval from the traditional zero states has to meet $(T_0/4 - T_{sh}/3) \geq 0$ in Figure 4.4(c). As a result, the available maximum shoot-through time interval of ZSVM4 in one switching period is limited to $(3/4) T_0$.

Figure 4.4(d) shows the ZSVM2 switching time sequence, and the switching times are calculated by

$$\begin{cases} T_{max+} = T_{max} \\ T_{max-} = T_{max} + \dfrac{T_{sh}}{4} \end{cases} \begin{cases} T_{mid+} = T_{mid} \\ T_{mid-} = T_{mid} \end{cases} \begin{cases} T_{min+} = T_{min} - \dfrac{T_{sh}}{4} \\ T_{min-} = T_{min} \end{cases} \tag{4.8}$$

This modulation method divides the desired total shoot-through time interval into four parts and it only needs to modify two switching times in (4.8). Under $(T_0/4 - T_{sh}/4) \geq 0$ in Figure 4.4(d), the maximum shoot-through duty ratio can be the entire traditional zero duty ratio shown in (4.3).

Figure 4.4(e) and (f) shows two types of ZSVM1, where the desired total shoot-through time interval is divided into two parts, and only one switching time, T_{max-} in Figure 4.4(e) or T_{min+} in Figure 4.4(f), needs to be modified, respectively, by

$$T_{min+} = T_{min} - \frac{T_{sh}}{2} \tag{4.9}$$

$$T_{max-} = T_{max} + \frac{T_{sh}}{2} \tag{4.10}$$

They can be named ZSVM1-I and the ZSVM1-II, respectively. From Figure 4.4(e) and (f), there is $(T_0/4 - T_{sh}/2) \geq 0$, so the maximum shoot-through time interval of ZSVM1 is reduced to $(1/2) T_0$.

4.3 Pulse-Width Amplitude Modulation

A PAM concept has been proposed for three-phase inverters by producing the 6ω voltage ripple on the dc-link bus, so that one of the three legs conducts switching based on PWM, and the upper or lower device in the other two legs keeps its on or off state [10, 11], which is the so-called PAM. As a result, an 87% switching loss reduction is achieved and a bulky dc-link capacitor becomes unnecessary [12–14]. The three-phase PAM-based inverter has a higher power density, higher efficiency, less volume, and lighter weight than the PWM-based system. The PAM technique was also applied to the quasi-Z-source matrix converter [15] and the quasi-Z-source cascaded multilevel inverter (qZS-CMI) [16]. Since the PAM is a kind of discontinuous PWM and it is detailed for quasi-Z-source matrix converter in Chapter 9 and for qZS-CMI in Chapter 11, the following section of this chapter compares the above continuous PWM methods of the three-phase ZSI/qZSI.

4.4 Comparison of All Modulation Methods

In this section, the aforementioned carrier-based PWM and SVM methods, which are continuous PWM methods, of the three-phase ZSI/qZSI are compared by performance, simulations, and experiments, based on the authors' publication of [9].

4.4.1 Performance Analysis

From the aforementioned analysis, the different ZSVMs provide different maximum shoot-through time intervals to the qZSI and they can be summarized as: (1) $T_{sh,max} = T_0$ for the ZSVM6 and ZSVM2, (2) $T_{sh,max} = 3T_0/4$ for the ZSVM4, (3) $T_{sh,max} = T_0/2$ for the ZSVM1. The maximum shoot-through duty ratio, the maximum voltage gain, and the maximum dc-link peak voltage ratio V_{dc}/V_{in} (or voltage stress ratio V_s/V_{in}) when the qZSI uses ZSVMx (x = 6, 4, 2, and 1) are shown in Table 4.1. It should be noted that the maximum voltage gain of ZSVM1 is not changed even if the modulation index changes, i.e. $G_{ZSVM1,max} = MB = 2\pi/3\sqrt{3} \approx 1.21$. Therefore, the maximum voltage stress ratio of ZSVM1 is inversely proportional to the modulation index M, as shown in Table 4.1. With the limited maximum voltage gain of 1.21, the maximum voltage stress will increase when the modulation index decreases.

For clear comparison of four ZSVMs, Figures 4.5(a) and (b) show the curves of maximum shoot-through duty ratio D_{max} versus the modulation index M, and the maximum voltage stress ratio versus the voltage gain G, respectively. It is possible to achieve ZSVM6 and ZSVM2 having a larger shoot-through duty ratio D_{max} than ZSVM4 and ZSVM1 at the same modulation index M, which implies higher voltage gain for the ZSVM6 and ZSVM2. Also, Figure 4.5(b) shows the lower voltage stress of ZSVM6 and ZSVM2 for the same voltage gain.

4.4.2 Simulation and Experimental Results

The four ZSVMs – ZSVM6, ZSVM4, ZSVM2, and ZSVM1 – are verified by using the designed qZSI experimental platform. The quasi-Z-source inductance and capacitance are $500\,\mu H$ and $470\,\mu F$, respectively. The three-phase load consists of a resistor of $10\,\Omega$ and an inductor of $12\,mH$ per phase. The same system and control parameters are used in all tests for both the simulation and experiments. They are: $V_{in} = 110\,V$, $D = 0.12$, $M = 0.7$, ac output frequency of $50\,Hz$, and the control cycle of $100\,\mu s$ [9].

Table 4.1 Maximum shoot-through duty ratio D_{max}, maximum voltage gain G_{max}, and maximum voltage stress V_s/V_{in} when using different ZSVMs (*Source:* Liu 2014 [9]. Reproduced with permission of IEEE)

	SBC	MBC	MCBC	ZSVM6	ZSVM4
D_{max}	$1-M$	$1-\dfrac{3\sqrt{3}}{2\pi}M$	$1-\dfrac{\sqrt{3}}{2}M$	$1-\dfrac{3\sqrt{3}}{2\pi}M$	$\dfrac{3}{4}\left(1-\dfrac{3\sqrt{3}}{2\pi}M\right)$
G_{max}	$\dfrac{M}{2M-1}$	$\dfrac{\pi M}{3\sqrt{3}M-\pi}$	$\dfrac{M}{\sqrt{3}M-1}$	$\dfrac{\pi M}{3\sqrt{3}M-\pi}$	$\dfrac{4\pi M}{9\sqrt{3}M-2\pi}$
V_s/V_{in}	$2G-1$	$\dfrac{3\sqrt{3}G}{\pi}-1$	$\sqrt{3}G-1$	$\dfrac{3\sqrt{3}G}{\pi}-1$	$\dfrac{9\sqrt{3}G}{2\pi}-2$

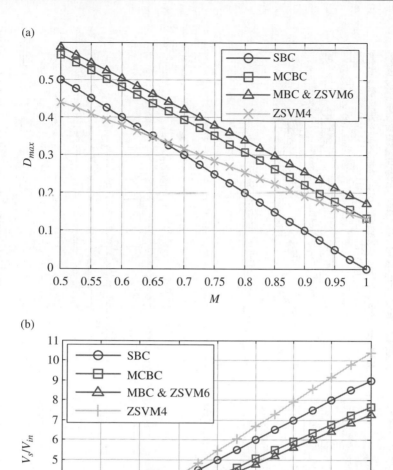

Figure 4.5 Maximum shoot-through duty ratio and voltage stress ratio of qZSI when using the different ZSVMs: (a) maximum shoot-through duty ratio versus the modulation index (b) switch's maximum voltage stress ratio versus the voltage gain (*Source*: Liu 2014 [9]. Reproduced with permission of IEEE).

The SBC is first tested in the simulation and experiment for the comparison of the ZSVMx (x = 6, 4, 2, and 1), and the results are shown in Figure 4.6. Figures 4.7–4.10 show the simulation and experimental results of the ZSVM6, ZSVM4, ZSVM2, and ZSVM1-I. Tables 4.2 and 4.3 summarize the average inductor current, inductor current ripples, and ac output voltage of the qZSI when using the simple boost control and the ZSVMx (x = 6, 4, 2, and 1).

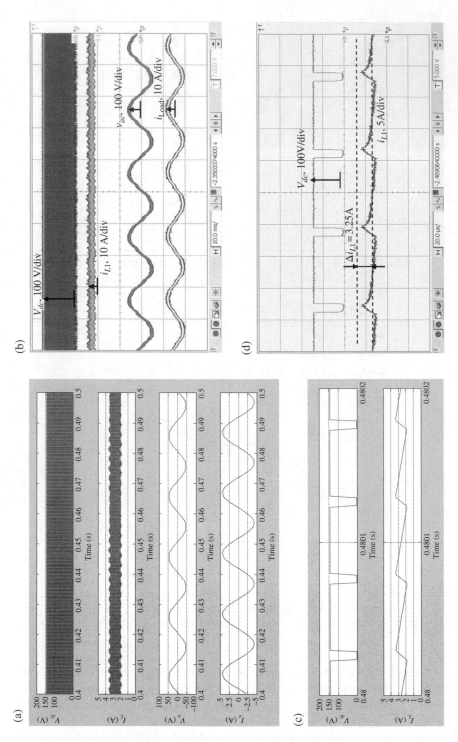

Figure 4.6 Simulation and experimental results of the qZSI system when using simple boost control: (a) simulated dc-link voltage, inductor L_1 current, ac load phase voltage and current, (b) experimental dc-link voltage, inductor L_1 current, ac load phase voltage and current, (c) simulated dc-link voltage and inductor L_1 current in two control cycles, (d) experimental dc-link voltage and inductor L_1 current shown in two control cycles (*Source:* Liu 2014 [9]). Reproduced with permission of IEEE).

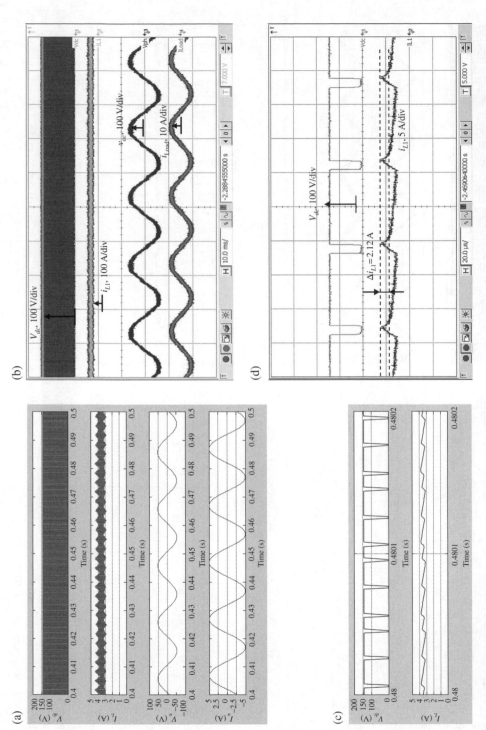

Figure 4.7 Simulation and experimental results of the qZSI system when using ZSVM6: (a) simulated dc-link voltage, inductor L_1 current, ac load phase voltage and current, (b) experimental dc-link voltage, inductor L_1 current, ac load phase voltage and current, (c) simulated dc-link voltage and inductor L_1 current in two control cycles, (d) experimental dc-link voltage and inductor L_1 current shown in two control cycles (*Source*: Liu 2014 [9]). Reproduced with permission of IEEE).

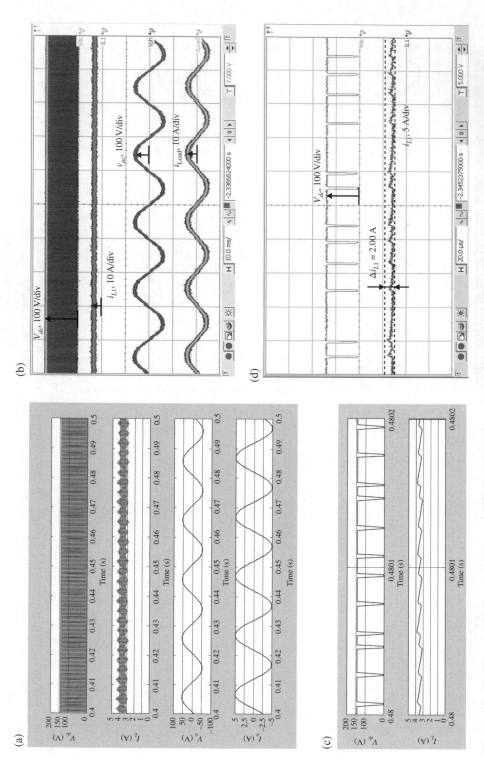

Figure 4.8 Simulation and experimental results of the qZSI system when using ZSVM4: (a) simulated dc-link voltage, inductor L_1 current, ac load phase voltage and current, inductor L_1 current, (b) experimental dc-link voltage, inductor L_1 current, ac load phase voltage and current, (c) simulated dc-link voltage and inductor L_1 current in two control cycles, (d) experimental dc-link voltage and inductor L_1 current shown in two control cycles (*Source*: Liu 2014 [9]). Reproduced with permission of IEEE).

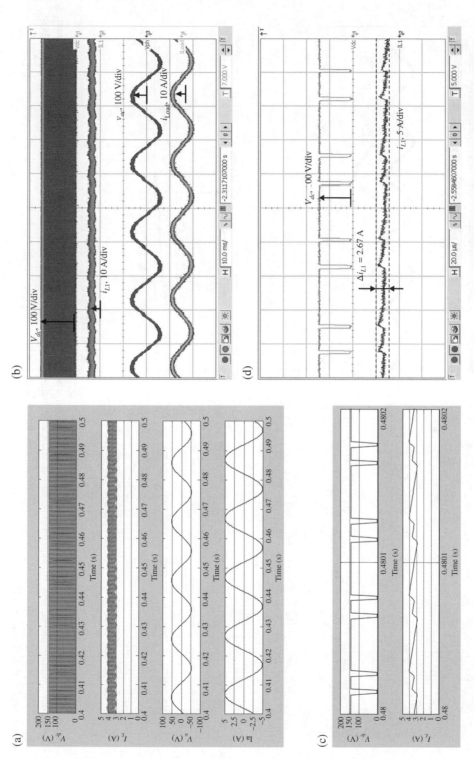

Figure 4.9 Simulation and experimental results of the qZSI system when using ZSVM2: (a) simulated dc-link voltage, inductor L_1 current, ac load phase voltage and current, (b) experimental dc-link voltage, inductor L_1 current, ac load phase voltage and current, (c) simulated dc-link voltage and inductor L_1 current in two control cycles, (d) experimental dc-link voltage and inductor L_1 current shown in two control cycles (*Source*: Liu 2014 [9]. Reproduced with permission of IEEE).

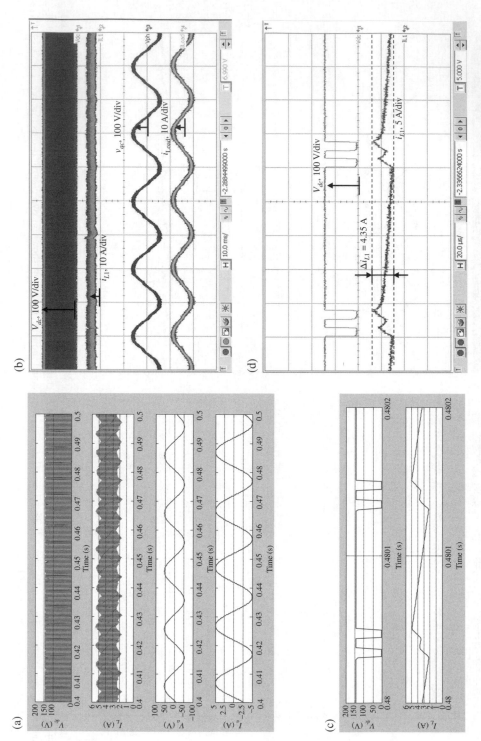

Figure 4.10 Simulation and experimental results of the qZSI system when using the ZSVM1-I: (a) simulated dc-link voltage, inductor L_1 current, ac load phase voltage and current, (b) experimental dc-link voltage, inductor L_1 current, ac load phase voltage and current, (c) simulated dc-link voltage and inductor L_1 current in two control cycles, (d) experimental dc-link voltage and inductor L_1 current shown in two control cycles (*Source*: Liu 2014 [9]. Reproduced with permission of IEEE).

Table 4.2 Summary of the simulated results (*Source:* Liu 2014 [9]. Reproduced with permission of IEEE)

	SBC	ZSVM6	ZSVM4	ZSVM2	ZSVM1
I_{L_av} (A)	2.69	3.54	3.54	3.57	3.58
ΔI_L (A)	1.85	1.26	1.23	1.42	3.11
V_{ac} (V)	32.23	37.58	37.57	37.58	37.44

Table 4.3 Summary of the experimental results (*Source*: Liu 2014 [9]. Reproduced with permission of IEEE)

	SBC	ZSVM6	ZSVM4	ZSVM2	ZSVM1
I_{L_av} (A)	2.46	3.12	3.13	3.21	3.02
ΔI_L (A)	3.25	2.12	2.00	2.67	4.35
V_{ac} (V)	36.92	42.8	42.4	43.07	42.5

For the 110 V dc source voltage and the shoot-through duty ratio of 0.12, the theoretical dc-link peak voltage will be 144.7 V for all of the simple boost control and the ZSVMx (x = 6, 4, 2, and 1), as shown in Figures 4.6–4.10. However, the output ac voltage, the inductor current ripple, and the qZSI efficiency will be different for the different PWM methods. The ZSVM6, ZSVM4, ZSVM2, and ZSVM1-I present the same dc-link voltage utilization, but the simple boost control causes a lower voltage utilization than the ZSVMs, as shown in the simulation and experimental ac output voltage waveforms of Figures 4.6–4.10 and/or in the ac output voltage values of Tables 4.2 and 4.3. In addition, Figures 4.6 and 4.10 show the two times of shoot-through behavior per control cycle in both the simple boost control and the ZSVM1-I; the ZSVM6 and ZSVM4 provide the six times shoot-through behavior per control cycle to the qZSI, as shown in both Figure 4.7 and 4.8; the four times shoot-through behavior per control cycle are achieved by the ZSVM2 as shown in Figure 4.9.

With the same shoot-through duty ratio, the lower shoot-through behavior per control cycle will cause higher inductor current ripples; for example, the simple boost control and the ZSVM1-I produce higher current ripple than the ZSVM6, ZSVM4, and ZSVM2, as shown in Figures 4.6–4.10 and Tables 4.2 and 4.3. Moreover, the two shoot-through behaviors of ZSVM1-I are close to each other, but two shoot-through behaviors of the simple boost control are dispersed into each other, which causes the ZSVM1-I to have higher current ripples than the simple boost control, as shown in Figures 4.6 and 4.10, and Tables 4.2 and 4.3. The six shoot-through behaviors of the ZSVM6 and ZSVM4 result in lower current ripples than the ZSVM2 of four shoot-through behaviors, as shown in Figures 4.7–4.9 and Tables 4.2 and 4.3. All of the experimental results are identical to the simulation results, which verifies the theoretical analysis.

Figure 4.11 shows the qZSI efficiency curves when using the different PWM methods. The ZSVM4 leads to higher efficiency than the ZSVM6, and both of them provide higher efficiency than the ZSVM1, ZSVM2, and simple boost control. The efficiency calculation will be introduced in Chapter 13.

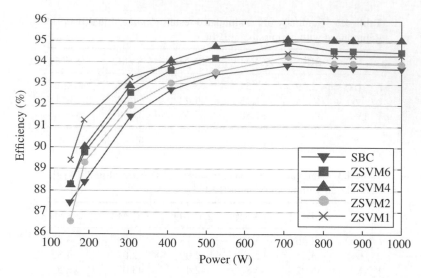

Figure 4.11 The qZSI efficiency curves when using the different PWM methods (*Source*: Liu 2014 [9]. Reproduced with permission of IEEE).

4.5 Conclusion

The PWM method is essential for operating impedance source inverters/converters. In the chapter, the three SPWM methods – the simple boost control, maximum boost control, and maximum constant boost control – as well as four SVM methods – the ZSVM6, ZSVM4, ZSVM2, and ZSVM1 – were illustrated for the three-phase two-level ZSI/qZSI, which are also applicable to most of the three-phase two-level inverters/converters coupling improved impedance source networks, to be analyzed in the later chapters of this book.

References

[1] F. Z. Peng, "Z-source inverter," *IEEE Trans. Ind. Appl.*, vol.39, pp.504–510, 2003.

[2] F. Z. Peng, M. Shen, Z. Qian, "Maximum boost control of the Z-source inverter," *IEEE Trans. Power Electron.*, vol.20, no.4, pp.833–838, July 2005.

[3] M. S. Shen, J. Wang, A. Joseph, F. Z. Peng, L. M. Tolbert, D. J. Adams, et al., "Maximum constant boost control of the Z-source inverter," in *Proc. 39th IAS Annual Meeting Conference Record of the 2004 IEEE Industry Applications Conference*, vol.1, 2004, pp.142–147.

[4] D. G. Holmes, T. A. Lipo, *Pulse Width Modulation for Power Converters: Principles and practice*, John Wiley, 2003.

[5] Y. Liu, B. Ge, F. J. T. E. Ferreira, A. T. de Almeida, H. Abu-Rub, "Modeling and SVM control of quasi-Z-source inverter," in *Proc. 2011 11th International Conference on Electrical Power Quality and Utilisation (EPQU)*, 2011, pp.1–7.

[6] J. Jin-Woo, A. Keyhani, "Control of a fuel cell based Z-source converter," *IEEE Trans. Energy Convers.*, vol.22, no.2, pp.467–476, June 2007.

[7] U. S. Ali, V. Kamaraj, "A novel space vector PWM for Z-source inverter," in *Proc. 1st International Conference on Electrical Energy Systems (ICEES)*, 2011, pp.82–85.

[8] Y. Liu, H. Abu-Rub, B. Ge, F. Z. Peng, "Analysis of space vector modulations for three-phase Z-source/quasi-Z-source inverter," in *Proc. 38th Annual Conference on IEEE Industrial Electronics Society (IECON)*, pp.5268–5273, 2012.

[9] Y. Liu, B. Ge, H. Abu-Rub, F. Z. Peng, "Overview of space vector modulations for three-phase Z-source/quasi-Z-source inverters," *IEEE Trans. Power Electron.*, vol.29, no.4, pp.2098–2108, April 2014.

[10] H. Fujita, "A three-phase voltage-source solar power conditioner using a single-phase PWM control method," in *Proc. IEEE Energy Conversion Congress and Exposition (ECCE)*, pp.3748–3754, 2009.

[11] H. Fujita, "Switching loss analysis of a three-phase solar power conditioner using a single-phase PWM control method," in *IEEE Energy Conversion Congress and Exposition (ECCE)*, pp.618–623, 2010.

[12] X. Yu, Q. Lei, F. Z. Peng, "Boost converter – Inverter system using PAM for HEV/EV motor drive," in *Proc. Twenty-Seventh Annual IEEE Applied Power Electronics Conference and Exposition (APEC)*, pp.946–950, 5, 2012.

[13] Q. Lei, F. Z. Peng, "Space vector pulse-width amplitude modulation for a buck–boost voltage/current source inverter," *IEEE Trans. Power Electron.*, vol.29, no.1, pp.266–274, Jan. 2014.

[14] Q. Lei, F. Z. Peng, B. Ge, "Pulse-width-amplitude-modulated voltage-fed quasi-Z-source direct matrix converter with maximum constant boost," in *Proc. 2012 Twenty Seventh Annual IEEE Applied Power Electronics Conference and Exposition (APEC)*, pp.641–646, 2012.

[15] B. Ge, Q. Lei, W. Qian, F. Z. Peng, "A family of Z-source matrix converters," *IEEE Trans. Ind. Electron.*, vol.59, no.1, pp.35–46, Jan. 2012.

[16] Y. Liu, B. Ge, H. Abu-Rub, F. Z. Peng, "Phase-shifted pulse-width-amplitude modulation for quasi-Z-source cascade multilevel inverter based photovoltaic power system," *IET Power Electron.*, vol.7, no.6, pp.1444–1456, June. 2014.

5

Control of Shoot-Through Duty Cycle: An Overview

Closed-loop control of power electronic systems is generally developed based on state-space averaging or small-signal modeling [1, 2]. In Chapter 2, the modeling of a voltage-fed qZSI is presented for the analysis of dynamic response of impedance network and closed-loop control of shoot-through duty cycle. Several feedback control methods of shoot-through duty cycle have been investigated to stabilize the dc-link voltage of the ZSI/qZSI. In this chapter, those methods are summarized and classified as single-loop, double-loop, linear, and non-linear methods, and an overview is presented in the following. The contents are based on the authors' publication of [1]. In the later chapters of this book, detailed control will be addressed for the entire system control based on the specific applications of the impedance source inverters/converters, such as photovoltaic (PV), distributed generation, and motor drive systems.

5.1 Summary of Closed-Loop Control Methods

Figure 5.1 shows a summary of control approaches used for the ZSI/qZSI shown in Figure 2.1 of Chapter 2. The dc-source voltage V_{in}, Z-source capacitor voltage v_C, Z-source inductor current i_L, or inverter dc-link voltage V_{PN} of the ZSI/qZSI are commonly used for the control through proportional-integral (PI) regulators or non-linear control algorithms, as shown in Figure 5.1(a) and (b) [2–16]. Those approaches are also applicable for all other derived ZSI/qZSI topologies, but with different controller parameters (for PI regulators) or models of non-linear control.

In the conventional PI regulators, as Figure 5.1(a) shows, the dc-source voltage V_{in}, Z-source capacitor voltage v_C, or measured/calculated inverter dc-link voltage V_{PN} is compared with the related reference value to obtain the command signal through a voltage controller, which is usually a PI regulator. The command signal is the shoot-through duty ratio D for single-loop

Impedance Source Power Electronic Converters, First Edition. Yushan Liu, Haitham Abu-Rub, Baoming Ge, Frede Blaabjerg, Omar Ellabban, and Poh Chiang Loh.
© 2016 John Wiley & Sons, Ltd. Published 2016 by John Wiley & Sons, Ltd.

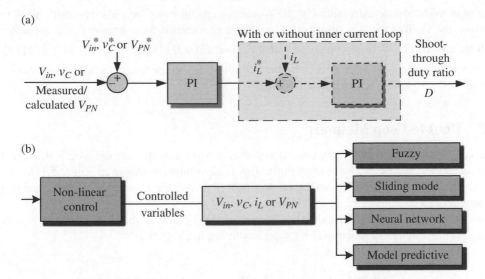

Figure 5.1   ZSI/qZSI's shoot-through duty cycle control by (a) PI regulators and (b) non-linear methods (*Source*: Liu 2014 [1]. Reproduced with permission of IEEE).

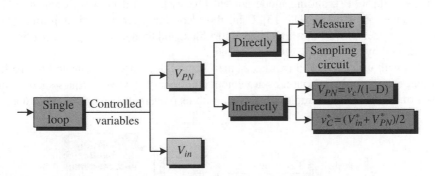

Figure 5.2   Single-loop control methods of shoot-through duty cycle (*Source*: Liu 2014 [1]. Reproduced with permission of IEEE).

controls, or the inductor current i_L^* for dual-loop controls. In non-linear control methods, fuzzy, sliding mode, neural network, and model predictive control have been developed, as Figure 5.1(b) shows.

5.2   Single-Loop Methods

The single-loop controls apply the feedback by the dc source voltage V_{in} [1], capacitor voltage v_C [3, 13], or the dc-link voltage V_{PN} [4, 7], and then the output is the command shoot-through duty ratio D. Figure 5.2 shows a summary of single-loop control methods. Note that the dc-link voltage of a ZSI is a pulse voltage waveform due to shoot-through zero states, which is inconvenient to measure. Therefore, the ability to control the dc-link voltage by measuring it directly [4] is improved by an assisted sampling circuit in [5], although this method results in complicated hardware. Alternatively, the Z-source capacitor voltage v_C is frequently used to control the dc-link voltage indirectly, based on the correlations between the capacitor voltage and the dc-link peak voltage such that $V_{PN} = v_C/(1-D)$ [3], or $v_C^* = (V_{in}^* + V_{PN}^*)/2$ [13]. However, the single-loop controls cause the dc-link peak voltage to change along with the dc source voltage variations and it is not constant.

5.3   Double-Loop Methods

Figure 5.3 shows the double-loop control system. An inner inductor current loop is dedicated to realize fast response and stabilizing the dc-link peak voltage, as shown in Figure 5.1. In [2], the inner current loop is performed by proportional (P) regulator. Later on, it is improved to a PI regulator in order to increase the stability margin [6].

As an example of design, the simulation investigations of the double-loop control by outer PI controller and inner P controller were demonstrated in Chapter 2.

5.4   Conventional Regulators and Advanced Control Methods

Figure 5.4 shows a block diagram of non-linear control algorithms applied to ZSI/qZSI, such as the fuzzy control [17], sliding mode control [10, 18], neural network control [11], and model predictive control (MPC) [12, 13]. Using those non-linear controls, the real-time variables V_{in}, v_C, i_L, or v_{dc} are employed for the related algorithms, and then switching control signals are directly obtained by tracking the reference values.

Compared with the classical PI regulators, the fuzzy, sliding mode, and neural network controls present fast dynamic response, even without additional PWM techniques such as sliding mode control. Nevertheless, complex control algorithms prevent their broad applications.

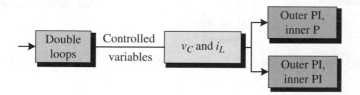

Figure 5.3   Double-loop control methods of shoot-through duty cycle (*Source*: Liu 2014 [1]. Reproduced with permission of IEEE).

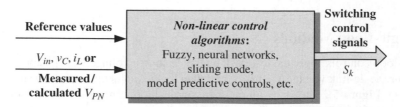

Figure 5.4   Non-linear control methods of shoot-through duty cycle (*Source*: Liu 2014 [1]. Reproduced with permission of IEEE).

Recently, the MPC has been proposed. It adjusts the quality of the output waveforms, stability of the impedance network, and robustness of transient response by a formulated Z-source network model and a minimized cost function [12, 13], which is much simpler than the other non-linear algorithms. However, to date, the reduction of switching states especially in multilevel topologies and the fixing of variable switching frequency are still under investigation for the MPC technique with ZSIs. In the later chapters of this book, the MPC for the three-phase, four-leg, and five-phase qZSI will be introduced.

References

[1] Y. Liu, H. Abu-Rub, B. Ge, "Z-source/quasi-Z-source inverters – derived networks, modulations, controls, and emerging applications to photovoltaic conversion," *IEEE Ind. Electron. Mag.*, vol.8, no.4, pp.32–44, Dec. 2014.

[2] S. Bacha, I. Munteanu, A. I. Bratcu, *Power Electronic Converters Modeling and Control: with Case Studies*, Wiley & Sons Ltd, 2013.

[3] A. Yazdani, R. Iravani, *Voltage-Sourced Converters in Power Systems: Modeling, Control, and Applications*, Wiley & Sons Ltd, 2010.

[4] Y. Li, S. Jiang, J. Cintron-Rivera, F. Z. Peng, "Modeling and control of quasi-Z-source inverter for distributed generation applications," *IEEE Trans. Ind. Electronics*, vol.60, no.4, pp.1532–1541, April 2013.

[5] C. J. Gajanayake, D. M. Vilathgamuwa, P. C. Loh, "Development of a comprehensive model and a multiloop controller for Z-source inverter DG systems," *IEEE Trans. Ind. Electron.*, vol.54, pp.2352–2359, Aug 2007.

[6] Q.-V. Tran, T.-W. Chun, H.-G. Kim, E.-C. Nho, "Minimization of voltage stress across switching devices in the Z-source inverter by capacitor voltage control," *J. Power Electron.*, vol.9, no.3, pp.335–342, May 2009.

[7] X. Ding, Z. Qian, S. Yang, B. Cui, F. Z. Peng, "A direct dc link boost voltage PID-like fuzzy control strategy in Z-source inverter," in *Proc. IEEE Power Electron. Spec. Conf. (PESC)*, 2008, pp.405–411.

[8] Y. Tang, J. Wei, S. Xie, "A new direct peak dc-link voltage control strategy of Z-source inverters," in *Proc. 2010 Twenty-Fifth Annual IEEE Applied Power Electronics Conference and Exposition (APEC)*, pp.867–872, 21–25 Feb. 2010.

[9] O. Ellabban, J. Van Mierlo, P. Lataire, "A DSP-based dual-loop peak dc-link voltage control strategy of the Z-source inverter," *IEEE Trans. Power Electron.*, vol.27, no.9, pp.4088–4097, Sept. 2012.

[10] A. H. Rajaei, S. Kaboli, A. Emadi, "Sliding-mode control of Z-source inverter," in *Proc. IEEE 34th Annu. Conf. Ind. Electron.*, 10–13 Nov. 2008, pp.947–952.

[11] H. Rostami, D. A. Khaburi, "Neural networks controlling for both the DC boost and AC output voltage of Z-source inverter," in *Proc. 1st Power Electron. Drive Syst. Technol. Conf.*, 17–18 Feb., 2010, pp.135–140.

[12] W. Mo, P. C. Loh, F. Blaabjerg, "Model predictive control for Z-source power converter," in *Proc. 8th Int. Conf. Power Electron.*, 30 May–3 Jun. 2011, pp.3022–3028.

[13] M. Mosa, H. Abu-Rub, J. Rodríguez, "High performance predictive control applied to three phase grid connected quasi-Z-source inverter," in *Proc. 39th Annual Conference of the IEEE Industrial Electronics Society (IECON)*, pp.5812–5817, 2013.

[14] S. Rajakaruna, L. Jayawickrama, "Steady-state analysis and designing impedance network of Z-source inverters," *IEEE Trans. Ind. Electron.*, vol.57, no.7, pp.2483–2491, Jul. 2010.

[15] J. Liu, J. Hu, L. Xu, "Dynamic modeling and analysis of Z-source converter – derivation of AC small signal model and design-oriented analysis," *IEEE Trans. Power Electron.*, vol.22, no.5, pp.1786–1796, Sep.2007.

[16] F. Guo, L. X. Fu, C. H. Lin, C. Li, W. Choi, J. Wang, "Development of an 85-kW bidirectional quasi-Z-source inverter with dc-link feed-forward compensation for electric vehicle applications," *IEEE Trans. Power Electron.*, vol.28, pp.5477–5488, Dec 2013.

[17] H. Abu-Rub, A. Iqbal, Sk. Moin Ahmed, F. Z. Peng, Y. Li, B. Ge, "Quasi-Z-source inverter-based photovoltaic generation system with maximum power tracking control using ANFIS," *IEEE Trans. Sustain. Energy*, vol.4, no.1, pp.11–20, Jan. 2013.

[18] J. Liu, S. Jiang, D. Cao, F. Z. Peng, "A digital current control of quasi-Z-source inverter with battery," *IEEE Trans. Ind. Informat.*, vol.9, no.2, pp.928–937, May 2013.

6

Z-Source Inverter: Topology Improvements Review

The Z-source inverter (ZSI) is an emerging topology for power electronics converters with very interesting properties such as buck–boost characteristics and single-stage conversion with potential for reduced cost, reduced volume, and higher efficiency due to a lower component number. In recent years, various ZSI topologies have been presented in numerous diversified studies. This chapter presents an overview of different dc/ac ZSI topology improvements and topology arrangements. There are many topology modifications either to overcome the drawbacks of the basic ZSI topology or to increase its voltage gain. In addition, the impedance source topological concept has been integrated with cascaded and multilevel concepts to improve the output quality and to reduce switch voltage and current stress. This chapter provides a comprehensive overview of most dc-ac ZSI topology improvements and arrangements, with some recommendations for the reader.

6.1 Introduction

One of the most promising power electronics converter topologies is the Z-source inverter (ZSI). The ZSI is an emerging topology for power electronics dc-ac converters with very interesting properties such as buck–boost characteristics and single-stage conversion. A two-port network, composed of two capacitors and two inductors connected in an X shape, is employed to provide an impedance source (Z-source) network, coupling the inverter main circuit to the dc input source. The ZSI advantageously utilizes the shoot-through (ST) state to boost the input voltage, which improves the inverter reliability and enlarges its application fields. In comparison with other power electronics converters, it provides an attractive single-stage dc-ac conversion with buck–boost capability with reduced cost, reduced volume, and higher efficiency due to a lower component count. For emerging power generation technologies, such as fuel cells, photovoltaic

Impedance Source Power Electronic Converters, First Edition. Yushan Liu, Haitham Abu-Rub, Baoming Ge, Frede Blaabjerg, Omar Ellabban, and Poh Chiang Loh.
© 2016 John Wiley & Sons, Ltd. Published 2016 by John Wiley & Sons, Ltd.

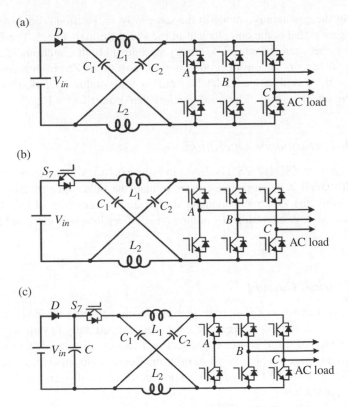

Figure 6.1 Z-source inverter topologies: (a) basic ZSI, (b) bidirectional ZSI, (c) high-performance ZSI.

arrays, and wind turbines, and for new power electronic applications, such as electric and hybrid vehicles, the ZSI is a very promising and competitive topology [1, 2].

The Z-source concept can be applied to all dc-to-ac, ac-to-dc, ac-to-ac, and dc-to-dc power conversion. In addition, it can be used as a voltage/current-fed Z-source inverter for two-level or multilevel configurations. This chapter will focus on the topology modifications of the two-level voltage-fed ZSI. The basic two-level voltage-fed ZSI topology, as shown in Figure 6.1(a), has some drawbacks, which resulted in decreasing the converter efficiency, such as: unidirectional power flow, light-load operation, high inrush current on starting, a discontinuous input current, higher Z-network capacitor voltage, and isolated source and inverter dc rail. In order to solve these drawbacks in basic ZSI topology, there have been many modifications of its structure. This chapter will presents a review of these topological modifications. The contents are based on the authors' publication of [3].

6.2 Basic Topology Improvements

6.2.1 Bidirectional Power Flow

The basic version of ZSI topology can be changed into a bidirectional ZSI (BZSI), as shown in Figure 6.1(b), by the replacement of input diode D with a bidirectional switch S_7, which operates during the regenerative mode in the same way as the diode did during the inverter mode, and its gate signal is the complement of the shoot-through signal. The BZSI is able to exchange energy between the ac and dc energy storages in both directions. There is no need for any additional sensors or control circuits, and the cost of the ZSI improvement is very low [4, 5]. Also, the BZSI is able to completely avoid the undesirable operation modes, when the ZSI is operating under a small inductance or a low load power factor [6].

6.2.2   High-Performance Operation

The high performance ZSI (HP-ZSI) is shown in Figure 6.1(c), which can operate over a wide load range with small Z-network inductor, eliminate the possibility of the dc-link voltage drops, and simplify the Z-network inductor design and system control [6]. The HP-ZSI is derived from the basic ZSI topology by adding an additional capacitor, C, and a bidirectional switch, S_7 [7, 8].

6.2.3   Low Inrush Current

Huge inrush current exists on ZSI startup. The initial voltage across the Z-source capacitors is zero, so a huge inrush current charges the capacitors immediately to half the input voltage. Then, the resonance of the Z-source capacitors and inductors starts, which results in large voltage and current surges, which might destroy the devices. Because of the inherent current path at startup in the basic ZSI topology, it cannot achieve the soft-start capability to suppress resonant current at startup.

The improved Z-source inverter (IZSI), as shown in Figure 6.2(a), is derived from the basic ZSI topology by exchanging the positions of the inverter bridge and the input diode, and their connection directions are reversed. So, the elements used are exactly the same as the basic ZSI topology. Although the IZSI produces the same voltage boost, its capacitor voltage stress can be reduced to a significant extent. In addition, it has inherent inrush-current limitation ability, because there is no current path at startup [9, 10]. In [11], the improved Z-source inverter has been combined with the HP-ZSI topology to produce the high performance improved ZSI (HP-IZSI), as shown in Figure 6.2(b). The HP-IZSI has two merits: the low voltage stresses of the Z-source network capacitors and the inherent limitation of the inrush current and voltage at startup.

The series Z-source inverter (SZSI), which is formed by connecting the Z-network in series between the dc source and the inverter bridge, as shown in Figure 6.2(c), has a lower Z-source capacitor voltage stress. Furthermore, the SZSI has lower inrush current because there is no current path on start–up [12, 13].

6.2.4   Soft-Switching

Quasi-resonant soft-switching ZSI (QRSSZSI) is formed by adding a quasi-resonant network with only one auxiliary switch to achieve the soft-switching, as shown in Figure 6.2(d). All switches in the inverter bridge are turned on and off under zero voltage switching conditions. The QRSSZSI has almost 10% overall efficiency increase compared with the hard switching one [14].

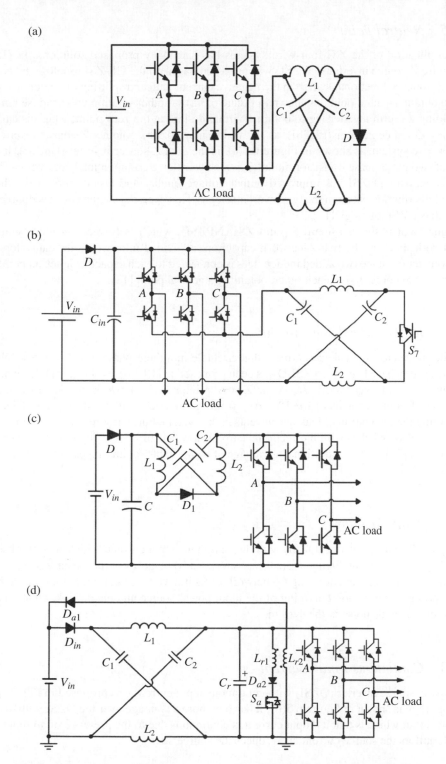

Figure 6.2 Improved ZSI topologies: (a) IZSI, (b) HP-IZSI, (c) SZSI, (d) QRSSZSI.

6.2.5   Neutral Point

The application of the ZSI four-wire systems has had many proposed solutions. In [15], a four-wire Z-source inverter (FWZSI) was proposed based on the HP-ZSI topology by adding an additional bidirectional switch in the negative current path and splitting the input dc capacitor to obtain a neutral point, as shown in Figure 6.3(a). Another solution was proposed in [16] by adding a fourth inverter leg to obtain a virtual dc-link at the zero point, which is called a four-leg Z-source inverter (FLZSI), as shown in Figure 6.3(b). Another Z-source inverter for a four-wire system, as shown in Figure 6.3(c), is the dual Z-source inverter (DuZSI). It consists of two three-phase transistor modules, which work on a common load, and with a single dc source. This DuZSI has improved output voltage quality and less current stress due to current distribution between the two inverter bridges. However, the switch count is double that of the basic ZSI topology [17].

Figure 6.3(d) shows a neutral point ZSI (NPZSI), which is formed by modifying the Z-network. Each of the two Z-network capacitors is divided into two series capacitors and one connection of the two added inductances is connected to each capacitor junction point, the other two are connected to each other, obtaining a neutral point [18].

6.2.6   Reduced Leakage Current

In order to reduce the leakage currents that circulate in a three-phase transformerless photovoltaic system based on ZSI, a ZSI-D has been proposed in [19, 20], as shown in Figure 6.4(a). Adding a fast recovery diode D_2 in the Z-source impedance network guarantees complete isolation of both terminals of the PV array from the inverter switches during the ST states, preventing the circulation of leakage currents in the system. Furthermore, a ZSI topology with two controlled switches with bidirectional current flow, called ZSI-S, Figure 6.4(b), has been proposed in [21] with the additional advantage of being stable for all ranges of the modulation indexes.

6.2.7   Joint Earthing

The quasi-Z-source inverter (QZSI) with discontinuous input current, which was proposed in [22, 23] and shown in Figure 6.5(a), has several advantage over the basic ZSI topology, including lower component rating (Z-network capacitor voltages are lower than in the basic ZSI topology) and the joint earthing of the input power source and the dc-link, which reduce the common-mode noise in the system.

6.2.8   Continuous Input Current

A quasi-Z-source inverter (QZSI) with continuous input current was proposed in [22, 23] and the topology is shown in Figure 6.5(b); it has one more advantage over the QZSI with discontinuous input which is that its input current is continuous due to the presence of an input coil, which buffers the source current and reduces the source stress.

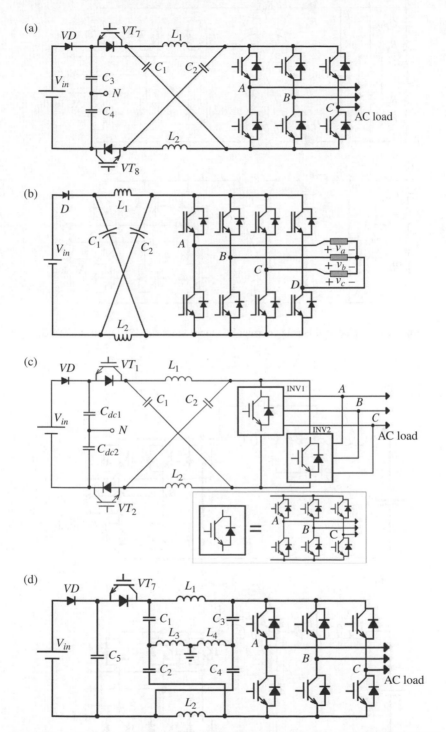

Figure 6.3 Neutral point ZSI topologies: (a) FWZSI, (b) FLZSI, (c) DuZSI, (d) NPZSI.

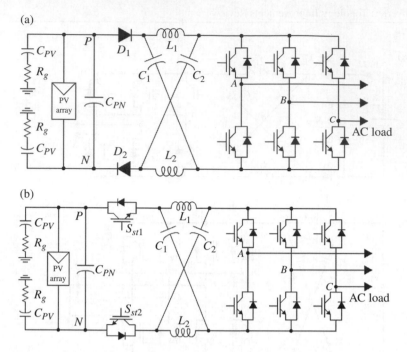

Figure 6.4 Reduced leakage current ZSI topologies: (a) ZSI-D and (b) ZSI-S.

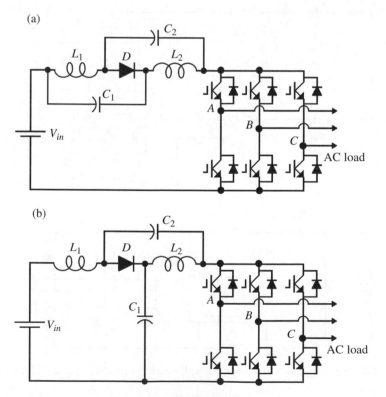

Figure 6.5 Joint earthing ZSI topologies: (a) QZSI with discontinuous input current, (b) QZSI with continuous input current.

6.2.9 Distributed Z-Network

Parasitic and distributed inductance and capacitance in high switching power converter circuits has been a major issue and a troublesome consideration for the physical layouts of power converters. The distributed ZZS (DiZSI), as shown in Figure 6.6, intentionally utilizes the parasitic and distributed inductance and capacitance for power conversion and at the same time for EMI attenuation [24, 25].

6.2.10 Embedded Source

The embedded Z-source inverter (EZSI) is formed by inserting two isolated dc sources into the Z-network with each connecting to an inductor (L_1 or L_2) in series, as shown in Figure 6.7. The filter function of the series inductors in the EZSI smooths the source current and ensures a continuously unidirectional operation, which makes it more suitable in PV and fuel cell applications [26, 27].

All the above circuit modifications try to improve the performance of the basic ZSI and avoid its drawbacks. Further, they do not have any effect on the voltage gain of the ZSI. Table 6.1 gives a summary of these improvements. It can be seen that the QZSI has continuous input current, reduced passive components rating, common earthing of the input power source,

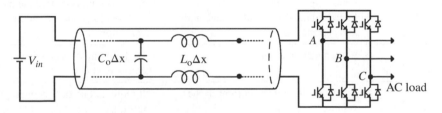

Figure 6.6 Distributed Z-source inverter (DiZSI) topology (*Source*: Jiang 2011 [25]. Reproduced with permission of IEEE).

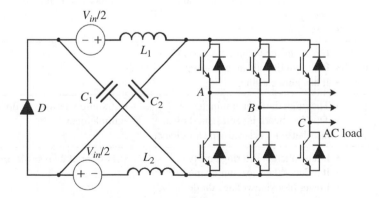

Figure 6.7 Embedded Z-source inverter (EZSI) topology.

Table 6.1 ZSI Topology Improvement Summary (*Source*: Ellabban 2016 [3]. Reproduced with permission of IEEE)

Topology Improvement	Advantages	Disadvantages
BZSI	• Bidirectional power flow	• Input diode is replaced by a bidirectional switch
HP-ZSI	• Wide load range operation • Simple Z-network inductor design	• Additional capacitor and a bidirectional switch
IZSI SZSI	• It can reduce the Z-source capacitor voltage stress • It has an inherent limitation to inrush current	• It has discontinuous input current
QRSSZSI	• Improved efficiency	• Additional quasi-resonant network and one auxiliary switch
FWZSI	• It has a neutral point • Better output quality	• Additional capacitor and two bidirectional switches
FLZSI	• It has a neutral point • Better output quality	• Additional two bidirectional switches
DuZSI	• It has a neutral point • Improved quality output voltage and less current stress	• Switch count is double the basic ZSI topology and two additional capacitors
NPZSI	• It has a neutral point	• Higher number of passive components
ZSI-D	• It reduces the leakage current in the system	• It requires one fast recovery diode more
SZI-S	• It reduces the leakage current and improves the system stability for wide range of modulation index	• It requires two bidirectional controlled switches more
QZSI with discontinuous input current	• It reduces the voltage stress on C_1 and C_2 • It has joint earthing	• It has discontinuous input current
QZSI with continuous input current	• It reduces the voltage stress on C_2 • It has continuous input current • It has joint earthing	
DiZSI	• The input diode is eliminated • The Z-network is represented by a distributed inductance and capacitance	• Z-source network is difficult to implement
EZSI	• It has inherent filtering ability • It reduces the capacitor sizing • Lower blocking voltage diode	• Two input dc sources are required

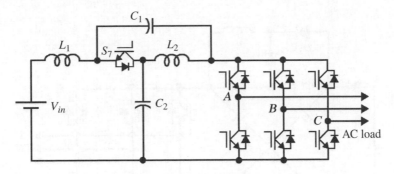

Figure 6.8 Bidirectional QZSI topology.

the dc-link, reduced leakage currents and is replacing the input diode by a bidirectional switch producing bidirectional QZSI (BQZSI), as shown in Figure 6.8, and will also allow bidirectional power flow [28]. Therefore, the BQZSI appears to be the most improved ZSI topology, which gives the best performance by just replacing the input diode by a bidirectional switch and rearranging the Z-network elements.

6.3 Extended Boost Topologies

6.3.1 Switched Inductor Z-Source Inverter

Here we replace one or two inductors of the Z-network with switched inductor (SL) circuits in ZSI/QZSI resulting in SL-ZSI or SL-QZSI to achieve high boost capability, which will also lead to size saving, high power density, and improved dependence between voltage gain and modulation index. There are different versions of combined ZSI and SL circuits. Figure 6.9(a) shows the SL-ZSI topology, where two inductors of the Z-network in the basic ZSI topology have been replaced by two SL cells [29, 30]. Figure 6.9(b) shows the SL-IZSI topology where two inductors of the Z-network in the IZSI topology have been replaced by two SL cells [31]. Figure 6.10(a) shows the switched inductor QZSI (SL-QZSI) topology, where one inductor of the Z-network in the QZSI topology has been replaced by an SL cell [32]. Figure 6.10(b) shows the improved switched-inductor QZSI (ISL-QZSI) topology where one diode in the SL cell has been replaced by a capacitor, a voltage-lifting unit is established, and, compared with the SL-QZSI, ISL-QZSI topologies can greatly improve the boost ability with improved efficiency [33, 34]. In [35, 36], a combination of switched-capacitor (SC) and a three-winding switched-coupled-inductor (SCL) is applied to the QZSI, and the topology obtained is termed SCL-QZSI, which is shown in Figure 6.10(c). The SCL-QZSI topology retains all of the advantages of the QZSI topology but, in addition, the integration of the SC with the SCL enhances the boost ability of the SCL-QZSI with a smaller component count and lower turn ratio. In [37], the active-switched capacitor/switched-inductor quasi-Z-source inverter (ASC/SL-qZSI) has been proposed, as shown in Figure 6.10(d), based on the QZSI topology. The ASC/SL-QZSI offers a small number of passive components such as capacitors and inductors in the impedance network, high boost ability, and low voltage stress across the switching devices. However, the ASC/SL-QZSI topology does not share the negative point between the dc input voltage source and the inverter. Figures 6.11(a) and (b) show the two switched-inductor QZSI topologies (rSL-QZSI and cSL-QZSI) where two inductors of the Z-network in the

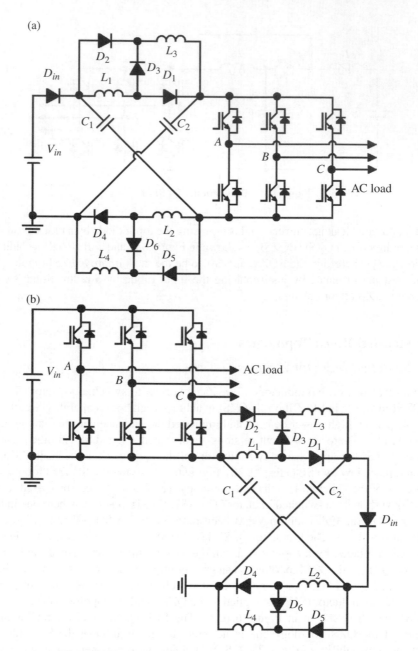

Figure 6.9 (a) SL-ZSI topology and (b) SL-IZSI topology.

QZSI topology have been replaced by two SL cells [38]; and Figure 6.11(c) shows the extended switched-inductor quasi-Z-source inverter (ESL-QZSI) topology where the SL-QZSI topology has been combined with the traditional boost converter [39].

Table 6.2 summarizes the different characteristics of these topologies, and Figure 6.12 shows the relationship of the boost factor with the duty cycle for the different topologies.

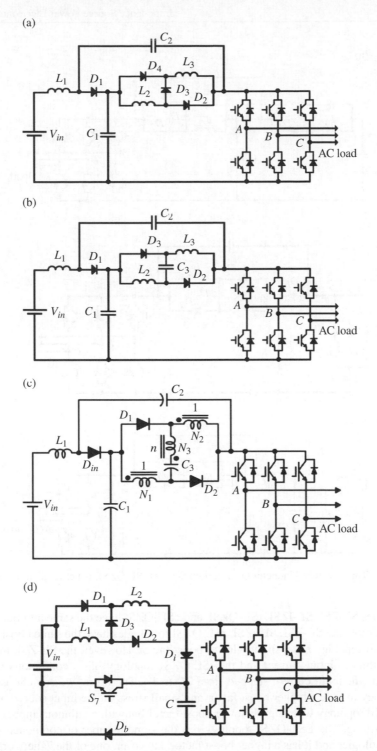

Figure 6.10 SL-QZSI different topologies: (a) SL-QZSI, (b) ISL-QZSI, (c) SCL-QZSI, (d) ASC/SL-QZSI.

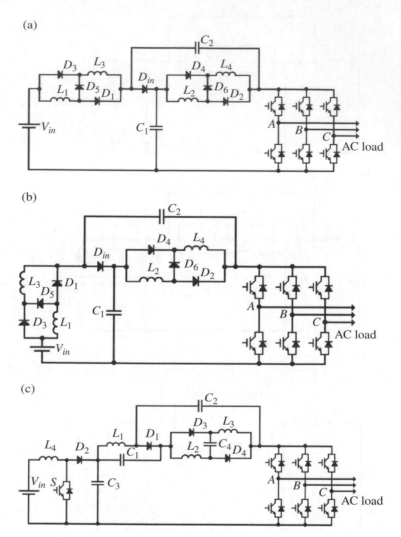

Figure 6.11   Three switched inductor QZSI topologies: (a) rSL-QZSI, (b) cSL-QZSI, (c) ESL-QZSI.

The topologies SL-ZSI, SL-IZSI, rSL-QZSI, and cSL-QZSI have the same number of components. In addition, the SL-ZSI, SL-IZSI, rSL-QZSI topologies have the same boost factor and the cSL-QZSI topology has a slightly lower boost factor. However, the SL-ZSI and SL-IZSI have discontinuous input current and the rSL-QZSI topology has a continuous but rippled input current (the input current stepped from $2I_L$ to I_L). Therefore, we can choose the rSL-QZSI topology to have a higher boost factor and small stress on the input source or choosing the cSL-QZSI topology with slightly lower boost factor but with continuous input current and common earthing. The ESL-QZSI topology has the same number of components as in the previous topologies and it has a higher boost factor. However, one of the Z-network elements is an active switch. The SL-QZSI, ISL-QZSI, and SCL-QZSI topologies have a lower component count than the previous topologies. Furthermore, the SL-QZSI topology has the lowest

Table 6.2 Characteristics of different switched inductor, ZSI combinations (*Source:* Ellabban 2016 [3]. Reproduced with permission of IEEE)

	ZSI	SL-ZSI	SL-IZSI	SL-QZSI	ISL-QZSI	SCL-QZSI	ASC/SL-QZSI	rSL-QZSI	cSL-QZSI	ESL-QZSI
No. of elements	2 inductors 2 capacitors 1 diode	4 inductors 2 capacitors 7 diodes	4 inductors 2 capacitors 7 diodes	3 inductors 2 capacitors 4 diodes	3 inductors 3 capacitors 3 diodes	1 inductor 1 coupled inductor 3 capacitors 3 diode	2 inductors 1 capacitor 5 diodes 1 switch	4 inductors 2 capacitors 7 diodes	4 inductors 2 capacitors 7 diodes	4 inductors 4 capacitors 4 diodes 1 switch
Continuous input current	No	No	No	Yes	Yes	Yes	Yes, rippled	Yes, rippled	Yes	Yes
Startup inrush current	Yes	Yes	No	No	No	No	No	No	No	No
Common Earth	No	No	No	Yes	Yes	Yes	No	Yes	Yes	Yes
Boost factor	$\dfrac{1}{1-2D_0}$ $0 \le D_0 \le 0.5$	$\dfrac{1+D_0}{1-3D_0}$ $0 \le D_0 \le 1/3$	$\dfrac{1+D_0}{1-3D_0}$ $0 \le D_0 \le 1/3$	$\dfrac{1+D_0}{1-2D_0-D_0^2}$ $0 \le 1-2D_0-D_0^2 \le 1$	$\dfrac{2}{1-3D_0}$ $0 \le D_0 \le 1$	$\dfrac{3}{1-4D_0}$ $0 \le D_0 \le 1/4$	$\dfrac{1+D_0}{1-3D_0}$ $0 \le D_0 \le 1/3$	$\dfrac{1+D_0}{1-3D_0}$ $0 \le\ \le 1/3$	$\dfrac{1}{1-3D_0}$ $0 \le D_0 \le 1/3$	$\dfrac{2}{(1-3D_0)(1-D_0)}$ $0 \le D_0 \le 1/3$

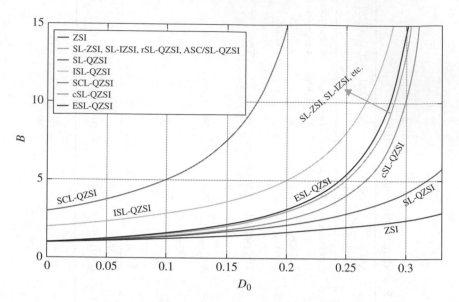

Figure 6.12 Relationship of the boost factor and duty ratio for different switched inductor ZSI topologies (*Source*: Ellabban 2013 [3]. Reproduced with permission of IEEE).

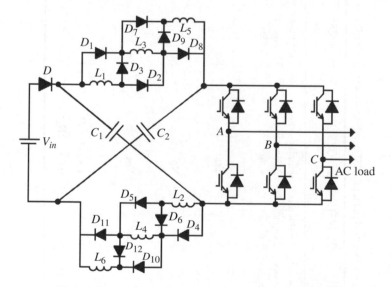

Figure 6.13 Multi-cell switched inductor ZSI topology.

boost factor (but higher than the classical ZSI topology), while the SCL-QZSI topology has the highest boost factor of the SL-ZSI different topologies. Therefore, the SCL-QZSI topology may be the best choice, if a high boost factor and lower component count with continuous input current, common earthing, and higher efficiency are demanded [35].

Furthermore, the SL-ZSI/QZSI topology can be generalized by adding extra cells to the original two inductors rather than replacing them. Each cell consists of one inductor and three diodes. These cells introduce additional inductors in parallel during ST charging and more inductors in series during the NST discharging, as shown in Figure 6.13 [40].

6.3.2 Tapped-Inductor Z-Source Inverter

A tapped inductor (TL) Z-source inverter (TL-ZSI), as shown in Figure 6.14(a), can be developed by replacing the two inductors in the classical Z-network with two two-terminal tapped inductor (TL) cells. The output voltage range of the TL-ZSI can be widely expanded by varying the turn ratio of the TL and by using much shorter shoot-through duration than the classical ZSI which in turn leads to a larger modulation index, and hence a better output waveform quality. Therefore, by using only five diodes and two TL cells, the TL-ZSI can produce a voltage gain that is much higher than many existing ZSI topologies with lower weight, size, complexity, and cost [41]. Furthermore, a tapped inductor (TL) quasi Z-source inverter (TL-QZSI), as shown in Figure 6.14(b), was proposed in [42]. It introduces an impedance network including a tapped inductor and two diodes to replace one inductor of the traditional QZSI. The TL-QZSI can improve the boost inversion ability of the traditional QZSI, together with a voltage-buck freedom. Table 6.3 summarizes the different characteristics of the TL-ZSI and TL-QZSI, topologies. As noted, when $N=1$, the TL network resembles the SL network and gives the same boost factor with lower component count. Furthermore, Figure 6.15 shows the relationship of the boost factor to the duty cycle for the different topologies, where the boost factor of the TL-QZSI is higher than that of the traditional ZSI/QZSI, and lower than the TL-ZSI.

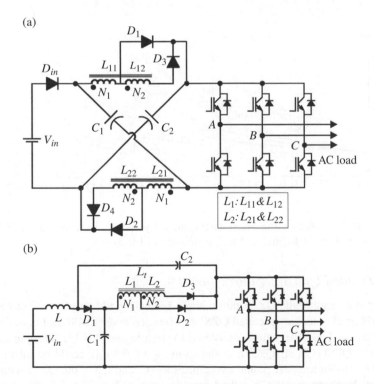

Figure 6.14 Different tapped inductor ZSI topologies: (a) tapped inductor ZSI, (b) tapped inductor QZSI.

Table 6.3 Characteristics of different tapped inductor ZSI topologies (*Source*: Ellabban 2016 [3]. Reproduced with permission of IEEE)

	QZSI	TL-ZSI	TL-QZSI
No. of elements	2 inductors	2 TL cells	1 inductor
	2 capacitors	3 capacitors	1 TL cell
	1 diode	5 diodes	2 capacitors
			3 diodes
Continuous input current	Yes	No	Yes
Startup inrush current	No	Yes	No
Common earth	Yes	No	Yes
Boost factor	$\dfrac{1}{1-2D_0}$	$\dfrac{1+ND_0}{1-2D_0-ND_0}$	$\dfrac{1+ND_0}{1-2D_0-ND_0^2}$
	$0 \le D_0 \le 0.5$	$D_0 < 1/(2+N)$	$D_0 < \left(\sqrt{N+1}-1\right)/N$

N is turns ratio of the tapped inductor

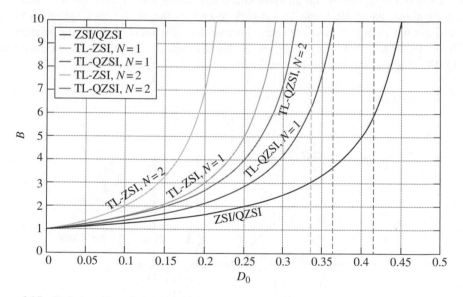

Figure 6.15 Relationship of the boost factor and duty ratio with different TL-ZSI topologies (*Source*: Ellabban 2016 [3]. Reproduced with permission of IEEE).

6.3.3 Cascaded Quasi-Z-Source Inverter

The concept of extending the QZSI gain without increasing the number of active switches is commonly referred to as the cascaded QZSI or extended boost QZSI and could be classified as capacitor assisted (CA) and diode assisted (DA) topologies [43–45]. The capacitor assisted extended boost QZSI (CAEB-QZSI), as shown in Figure 6.16(a), could be formed by adding one diode (D_2), one inductor (L_3), and two capacitors (C_3 and C_4) to the QZSI with continuous input current. Furthermore, the modified capacitor assisted cascaded QZSI (MCAC-QZSI) could be derived from the CAC-QZSI simply by changing the connection points of

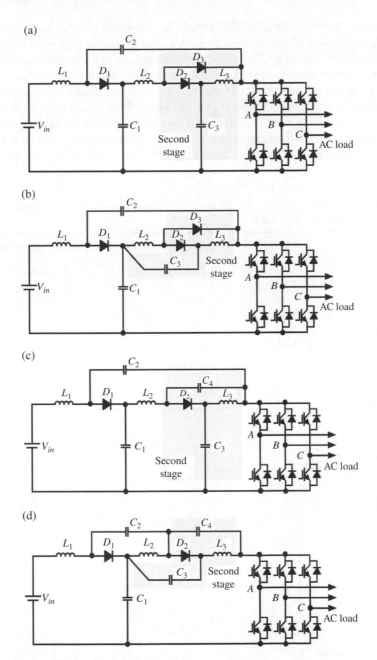

Figure 6.16 Extended boost quasi-Z-source inverter topologies: (a) CAC-QZSI, (b) MCAC-QZSI, (c) DAC-QZSI, (d) MDAC-QZSI.

capacitors C_2 and C_3, as shown in Figure 6.16(b). Moreover, the diode assisted extended boost QZSI (DAEB-QZSI), as shown in Figure 6.16(c), could be derived by adding one capacitor (C_3), one inductor (L_3), and two diodes (D_2 and D_3) to the continuous input current QZSI topology. In addition, the modified diode assisted extended boost QZSI (MDAEB-QZSI) could be derived from the DAEB-QZSI simply by changing the connection points of the capacitor C_3, as shown in Figure 6.16(d). Table 6.4 summarizes the different characteristics of these cascaded QZSI topologies. The capacitor-assisted topologies would produce high boost with smaller shoot-through, as shown in Figure 6.17, and would apply lower voltage stress on the devices. Furthermore, two topology families have the expandability to achieve even higher boost by adding more stages on the front end, and also hybrid extended-boost topologies can be achieved by combining both diode-assisted and capacitor-assisted techniques. However, these extended-boost QZSI topologies have obvious shortcomings, such as a small boost effect, a more complicated structure, complicated passive element design, and also large volume.

Table 6.4 Characteristics of different cascaded QZSI topologies (*Source*: Ellabban 2016 [3]. Reproduced with permission of IEEE)

	QZSI	CAC-QZSI/MCAC-QZSI	DAC-QZSI/MDAC-QZSI
No. of elements	2 inductors	3 inductors	3 inductors
	2 capacitors	3 capacitors	4 capacitors
	1 diode	3 diode	2 diode
Continuous input current	Yes	No	Yes
Startup inrush current	No	No	No
Common earth	Yes	Yes	Yes
Boost factor	$\dfrac{1}{1-2D_0}$	$\dfrac{1}{1-3D_0}$	$\dfrac{1}{\left(1-D_0\right)\left(1-2D_0\right)}$
	$0 \le D_0 \le 0.5$	$0 \le D_0 \le 1/3$	$0 \le D_0 \le 0.5$

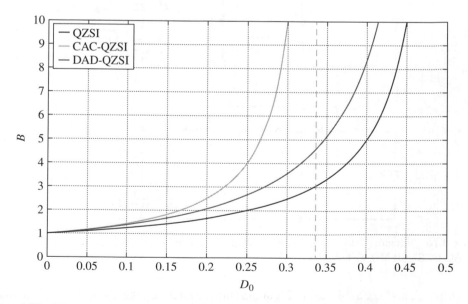

Figure 6.17 Voltage boosting capability of different extended boost QZSI topologies (*Source*: Ellabban 2016 [3]. Reproduced with permission of IEEE).

6.3.4 Transformer-Based Z-Source Inverter

Magnetically coupled inductors and transformers find a niche in impedance networks to improve the voltage boost capability as well as the modulation index. In addition, they reduce the number of passive components needed in the network, which improves the power density and reduces the cost of the system [46]. The T-source inverter (TSI) and the trans-Z-source inverter (Trans-ZSI) are two modified forms of the basic ZSI topology, which are achieved by reducing the size of the Z-network. The two inductors are built together on one core to form a coupled inductor (transformer) with low leakage inductance and instead of having two capacitors only one capacitor is used in the TSI/Trans-ZSI. In addition, the TSI/Trans-ZSI have the feature of the common dc rail used between the dc source and the inverter bridge. But the most significant advantage of the TSI/Trans-ZSI is the extended possibility of manipulation of inverter output voltage and ST duty ratio using transformer turns ratio greater than 1 and reducing the voltage stress on the inverter switches. Figure 6.18(a) shows the TSI topology [47, 48] and Figure 6.18(b) shows the trans-QZSI topology [49].

Figure 6.19(a) shows an improved trans-QZSI with continuous input current and high boost inversion capability. It consists of one inductor (L_3), one transformer, two capacitors (C_1 and C_2), and one diode (D). Its main characteristics are: the input dc current is continuous with lower ripples and lower current stress flow on the transformer windings and input diode compared with the trans-QZSI topology; it provides resonant current suppression, unlike the trans-QZSI topology in Figure 6.18(b), because no current flows to the main

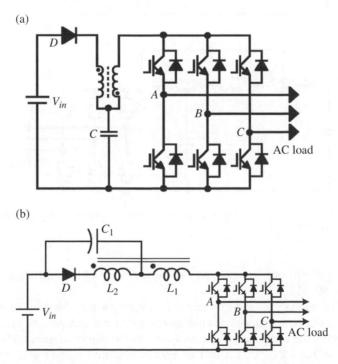

Figure 6.18 Transformer based Z-source inverter topologies: (a) T-source inverter (TSI)/trans-Z-source inverter (trans-ZSI), (b) trans-QZSI.

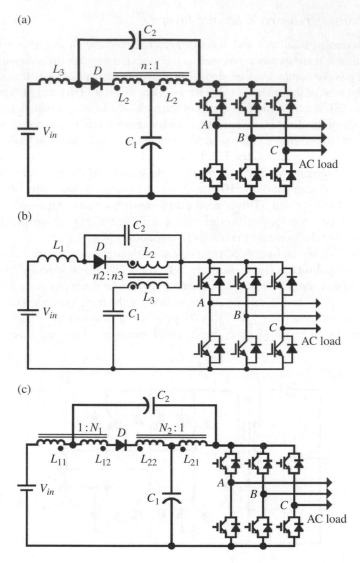

Figure 6.19 Improved transformer-based Z-source inverter topologies: (a) Improved trans-QZSI, (b) LCCT-ZSI, (c) TQZSI.

circuit at startup and only one inductor and one capacitor are added, and a higher boost factor can be obtained [50, 51].

Figure 6.19(b) shows an inductor-capacitor-capacitor-transformer Z-source inverter (LCCT-ZSI) using one more inductor and one more capacitor in comparison with the trans-ZSI/ trans-QZSI and the same component count compared to the improved trans-QZSI [52, 53]. The unique property of the LCCT-ZSI is that no energy is stored in the transformer windings. The two DC-current blocking capacitors connected in series with the transformer also prevent the transformer core from saturating.

By replacing two inductors in the classical ZSI topology with two transformers, the TZ-source inverter (TZSI) produces a very high boost voltage gain when the turns ratio of the transformers is larger than 1. Compared to the trans-ZSI, the TZSI uses a lower transformer turns ratio, which reduces the transformer size and weight, while producing the same output voltage gain because the power rating and turns ratio of each transformer in the TZSI are smaller than those of the transformer in the trans-ZSI under the same output power and voltage gain conditions; Figure 6.19(c) shows the TQZSI topology [54].

The Γ-Z-source inverter (Γ-ZSI), as shown in Figure 6.20(a), uses a unique Γ-shaped impedance network, composed of a transformer and capacitor, for boosting its output voltage and producing flexible enhanced voltage gain. However, it has discontinuous input current [55]. The asymmetrical Γ Z Source inverter (Asym Γ ZSI), shown in Figure 6.20(b), which has a continuous input current, uses one coupled transformer, one inductor, and two capacitors, which are the same as the improved trans-QZSI [56]. Unlike other transformer-based topologies, the voltage gain of both Γ-ZSI and Asym-Γ-ZSI is raised by lowering their transformer turns ratio, rather than increasing it.

By reconfiguring the TZ-source network, a sigma-source network (Σ-source network) which consists of dual Γ-source networks is realized. The sigma-Z-source inverter (ΣZSI) topology, as shown in Figure 6.21(a), uses the same number of components as the TZSI,

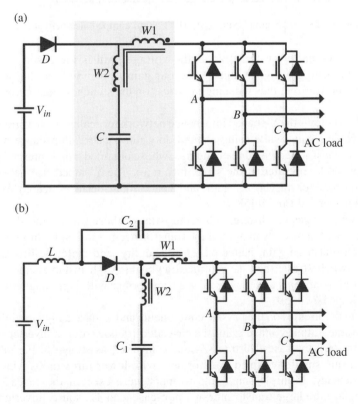

Figure 6.20   (a) Γ-Z-source inverter (Γ-ZSI); (b) asymmetrical Γ-Z-source inverter (Asym-Γ-ZSI).

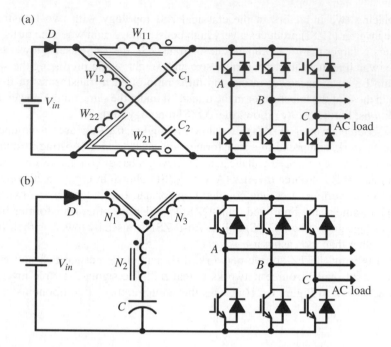

Figure 6.21   (a) Sigma-ZSI (ΣZSI), (b) Y-source impedance network (YSIN).

but it achieves a smaller system size as the voltage gain is increased by reducing the turns ratio. By having a higher input-to-output gain ratio, the modulation index of the inverter can be maximized, thus reducing the stress on the switches and improving the power quality [57].

Figure 6.21(b) shows the Y-shaped impedance network for realizing converters that demand a very high voltage gain, while using a small duty ratio. The Y-impedance network uses a tightly coupled transformer with three windings, whose obtained gain is presently not matched by existing networks operated at the same duty ratio. The Y impedance network has more degrees of freedom for varying its gain, and hence more design freedom for meeting the requirements demanded from it [58].

Table 6.5 summarizes the different characteristics of these transformer-based ZSI topologies. In spite of the many advantages of the transformer-based ZSI topologies such as: high modulation ratio at high gain, better waveform quality, and better dc-link utilization and hence lower switch stress, they have some drawbacks: high instantaneous shoot-through and magnetization currents and they require a transformer with high magnetic coupling and low leakage impedance [59].

To distribute the stresses on the passive components and to have a high modulation index to produce the same output voltage gain, the cascaded trans-Z-source inverter was proposed in [60, 61]. In [60], the cascaded trans-Z-source inverter, as shown in Figure 6.22(a), was proposed by using smaller coupled transformers with lower turns ratios which leads to an improved efficiency. Multiple small transformers were used in the cascaded TZ-source inverter to replace the large transformer in a non-cascaded TZ-source inverter, as shown in Figure 6.22(b) [61]. Since the cascaded TZ-source inverter has a large number of components,

Table 6.5 Characteristics of different transformer-based ZSI topologies (*Source*: Ellabban 2016 [3]. Reproduced with permission of IEEE)

	TSI/Trans-ZSI	Trans-QZSI	Improved trans-QZSI	LCCT-ZSI	TQZSI	Γ-ZSI	ΣZSI	Asym-Γ-ZSI	YSIN
No. of elements	1 transformer 1 capacitors 1 diode	1 transformer 1 capacitors 1 diode	1 transformer 1 inductor 2 capacitors 1 diode	1 transformer 1 inductor 2 capacitors 1 diode	2 transformer 2 capacitors 1 diode	1 transformer 1 capacitors 1 diode	2 transformer 2 capacitors 1 diode	1 transformer 1 inductor 2 capacitors 1 diode	1 3-winding transformer 1 capacitors 1 diode
Continuous input current	No	Yes, high rippled	Yes	Yes	Yes	No	No	Yes	No
Startup inrush current	Yes	Yes	No	No	No	Yes	Yes	No	Yes
Common earth	Yes	Yes	Yes	Yes	Yes	Yes	No	Yes	Yes
Boost factor	$\dfrac{1}{1-(n+1)D_0}$ $0 \le D_0 \le (n+1)^{-1}$ n is transformer turns ratio	$\dfrac{1}{1-(n+1)D_0}$ $0 \le D_0 \le (n+1)^{-1}$ n is transformer turns ratio	$\dfrac{1}{1-(n+2)D_0}$ $0 \le D_0 \le (n+2)^{-1}$ n is transformer turns ratio	$\dfrac{1}{1-(n+1)D_0}$ $0 \le D_0 \le (n+1)^{-1}$ n is transformer turns ratio	$\dfrac{1}{1-(n_1+n_2+2)D_0}$ $0 \le D_0 \le \dfrac{1}{n_1+n_2+2}$ n_1 is T_1 turns ratio n_1 is T_2 turns ratio	$\dfrac{1}{1-\left(1+\dfrac{1}{\gamma_{\mathrm{rz}}-1}\right)D_0}$ $0 \le D_0 \le \dfrac{1}{1+\dfrac{1}{\gamma_{\mathrm{rz}}-1}}$ γ_{rz} is transformer turns ratio	$\dfrac{1}{1-\gamma D_0}$ $\gamma = 2+\dfrac{1}{n_{\mathrm{T1}}-1}+\dfrac{1}{n_{\mathrm{T2}}-1}$ $0 \le D_0 \le \dfrac{1}{\gamma}$ $n_{\mathrm{T1}}, n_{\mathrm{T2}}$ transformers turns ratio	$\dfrac{1}{1-\left(2+\dfrac{1}{\gamma_{\mathrm{rz}}-1}\right)D_0}$ $0 \le D_0 \le \dfrac{1}{2+\dfrac{1}{\gamma_{\mathrm{rz}}-1}}$ γ_{rz} is transformer turns ratio	$\dfrac{1}{1-kD_0}$ $0 \le D_0 \le k^{-1}$ $k = \dfrac{N_3+N_1}{N_3-N_2}, k \ge 2$ N_1, N_2, N_3 3-winding transformer turns ratio

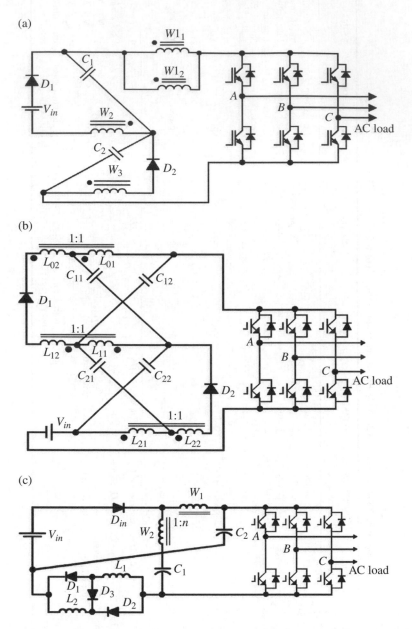

Figure 6.22 Different cascaded and combined transformer-based ZSI topologies: (a) cascaded trans-Z-source inverter, (b) cascaded TZ-source inverter, (c) switched inductor Γ-source inverter.

it should only be considered in a design where the use of a few higher-rated components is not feasible, or when the use of multiple lower-rated components is desirable.

Furthermore, in [62], a high step-up inverter with combined Γ-Z-source inverter and switched inductor voltage multiplier cell was presented, as shown in Figure 6.22(c), which resulted in a higher voltage gain than the Γ-Z-source inverter at any transformer turn ratio.

6.3.5　High Frequency Transformer Isolated Z-Source Inverter

The high frequency transformer isolated Z-source inverter (HFTI-ZSI) carefully integrates the Z-source network concept into a two-stage dc-dc-ac topology by employing a high frequency transformer. In addition to inheriting the merits of basic ZSI, the HFTI-ZSI can achieve electrical isolation and a higher voltage boost ratio with reduced total device stresses, abilities which are lacking in the single-stage ZSI [63, 64]. Figure 6.23 shows the high frequency transformer isolated quasi Z-source inverter (HFTI-QZSI) as an example for the HFTI-ZSI.

6.4　L-Z-Source Inverter

The L-Z-source inverter (L-ZSI), as shown in Figure 6.24, contains inductors and diodes in the Z-network. The L-ZSI uses a unique inductor and diode network for boosting its output voltage, and provides a common ground for the dc source and inverter. Furthermore, it avoids the disadvantage caused by capacitors in most ZSI topologies, especially in prohibiting the inrush current at startup and the resonance of Z-source capacitors and inductors, in addition to improving the converter efficiency [65]. Figure 6.25 shows the boost factor of the L-ZSI compared to the classical ZSI and the SL-ZSI topologies, and by adjusting the number of inductors, (a) the boost factor of the L-ZSI can be larger than that of the SL-ZSI and the classical ZSI with short shoot-through duty ratio, (b) the boost factor of L-ZSI is increased

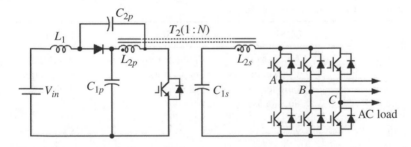

Figure 6.23　High frequency transformer isolated quasi Z-source inverter (HFTI-QZSI).

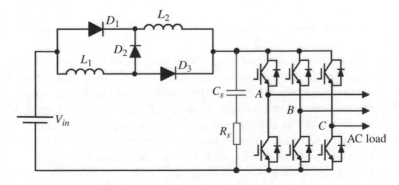

Figure 6.24　L-Z-source inverter topology.

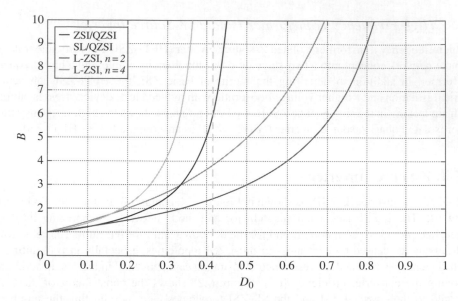

Figure 6.25 Boost factor comparison for the L-ZSI, SL-ZSI, and ZSI topologies (*Source*: Ellabban 2016 [3]. Reproduced with permission of IEEE).

Table 6.6 Characteristics of the L-ZSI topology compared to other ZSI topologies (*Source*: Ellabban 2016 [3]. Reproduced with permission of IEEE)

	QZSI	SL-QZSI	L-ZSI
No. of elements	2 inductors	3 inductors	2 inductors ($n=2$)
	2 capacitors	2 capacitors	0 capacitors
	1 diode	4 diode	3 diode
Continuous input current	No, rippled	Yes	Yes
Startup inrush current	No	No	No
Common earth	Yes	Yes	Yes
Boost factor	$\dfrac{1}{1-2D_0}$	$\dfrac{1+D_0}{1-2D_0-D_0^2}$	$\dfrac{1+(n-1)D_0}{(1-D_0)}$
	$0 \le D_0 \le 0.5$	$0 \le 1-2D_0-D_0^2 \le 1$	$0 \le D_0 \le 1$
			n is the no. of inductors

with the increasing number of inductors, and (c) the number of inductors is not limited and the range of D_0 is [0, 1) (Table 6.6). The average value of the boost factor change rate of the L-ZSI is far less than that of the SL-ZSI and the classical ZSI topologies, which makes the control of L-ZSI easier than other topologies. However, if the L-Z-source impedance network works in DCM mode, the dc-link voltage increases infinitely, the output voltage will be uncontrollable, and the system is unstable. In order to avoid the problem caused by the DCM mode, a snubber circuit (R_s and C_s, as shown in Figure 6.24) is needed. In addition, The L-ZSI operates at large ST duty cycles to obtain higher voltage gains, which reduces output power quality.

6.5   Changing the ZSI Topology Arrangement

In this section, the basic ZSI structure has been used to extend its benefits to multilevel, multi-output and multi-input/multi-output applications, as listed in Table 6.7.

The three-level neutral-point-clamped (3 L-NPC) inverter has many advantages over the two-level VSI: better than halving the semiconductor voltage stress, higher switching frequency because of lower switching losses, decreased dv/dt and better harmonic performance. However, the 3 L-NPC can only perform the voltage buck function. By integrating the Z-source topological concept to the basic neutral-point clamped (NPC) inverter topology, a family of single-stage buck–boost multilevel ZSI had been proposed [66–71]. Figure 6.26 shows different topologies of Z-source three-level NPC inverter and Figure 6.27 shows a five-level Z-source diode clamped NPC inverter.

A four-level cascaded Z-source inverter was proposed in [71], as shown in Figure 6.28(a), the four-level cascaded Z-source inverter gives a higher output voltage gain compared to the conventional four-level cascaded inverter with similar harmonic distortion and with less complex circuitry and control logic. A combination of a QZSI into cascaded H-bridge (CHB) has been proposed in [72], as shown in Figure 6.28(b), the cascaded structure reduced the voltage gain and improved the system reliability, due to the QZSI ST capability.

A dual-output Z-source inverter using nine switches was proposed in [73, 74] to independently supply two different ac loads with buck–boost capability, as shown in Figure 6.29. This topology will be advantageous for high-power inverter applications, where cost and efficiency are key decision factors. A dual-input/dual-output Z-source inverter with two ac outputs and two dc inputs was proposed in [75] and is shown in Figure 6.30. The dual-input/dual-output ZSI can control amplitude, frequency, and phase of both ac outputs and also control current of both dc inputs.

Table 6.7   Changing the ZSI topology arrangement overview (*Source*: Ellabban 2016 [3]. Reproduced with permission of IEEE)

Changing the ZSI Topology Arrangement	Comments
Multilevel Z-source neutral-point-clamped (NPC) inverter	Advantages: • Minimum harmonic distortion • Eliminated common-mode voltage • Lower individual semiconductor stress
Cascaded-multilevel (CML) Z-source inverter	• Modular • High voltage/high power grid-tied without transformer • Requires fewer components in each level
Nine switch ZSI	• A one-input two-output converter, and it has higher output voltage when compared with NS-VSI
Dual-input-dual-output ZSI	• A modification of the above topology to use two-isolated input as a source and produce two-ac output

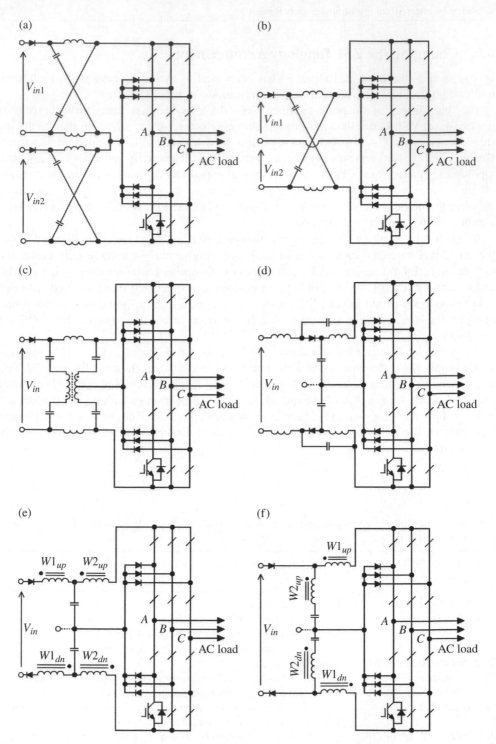

Figure 6.26 Three-level Z-source inverter topologies: (a) Z-source NPC inverter with two impedance networks and two DC sources, (b) Z-source NPC inverter with a single impedance network and two DC sources, (c) Z-source NPC inverter with a single impedance network and a single DC source, (d) quasi-Z-source NPC inverter, (e) Γ-Z-source NPC inverter, (f) trans-Z-source NPC inverter.

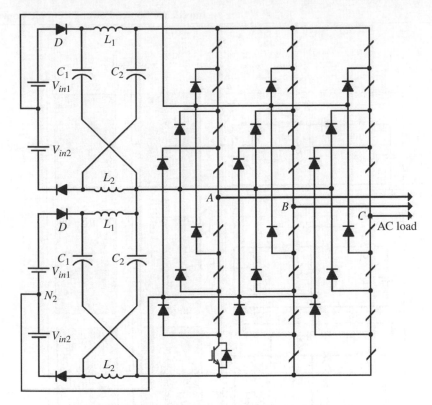

Figure 6.27 Five-level Z-source diode-clamped inverter.

(a)

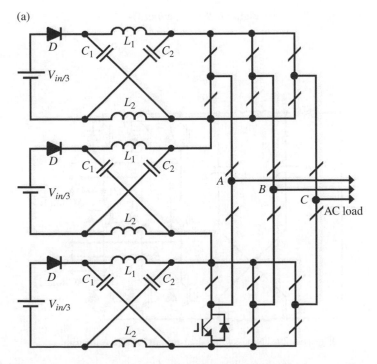

Figure 6.28 Cascaded multilevel ZSI: (a) Cascaded four-level ZSI, (b) quasi-Z-source cascade *n* multilevel inverter.

(b)

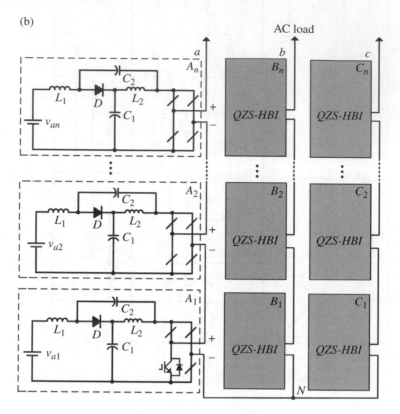

Figure 6.28 (*Continued*)

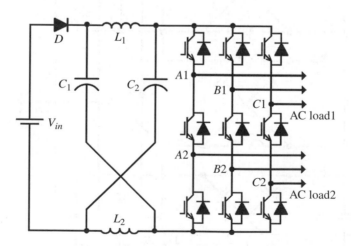

Figure 6.29   Nine-switch dual-output Z-source inverter.

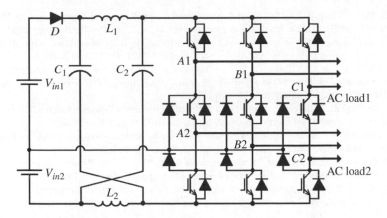

Figure 6.30 Dual-input/dual-output Z-source inverter.

6.6 Conclusion

The Z-source inverter is an innovative power electronics converter topology presented recently with very interesting properties such as buck–boost characteristics and also single-stage conversion. It employs a unique impedance network to couple the main circuit of the converter to the power source and utilizes the shoot-through state to boost the input voltage, which improves the inverter reliability and enlarges its application fields. Therefore, it has attracted more attention than other converters over the past decade and has already been tried with many applications. Various Z-source inverter topologies have been reported in the literature. This chapter has presented an intensive overview of the Z-source inverter's different topology improvements and different arrangements. However, this study is only limited to voltage-type dc/ac Z-source inverter configuration to present more comparative details. This overview will help power electronics researchers and engineers to understand and identify the pros and cons of each topology and choose the most suitable topology for their applications.

References

[1] F. Z. Peng, "Z-source inverter," *Proc. Ind. Appl. Conf.*, Oct. 13–18, 2002, vol.2, pp.775–781.

[2] F. Z. Peng, "Z-source inverter," *IEEE Trans. Ind. Appl.*, vol.39, no.2, pp.504–510, Mar/Apr. 2003.

[3] O. Ellabban, and H. Abu-Rub, "Z-Source Inverter: Topology Improvements Review," in *IEEE Industrial Electronics Magazine*, vol.10, no.1, pp. 6–24, Spring 2016.

[4] J. Rabkowski, "The bidirectional Z-source inverter as an energy storage/grid interface," *Proc. of the International Conference on "Computer as a Tool," EUROCON*, pp.1629–1635, 9–12 Sept. 2007.

[5] J. Rabkowski, "The bidirectional Z-source inverter for energy storage application," in *Proc. The European Conference on Power Electronics and Applications*, 2007, pp.1–10, 2–5 Sept. 2007.

[6] M. Shen and F. Z. Peng, "Operation modes and characteristics of the Z-source inverter with small inductance or low power factor," *IEEE Trans. Ind. Electron.*, vol.55, no.1, pp.89–96, January 2008.

[7] X. Ding, Z. Qian, S. Yang, B. Cui, F. Z. Peng, "A high-performance Z-source inverter operating with small inductor at wide-range load," in *Proc. The Twenty-Second Annual IEEE Applied Power Electronics Conference (APEC)*, 2007, pp.615–620.

[8] O. Ellabban, J. V. Mierlo, P. Lataire, "Voltage mode and current mode control for a 30 kW high-performance Z-source inverter," in *Proc. the IEEE Electrical Power & Energy Conference (EPEC)*, pp.1–6, 22–23 Oct. 2009.

[9] Y. Tang, S. Xie, C. Zhang, Z. Xu, "Improved Z-Source Inverter with Reduced Z-Source Capacitor Voltage Stress and Soft-Start Capability," *IEEE Trans. Power Electron.*, vol.24, no.2, pp.409–415 February 2009.

[10] K. Yu, F. Luo, M. Zhu, "Study of an improved Z-source inverter: small signal analysis," *Proc. of the 5th IEEE Conference on Industrial Electronics and Applications*, pp.2169–2174, 15–17 June 2010.

[11] L. Yang, D. Qiu, B. Zhang, G. Zhang, "A high-performance Z-source inverter with low capacitor voltage stress and small inductance," in *Proc. of the Twenty-Ninth Annual IEEE Applied Power Electronics Conference and Exposition (APEC)*, pp.2331–2337, 16–20 March 2014.

[12] J. Wei, Y. Tang, S. Xie, "Grid-connected PV System Based on the Series Z-Source Inverter," in *Proc. of the 5th IEEE Conference on Industrial Electronics and Applications (ICIEA)*, pp.532–537, 2010.

[13] J. Wei, Y. Tang, S. Xie, "Grid-connected PV System Based on the Series Z-Source Inverter," in *Proc. of the 5th IEEE Conference on Industrial Electronics and Applications*, pp.532–537, 2010.

[14] Y. Zhu, M. Chen, X. Lee, Y. Tsutomu, "A novel quasi-resonant soft-switching Z-source inverter," in *Proc. of the IEEE International Conference on Power and Energy (PECon)*, pp.292–297, 2–5 Dec. 2012.

[15] A. S. Khlebnikov, S. A. Kharitonov, "Application of the Z-source converter for aircraft power generation systems," in *Proc. of the 9th International Workshop and Tutorials on Electron Devices and Materials (EDM)*, pp.211–215, 1–5 July 2008.

[16] E. C. dos Santos, J. H. G. Muniz, E. P. X. P. Filho, E. R. C. Da Silva, "Dc-ac three-phase four-wire Z-source converter with hybrid PWM strategy," in *Proc. of the 36th Annual Conference on IEEE Industrial Electronics Society (IECON)*, pp.409–414, 7–10 Nov. 2010.

[17] A. S. Khlebnikov, S. A. Kharitonov, P. A. Bachurin, A. V. Geist, D. V. Makarov, "Modeling of dual Z-source inverter for aircraft power generation," in *Proc. of the International Conference and Seminar of Young Specialists on Micro/Nanotechnologies and Electron Devices (EDM)*, pp.373–376, 30 June 4 July 2011.

[18] P. A. Bachurin, D. V. Makarov, A. V. Geist, M. V. Balagurov, D. A. Shtein, "Z-source inverter with neutral point," in *Proc. of the 14th International Conference of Young Specialists on Micro/Nanotechnologies and Electron Devices (EDM)*, pp.255–258, 1–5 July 2013.

[19] F. Bradaschia, M. C. Cavalcanti, P. E. P. Ferraz, F. A. S. Neves, E. C. dos Santos, J. H. G. M. da Silva, "Modulation for three-phase transformerless Z-source inverter to reduce leakage currents in photovoltaic systems," *IEEE Trans. Ind. Electron.*, vol.58, no.12, pp.5385–5395, 2011.

[20] V. Erginer, M. H. Sarul, "Modified reduced common mode current modulation techniques for Z-Source inverter used in photovoltaic systems," *Proc. of the 4th Power Electronics, Drive Systems and Technologies Conference (PEDSTC)*, pp.459,464, 13–14 Feb. 2013.

[21] P. E. P. Ferraz, F. Bradaschia, M. C. Cavalcanti, F. A. S. Neves, G. M. S. Azevedo, "A modified Z-source inverter topology for stable operation of transformerless photovoltaic systems with reduced leakage currents," *Proc. of the 2011 Brazilian Power Electronics Conference (COBEP)*, pp.615–622, 11–15 Sept. 2011.

[22] J. Anderson, F. Peng, "A Class of Quasi-Z-Source Inverters," *IEEE Industry Applications Society Annual Meeting*, pp.1–7, 5–9 Oct. 2008.

[23] J. Anderson, F. Z. Peng, "Four quasi-Z-source inverters," *IEEE Power Electronics Specialists Conference (PESC)*, pp.2743–2749, 15–19 June 2008.

[24] F. Z. Peng, "Z-source networks for power conversion," *Proc. of the Twenty-Third IEEE Applied Power Electronics Conference and Exposition (APEC)*, pp.1258–1265, 24–28 Feb. 2008.

[25] S. Jiang, F. Z. Peng, "Transmission-line theory based distributed Z-source networks for power conversion," *Proc. of the Twenty-Sixth Annual IEEE Applied Power Electronics Conference and Exposition (APEC)*, pp.1138–1145, 6–11 March 2011.

[26] F. Gao, P. C. Loh, F. Blaabjerg, C. J. Gajanayake, "Operational analysis and comparative evaluation of embedded Z-source inverters," *Proc. of the IEEE Power Electronics Specialists Conference (PESC)*, pp.2757–2763, 15–19 June 2008.

[27] S. Khani, L. Mohammadian, S. H. Hosseini, K. R. Milani, "Application of embedded Z-Source inverters in grid-connected photovoltaic systems," *Proc. of the 18th Conference on Electrical Power Distribution Networks (EPDC)*, pp.1,5, 30 April–1 May 2013.

[28] F. Guo, L. Fu, C. Lin, C. Li, W. Choi, J. Wang, "Development of an 85-kW bidirectional quasi-Z-source inverter with dc-link Feed-Forward Compensation for Electric Vehicle Applications," *IEEE Trans. Power Electron.*, vol.28, no.12, pp.5477–5488, Dec. 2013.

[29] M. Zhu, K. Yu, F. Luo, "Topology analysis of a switched-inductor Z-source inverter," in *Proc. of the 5th IEEE Conference on Industrial Electronics and Applications*, pp.364–369, 15–17 June 2010.

[30] M. Zhu, K. Yu, F. Luo, "Switched inductor Z-source inverter," *IEEE Trans. Power Electron.*, vol.25, no.8, pp.2150–2158, Aug. 2010.

[31] M. Ismeil, M. Orabi, R. Kennel, O. Ellabban, H. Abu-Rub, "Experimental studies on a three phase improved switched Z-source inverter," in *Proc. of the Applied Power Electronics Conference and Exposition (APEC)*, pp.1248–1254, 16–20 March 2014.

[32] M.-K. Nguyen, Y. Lim, G. Cho, "Switched-inductor quasi-Z-source inverter," *IEEE Trans. Power Electron.*, vol.26, no.11, pp.3183–3191, Nov. 2011.

[33] K. Deng, J. Zheng, J. Mei, "Novel switched-inductor quasi-Z-source inverter," *Journal of Power Electronics*, vol.14, no.1, pp.11–21, January 2014.

[34] L. Li, Y. Tang, "A high set-up quasi-Z-source inverter based on voltage-lifting unit," in *Proc. of the IEEE Energy Conversion Congress and Exposition (ECCE)*, pp.1880–1886, 14–18 Sept. 2014.

[35] F. Ahmed, H. Cha, S. Kim, H. Kim, "Switched-coupled-inductor quasi-Z-source inverter," to appear in *IEEE Trans. Power Electron.*, doi: 10.1109/TPEL.2015.2414971.

[36] F. Ahmed, H. Cha, S. Kim, H. Kim, "A high voltage gain switched-coupled-inductor quasi-Z-source inverter," in *Proc. of the International Power Electronics Conference (IPEC-Hiroshima 2014 – ECCE-ASIA)*, pp.480,484, 18–21 May 2014.

[37] A. Ho, T. Chun, H. T. Kim, "Extended boost active-switched-capacitor/switched-inductor quasi-Z-source inverters," *IEEE Trans. Power Electron.*, vol.30, no.10, pp.568- 5690, May 2015.

[38] M.-K. Nguyen, Y.-C. Lim, J.-H. Choi, "Two switched-inductor quasi-Z-source inverters," *IET Power Electron.*, vol.5, no.7, pp.1017–1025, August 2012.

[39] K. Deng, F. Mei, J. Mei, J. Zheng, G. Fu, "An extended switched-inductor quasi-Z-source inverter," *Journal of Electron. Eng. Technol.*, vol.8, pp.742–750, 2013.

[40] D. Li, P. C. Loh, M. Zhu, F. Gao, F. Blaabjerg, "Generalized multi-cell switched-inductor and switched-capacitor Z-source inverters," *IEEE Trans. Power Electron.*, vol.28, no.2, pp.837–848, Feb. 2013.

[41] M. Zhu, D. Li, P. C. Loh, F. Blaabjerg, "Tapped-inductor Z-Source inverters with enhanced voltage boost inversion abilities," in *Proc. of the 2nd IEEE Inter. Conf. on Sustainable Energy Technologies (ICSET)*, 6–9 Dec 2010.

[42] Y. Zhou, W. Huang, J. Zhao, P. Zhao, "Tapped inductor quasi-Z-source inverter," in *Proc. of the Twenty-Seventh Annual IEEE Applied Power Electronics Conference and Exposition (APEC)*, pp.1625–1630, 5–9 Feb. 2012.

[43] C. J. Gajanayake, F. Luo, H. Gooi, P. So, L. Siow, "Extended-boost Z-source inverters," *IEEE Trans. Power Electron.*, vol.25, no.10, pp.2642–2652, Oct. 2010.

[44] D. Vinnikov, I. Roasto, T. Jalakas, R. Strzelecki, M. Adamowicz, "Analytical comparison between capacitor assisted and diode assisted cascaded quasi-Z-source inverters," *Electrical Review*, vol.88, (1a/2012), pp.212–217.

[45] D. Vinnikov, I. Roasto, T. Jalakas, S. Ott, "Extended boost quasi-Z-source inverters: possibilities and challenges," *Electron. Elect. Eng.*, vol.112, no.6, pp.51–56, 2011.

[46] Y. P. Siwakoti, F. Z. Peng, F. Blaabjerg, P. C. Loh, G. E. Town, "Impedance-source networks for electric power conversion part I: A topological review," *IEEE Trans. Power Electron.*, vol.30, no.2, pp.699–716, Feb. 2015.

[47] M. Adamowicz, N. Strzelecka, "T-source inverter," *Electrical Review*, R. 85 NR 10/2009, pp.233–238.

[48] R. Strzelecki, M. Adamowicz, N. Strzelecka, W. Bury, "New type T-source inverter," in *Proc. of the Compatibility and Power Electronics (CPE '09)*, pp.191–195, 20–22 May 2009.

[49] W. Qian, F. Z. Peng, H. Cha, "Trans-Z-source inverters," *IEEE Trans. Power Electron.*, vol.26, no.12, pp.3453–3463, Dec. 2011.

[50] M. Nguyen, Y. Lim, S. Park, "Improved trans-Z-source inverter with continuous input current and boost inversion capability," *IEEE Trans. Power Electron.*, vol.28, no.10, pp.4500–4510, Oct. 2013.

[51] M. Nguyen, Q. Phan, Y. Lim, S. Park, "Transformer-based quasi-Z-source inverters with high boost ability," in *Proc. of the IEEE International Symposium on Industrial Electronics (ISIE)*, 28–31 May 2013.

[52] M. Adamowicz, "LCCT-Z-source inverters," in *Proc. of the 10th International Conference on Environment and Electrical Engineering (EEEIC)*, 8–11 May 2011.

[53] M. Adamowicz, R. Strzelecki, F. Z. Peng, J. Guzinski, H. Abu-Rub, "New type LCCT-Z-source inverters," in *Proc. of the 2011–14th European Conference on Power Electronics and Applications (EPE)*, 30 Aug.–1 Sept. 2011.

[54] M.-K. Nguyen, Y.-C. Lim, Y.-G. Kim, "TZ-source inverters," *IEEE Trans. Ind. Electron.*, vol.60, no.12, pp.5686–5695, Dec. 2013

[55] P. C. Loh, D. Li, F. Blaabjerg, "Γ-Z-source inverters," *IEEE Trans. Power Electron.*, vol.28, no.11, pp.4880–4884, Nov. 2013.

[56] W. Mo, P. C. Loh, F. Blaabjerg, "Asymmetrical Γ-source inverters," *IEEE Trans. Ind. Electron.*, vol.61, no.2, pp.637–647, Feb. 2014.

[57] J. Soon, K. Low, "Sigma-Z-source inverters," *IET Power Electron.*, vol.8, no.5, pp.715–723, 2015.

[58] Y. P. Siwakoti, Poh Chiang Loh, F. Blaabjerg, G. E. Town, "Y-source impedance network," *IEEE Trans. Power Electron.*, vol.29, no.7, pp.3250–3254, July 2014.

[59] P. C. Loh, F. Blaabjerg, "Magnetically coupled impedance-source inverters," *IEEE Trans. Ind. Appl.*, vol.49, no.5, pp.2177–2187, Sept.–Oct. 2013.

[60] D. Li, P. C. Loh, M. Zhu, F. Gao, F. Blaabjerg, "Cascaded multi-cell trans-Z-source inverters," *IEEE Trans. Power Electron.*, vol.28, no.2, pp.826–836, February 2013.

[61] M.-K. Nguyen, Y.-C.l Lim, S.-J. Park, Y.-G. Jung, "Cascaded TZ-source inverters," *IET Power Electron.*, vol.7, no.8, pp.2068–2080, August 2014.

[62] H. Zeinali, A. Mostaan, A. Baghramian, "Switched inductor Γ source inverter," in *Proc. of the 5th Power Electronics, Drive Systems and Technologies Conference (PEDSTC)*, pp.187–192, 5–6 Feb. 2014.

[63] S. Jiang, D. Cao, F. Z. Peng, "High frequency transformer isolated Z-source inverters," in *Proc. of the Twenty-Sixth Annual IEEE Applied Power Electronics Conference and Exposition (APEC)*, 2011, pp.442–449.

[64] Y. Ding, L. Li, "Research and application of high frequency isolated quasi-Z-source inverter," in *Proc. of the 38th Annual Conference on IEEE Industrial Electronics Society (IECON)*, pp.714–718, 25–28 Oct. 2012.

[65] L. Pan, "L-Z-source inverter," *IEEE Trans. Power Electron.*, vol.29, no.12, pp.6534–6543, Dec. 2014.

[66] P. C. Loh, F. Blaabjerg and C. P. Wong, "Comparative evaluation of pulse width modulation strategies for Z-source neutral-point-clamped inverter," *IEEE Trans. Power Electron.*, vol.22, no.3, pp.1005–1013, May 2007.

[67] P. C. Loh, S. W. Lim, F. G. and F. Blaabjerg, "Three-level Z-Source inverters using a single LC impedance network," *IEEE Trans. Power Electron.*, vol.22, No. 2, pp.706–711, March 2007.

[68] O. Husev, C. Roncero-Clemente, E. Romero-Cadaval, D. Vinnikov, S. Stepenko, "Single phase three-level neutral-point-clamped quasi-Z-source inverter," *IET Power Electron.*, vol.8, no.1, pp.1–10, 2015.

[69] W. Mo, P. C. Loh, F. Blaabjerg, P. Wang, "Trans-Z-source and Γ-Z-source neutral-point-clamped inverters," *IET Power Electron.*, vol.8, no.3, pp.371–377, 2015.

[70] F. Gao, P. C. Loh, F. Blaabjerg, R. Teodorescu and D. M. Vilathgamuwa, "Five-level Z-source diode-clamped inverter," *IET Power Electron.*, vol.3, no.4, pp.500–510, 2010.

[71] B. K. Chaithanya, A. Kirubakaran, "A novel four level cascaded Z-source inverter," in *Proc. of the IEEE International Conference on Power Electronics, Drives and Energy Systems (PEDES)*, pp.1–5, 16–19 Dec. 2014.

[72] Y. Liu, B. Ge, H. Abu-Rub, F. Z. Peng, "An effective control method for quasi-Z-source cascade multilevel inverter-based grid-tie single-phase photovoltaic power system," *IEEE Trans. Ind. Informat.*, pp.399–407, vol.10, no.1, Feb. 2014

[73] S. M. Dehghan, M. Mohamadian, A. Yazdian, F. Ashrafzadeh, "Space vectors modulation for nine-switch converters," *IEEE Trans. Power Electron.*, vol.25, no.6, pp.1488–1496, June 2010.

[74] S. M. Dehghan, M. Mohamadian, A. Yazdian, "Hybrid electric vehicle based on bidirectional Z-source nine-switch inverter," *IEEE Trans. Veh. Technol.*, vol.59, no.6, pp.2641–2653, July 2010.

[75] S. M. Dehghan, M. Mohamadian, A. Yazdian, F. Ashrafzadeh, "A dual-input–dual-output Z-source inverter," *IEEE Trans. Power Electron.*, vol.25, no.2, pp.360–368, February 2010.

7

Typical Transformer-Based Z-Source/Quasi-Z-Source Inverters

Poh Chiang Loh[1], Yushan Liu[2], and Haitham Abu-Rub[2]

[1] *Department of Energy Technology, Aalborg University, Aalborg East, Denmark*
[2] *Electrical and Computer Engineering Program, Texas A&M University at Qatar, Qatar Foundation, Doha, Qatar*

As the overview of derived networks has shown in Chapter 4, there are several transformer-based Z-source/quasi-Z-source inverters (ZSIs/qZSIs). In this chapter, the typical trans and LCCT ZSIs/qZSIs are discussed. Their configurations, working operations, voltage and current principles, and simulation results are presented, providing a demonstration of the investigation of other extended transformer-based ZSIs/qZSIs.

7.1 Fundamentals of Trans-ZSI

7.1.1 Configuration of Current-Fed and Voltage-Fed Trans-ZSI

The original Z-source inverters have been shown to demonstrate output voltage or current buck–boost ability. Despite their improved flexibility, these existing Z-source inverters still have some limitations, mostly linked to their requirement for low modulation ratio at high input-to-output gain, and the presence of an impedance network. The former means a high dc-link voltage, which can stress the semiconductor switches unnecessarily. The latter leads to increases in cost and size, which similarly are undesirable. To lessen these concerns, an interesting approach is to use magnetically coupled transformers or inductors to raise the gain and modulation ratio simultaneously, while reducing the number of passive components needed. One possible family of topologies that uses such magnetic components is known as the trans-Z-source inverter (trans-ZSI), whose variants and operating principles are described as follows.

Impedance Source Power Electronic Converters, First Edition. Yushan Liu, Haitham Abu-Rub, Baoming Ge, Frede Blaabjerg, Omar Ellabban, and Poh Chiang Loh.
© 2016 John Wiley & Sons, Ltd. Published 2016 by John Wiley & Sons, Ltd.

Earlier, the gain expression derived for the traditional Z-source inverters was expressed as

$$\hat{v}_{ac} = M_t \frac{\hat{v}_i}{2} = \frac{1}{1-2d_t} \cdot \frac{1}{2} M_t V_{dc} \qquad (7.1)$$

where d_t is the fractional shoot-through time per switching period (or open-circuit time for the current-type inverter), M_t is the modulation ratio, and V_{dc}, $V_c = V_{C1} = V_{C2}$, $\hat{v}_i$, and $\hat{v}_{ac}$ are the input, capacitor, peak dc-link, and peak ac output voltages, respectively.

By setting the denominator of (7.1) to be greater than zero, limits of d_t can promptly be determined as $0 \le d_t < 0.5$. Noting next that the shoot-through state can only be inserted within the null interval to avoid introducing volt-sec error, the limit imposed on M_t can be written as $0 \le M_t \le 1.15(1-d_t)$, where the factor of 1.15 is introduced by the triplen offset added to the inverter modulation. In addition to the voltages in (7.1), the shoot-through instantaneous current i_i might also be of interest since it is usually high, even though it will not cause damage because of its short duration. For the traditional Z-source inverter, this value would be:

$$i_i = 2I_L = 2I_{dc} \qquad (7.2)$$

where I_{dc} is the average input current drawn from the dc source. The same operating characteristics are applied to the current-type Z-source inverter except for the replacement of shoot-through state by open-circuit state for current boosting.

The limitation faced by the traditional Z-source inverters can thus be summarized as their limited maximum $M_t \le 1.15(1-d_t)$ as d_t increases for gain boosting. Lower M_t is generally not encouraged since it leads to a poorer spectral performance and a higher dc-link stress for certain considered ac amplitudes. A few suggestions for solving these concerns have since emerged, with most focusing on replacing the two inductors or two capacitors in the original X-shaped network with other passive components or circuits. Particularly, in [1–3], the inductors of the original network have been replaced by a single two-winding transformer, whose turns ratio can be increased for gain boosting. The resulting circuits are called T-source inverters in [1] before being renamed as trans-ZSIs in [3].

The T-ZSI or trans-ZSI can topologically be derived from the original Z-source network, with its two inductors first replaced by the magnetizing inductances of two transformers shown in Figure 7.1(a). Coupled windings of the two transformers are shown as (W_{1p}, W_{1s}) and (W_{2p}, W_{2s}), and their turn ratios are hence expressed as

$$\gamma_1 = W_{1p}/W_{1s}, \quad \gamma_2 = W_{2p}/W_{2s} \qquad (7.3)$$

Figure 7.1(a) is no doubt incomplete since its inner X-shaped network, consisting of W_{1s}, W_{2s}, C_1, and C_2, is still floating. A way to fix its potential is shown in Figure 7.1(b), where the extra connections have been drawn with thicker lines. Two parallel branches are thus formed with the first consisting of C_1 and the second consisting of W_{1s}, W_{2s}, and C_2 in series. If the winding turns of W_{1s} and W_{2s} are further made equal, their combined effect will be nullified since they are of opposite polarities. The two shunt branches therefore appear similar, and can hence be combined. Another feature noted is with reference to the turns ratios γ_1 and γ_2, which need not be equal since any mismatch in currents through W_{1s} and W_{2s} will flow through C_1. Windings W_{1s} and W_{2s} can also be coupled together since they are already magnetically tied by

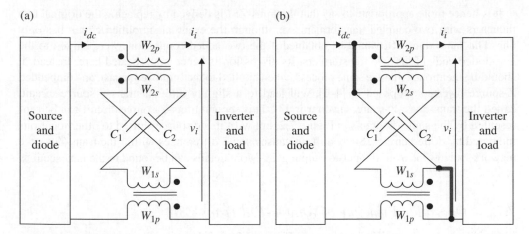

Figure 7.1 Illustration of (a) magnetically coupled transformers and (b) trans-ZSI (*Source*: Qian 2011 [3]. Reproduced with permission of IEEE).

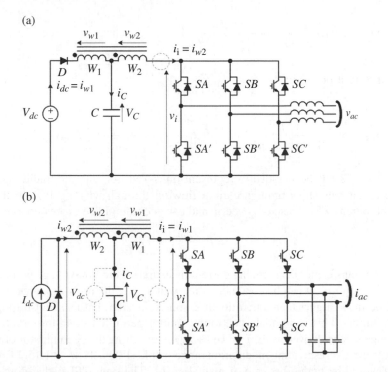

Figure 7.2 Topologies of (a) voltage-type and (b) current-type trans-ZSIs (*Source*: Qian 2011 [3]. Reproduced with permission of IEEE).

W_{1s} and W_{2s}, which have the same turns and carry the same current. With these integrations introduced, the inverter in Figure 7.1(b) is simplified to either of the trans-ZSIs shown in Figure 7.2, whose turns ratio is more appropriately expressed as

$$\gamma_{TZ} = W_1/W_2 = W_{1p}/W_{2p} = \gamma_1/\gamma_2, \quad W_{1s} = W_{2s} \tag{7.4}$$

It is hence more appropriate to say that the trans-ZSI is derived by replacing the original two inductors with two coupled transformers even though the eventual simplified circuit has only one. The same trans-ZSI would be obtained if the two added connections were shifted to the non-dotted ends of the two transformers. Its discussion is hence not repeated here. Instead, it should be mentioned that the same process, when applied to the improved, quasi, and embedded Z-source inverters proposed in [4–6], will lead to a slightly different trans-Z-source circuit, called the trans-quasi-Z-source inverter in [3]. This second trans-Z-source circuit can be represented by the same diagrams in Figure 7.2, but with its dc source shifted to other locations indicated by dotted circles. Such shifting causes some differences within the trans-Z-source network, but not the overall input-to-output gain. More details can be found in the next section.

7.1.2   Operating Principle of Voltage-Fed Trans-ZSI

The voltage-type trans-ZSI is shown in Figure 7.2(a), where an alternative dotted source position is indicated. Shifting the source to that position will only cause the capacitor voltage V_C to be smaller, but not the other voltage expressions stated as follows [3]:

Source in series with D: $V_C = \dfrac{1-d_{TZ}}{1-(\gamma_{TZ}+1)d_{TZ}}V_{dc}$

Source at dotted position: $V_C = \dfrac{\gamma_{TZ}d_{TZ}}{1-(\gamma_{TZ}+1)d_{TZ}}V_{dc}$

$$\hat{v}_i = \frac{1}{1-(\gamma_{TZ}+1)d_{TZ}}V_{dc}, \quad \hat{v}_{ac} = \frac{1}{1-(\gamma_{TZ}+1)d_{TZ}}\cdot\frac{1}{2}M_{TZ}V_{dc} \qquad (7.5)$$

where subscript TZ has been added to represent trans-ZSI. As with the traditional Z-source inverter, instantaneous shoot-through current flowing through winding W_2 and through the switches of the trans-ZSI is also of concern, and can accordingly be expressed as

$$i_i = i_{W2} = I_m = (\gamma_{TZ}+1)I_{dc} \qquad (7.6)$$

where I_m is the transformer magnetizing current measured at the low-voltage winding W_2.

From (7.5) and (7.6), it can be concluded that with $\gamma_{TZ} \geq 1$, the trans-Z-source voltage gain can indeed be made higher than a traditional Z-source inverter, but its shoot-through current will also be higher. This is no doubt due to the shorter permitted shoot-through time whose variation range can be derived as $0 \leq d_{TZ} < 1/(1+\gamma_{TZ})$ after setting the denominator of (7.5) to be greater than zero. Corresponding modulation range would still be $0 \leq M_{TZ} \leq 1.15(1-d_{TZ})$, but its upper limit would be higher because of a smaller d_{TZ}. The trans-ZSI is therefore proven to produce high gain and modulation ratio, but its winding turns can become excessive at high gain because of an increasing γ_{TZ}.

The same trans-Z-source network can also be used with a current-type inverter after swapping their winding positions. The resulting circuit is shown in Figure 7.2(b) with its other possible source locations also marked with dotted circles. Its governing expressions are similar to those in (7.5) and (7.6), and for easier reference, they are listed explicitly in (7.7) to (7.10).

$$\text{Source in parallel with } D: I_m = \frac{(1-d_{TZ})(\gamma_{TZ}+1)}{1-(\gamma_{TZ}+1)d_{TZ}}I_{dc} \tag{7.7}$$

$$\text{Source in parallel with } C: I_m = \frac{\gamma_{TZ}}{1-(\gamma_{TZ}+1)d_{TZ}}I_{dc} \tag{7.8}$$

$$\text{Source in parallel with } v_i: I_m = \frac{(\gamma_{TZ}+1)\gamma_{TZ}d_{TZ}}{1-(\gamma_{TZ}+1)d_{TZ}}I_{dc} \tag{7.9}$$

$$i_i = \frac{I_{dc}}{1\ (\gamma_{TZ}+1)d_{TZ}}, \quad \hat{i}_{ac} = \frac{M_{TZ}I_{dc}}{1-(\gamma_{TZ}+1)d_{TZ}}, \quad \hat{v}_i = (\gamma_{TZ}+1)V_{dc} \tag{7.10}$$

7.1.3 Steady-State Model

As in Section 2.2 for a traditional Z-source inverter, differential equations related to the trans-Z-source inverter can be written, from which the steady-state model can be derived by making all derivatives of state variables equal to zero. Among the steady-state functions that can be derived, the most widely used is probably the input-to-output gain function, which was shown in (7.5).

Even though relatively straightforward, as an imposed requirement, the necessary procedure is provided beginning with differential expressions for representing the shoot-through state of the voltage-type trans-Z-source inverter chosen as an example. Analogous expressions for the current-type trans-Z-source inverter can similarly be derived, but will not be explicitly shown here.

$$L_m \frac{di_m}{dt} - RC \frac{dv_c}{dt} = v_c - ri_m \tag{7.11}$$

$$C \frac{dv_c}{dt} = -i_m \tag{7.12}$$

Variables used in the expressions are shown in Figure 7.3(a), where lowercase variables are used for representing instantaneous voltages and currents. Parameters R and r have also been included to represent parasitic resistances of capacitance C and magnetizing inductance L_m, respectively. Equations (7.11) and (7.12) can further be written in matrix form:

$$\begin{bmatrix} L_m & -RC \\ 0 & C \end{bmatrix} \begin{bmatrix} di_m/dt \\ dv_c/dt \end{bmatrix} = \begin{bmatrix} -r & 1 \\ -1 & 0 \end{bmatrix} \begin{bmatrix} i_m \\ v_c \end{bmatrix} \tag{7.13}$$

Next, when in its non-shoot-through state represented by Figure 7.3(b), differential equations of the voltage-type trans-Z-source inverter become

$$\gamma_{TZ}L_m \frac{di_m}{dt} + RC \frac{dv_c}{dt} = -v_c - \gamma_{TZ}ri_m + V_{dc} \tag{7.14}$$

$$\gamma_{TZ}C \frac{dv_c}{dt} = i_m - (1+\gamma_{TZ})i_o \tag{7.15}$$

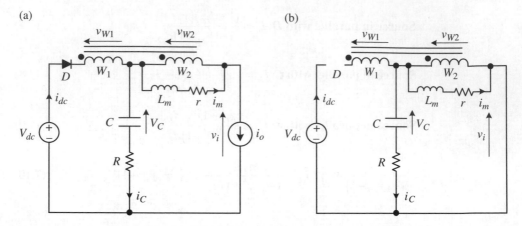

Figure 7.3 Equivalent circuits of trans-Z-source inverter when in (a) shoot-through and (b) non-shoot-through states (*Source*: Li 2013 [6]. Reproduced with permission of IEEE).

and in matrix form

$$
\begin{bmatrix} L_m & RC/\gamma_{TZ} \\ 0 & C \end{bmatrix}\begin{bmatrix} di_m/dt \\ dv_c/dt \end{bmatrix} = \begin{bmatrix} -r & -1/\gamma_{TZ} \\ 1/\gamma_{TZ} & 0 \end{bmatrix}\begin{bmatrix} i_m \\ v_c \end{bmatrix} + \begin{bmatrix} 1/\gamma_{TZ} & 0 \\ 0 & -\dfrac{1+\gamma_{TZ}}{\gamma_{TZ}} \end{bmatrix}\begin{bmatrix} V_{dc} \\ i_o \end{bmatrix} \tag{7.16}
$$

The left-hand coefficient matrices of (7.13) and (7.16) will be exactly the same when $R \approx 0$, which will usually be the case in practice. State-space averaging applied to (7.13) and (7.16) then results in

$$
\begin{bmatrix} L_m & 0 \\ 0 & C \end{bmatrix}\begin{bmatrix} di_m/dt \\ dv_c/dt \end{bmatrix} = \begin{bmatrix} -r & d-\dfrac{1-d}{\gamma_{TZ}} \\ -d+\dfrac{1-d}{\gamma_{TZ}} & 0 \end{bmatrix}\begin{bmatrix} i_m \\ v_c \end{bmatrix} + \begin{bmatrix} \dfrac{1-d}{\gamma_{TZ}} & 0 \\ 0 & -\dfrac{(1+\gamma_{TZ})(1-d)}{\gamma_{TZ}} \end{bmatrix}\begin{bmatrix} V_{dc} \\ i_o \end{bmatrix}
$$
$$\tag{7.17}$$

where the left-hand terms represent the average voltage across magnetizing inductance L_m and average current through capacitance C. These terms will thus be zero in the steady state. The steady-state expression can hence be expressed as

$$
\begin{bmatrix} -r & D-\dfrac{1-D}{\gamma_{TZ}} \\ -D+\dfrac{1-D}{\gamma_{TZ}} & 0 \end{bmatrix}\begin{bmatrix} I_m \\ V_c \end{bmatrix} + \begin{bmatrix} \dfrac{1-D}{\gamma_{TZ}} & 0 \\ 0 & -\dfrac{(1+\gamma_{TZ})(1-D)}{\gamma_{TZ}} \end{bmatrix}\begin{bmatrix} V_{dc} \\ i_o \end{bmatrix} = \begin{bmatrix} 0 \\ 0 \end{bmatrix} \tag{7.18}
$$

where state variables and control signal have been changed to uppercase, such as I_m, V_c, and D, to represent their respective steady-state values. Note however that i_o is not re-expressed in uppercase because it only represents the output current value when in the non-shoot-through state. The first row of (7.18) can then be separately written as

$$-rI_m + \left[D - \frac{1}{\gamma_{TZ}}(1-D)\right]V_c + \left[\frac{1}{\gamma_{TZ}}(1-D)\right]V_{dc} \tag{7.19}$$

Assuming $r=0$, (7.19) will lead to an expression for V_c:

$$V_c = \frac{(1-D)V_{dc}}{1-D(1+\gamma_{TZ})} \tag{7.20}$$

which is the same as the first expression derived in Subsection 7.1.2, because they are derived from the same circuit principles. The second row of (7.18) can similarly be expressed as

$$\left[-D + \frac{1}{\gamma_{TZ}}(1-D)\right]I_m = \left(1 + \frac{1}{\gamma_{TZ}}\right)(1-D)\,i_o \;\Rightarrow\; I_m = \frac{(\gamma_{TZ}+1)(1-D)}{1-D(1+\gamma_{TZ})}\,i_o \tag{7.21}$$

Further noting that input and output powers are equal for a lossless converter, we have

$$V_{dc}\int i_{dc}dt = (1-D)\times \hat{v}_i \times i_o$$

$$I_{dc} = \frac{(1-D)i_o}{1-(\gamma_{TZ}+1)D} \tag{7.22}$$

Substituting (7.22) into (7.21) results in

$$I_m = (\gamma_{TZ}+1)I_{dc} \tag{7.23}$$

This expression is again similar to (7.6) since they are derived from the same circuit principles.

7.1.4 Dynamic Model

Small-signal analysis with appropriate perturbations introduced to state and control variables can be performed with the trans-Z-source network to demonstrate its non-minimum-phase response. This response is also demonstrated by the conventional Z-source inverter, which is as expected since their overall operating principles remain unchanged. The only difference is the charging/discharging of magnetizing inductance in the trans-Z-source inverter, rather than discrete inductances found in the conventional Z-source inverter. To demonstrate this, small-signal perturbations $\hat{i}_m$, $\hat{v}_c$, and $\hat{d}$ can be introduced into (7.17), resulting in the following expressions in the Laplace domain:

$$sL_m\hat{i}_m = -r\hat{i}_m - \frac{1}{\gamma_{TZ}}\hat{v}_c + \left(1+\frac{1}{\gamma_{TZ}}\right)\left(V_c\hat{d}+D\hat{v}_c\right) - \frac{V_{dc}}{\gamma_{TZ}}\hat{d} \tag{7.24}$$

$$sC\hat{v}_c = \frac{1}{\gamma_{TZ}}\hat{i}_m - \left(1+\frac{1}{\gamma_{TZ}}\right)\left(I_m\hat{d}+D\hat{i}_m\right) + \frac{\left(1+\gamma_{TZ}\right)i_o}{\gamma_{TZ}}\hat{d} \tag{7.25}$$

Assuming $r=0$, (7.24) and (7.25) can be simplified as

$$sL_m\hat{i}_m = \left[\left(1+\frac{1}{\gamma_{TZ}}\right)D-\frac{1}{\gamma_{TZ}}\right]\hat{v}_c + \left[\left(1+\frac{1}{\gamma_{TZ}}\right)V_c - \frac{V_{dc}}{\gamma_{TZ}}\right]\hat{d} \tag{7.26}$$

$$sC\hat{v}_c = \left[\frac{1}{\gamma_{TZ}}-\left(1+\frac{1}{\gamma_{TZ}}\right)D\right]\hat{i}_m + \left[\frac{\left(1+\gamma_{TZ}\right)i_o}{\gamma_{TZ}}-\left(1+\frac{1}{\gamma_{TZ}}\right)I_m\right]\hat{d} \tag{7.27}$$

Substituting (7.27) into (7.26) results in

$$s^2L_mC\hat{v}_c = \left(\frac{1}{\gamma_{TZ}}-\left(1+\frac{1}{\gamma_{TZ}}\right)D\right)\left[\left(\left(1+\frac{1}{\gamma_{TZ}}\right)D-\frac{1}{\gamma_{TZ}}\right)\hat{v}_c + \left(\left(1+\frac{1}{\gamma_{TZ}}\right)V_c - \frac{V_{dc}}{\gamma_{TZ}}\right)\hat{d}\right]$$

$$+\left(1+\frac{1}{\gamma_{TZ}}\right)\left(i_o - I_m\right)sL_m\hat{d}$$

$$\Rightarrow \left[s^2L_mC+\left(D-\frac{1}{\gamma_{TZ}}(1-D)\right)^2\right]\hat{v}_c = \left[\left(\frac{1}{\gamma_{TZ}}(1-D)-D\right)\left(\left(1+\frac{1}{\gamma_{TZ}}\right)V_c - \frac{V_{dc}}{\gamma_{TZ}}\right)+\left(1+\frac{1}{\gamma_{TZ}}\right)\left(i_o - I_m\right)sL_m\right]\hat{d}$$

$$\tag{7.28}$$

Coefficients of $\hat{v}_c$ on the left of (7.28) are obviously all positive, meaning that there will be no unstable poles accompanying the transfer function $\hat{v}_c/\hat{d}$. Coefficients of $\hat{d}$ on the right of (7.28), however, require more reasoning, beginning with (7.22), which immediately conveys that $I_{dc} \geq i_o$. Expression (7.23) then further shows that $I_m > I_{dc} \geq i_o$. The coefficient $\left(1+\frac{1}{\gamma_{TZ}}\right)\left(i_o - I_m\right)$ on the right of (7.28) will hence be negative. Then for the second coefficient found on the right of (7.28), its polarity can be determined by first noting

$$\left(\frac{1}{\gamma_{TZ}}(1-D)-D\right)=\left(\frac{1}{\gamma_{TZ}}-\left(\frac{\gamma_{TZ}+1}{\gamma_{TZ}}\right)D\right)>0 \tag{7.29}$$

since the maximum of D is $1/(1+\gamma_{TZ})$, as explained in Subsection 7.1.2. Moreover,

$$\left(\left(1+\frac{1}{\gamma_{TZ}}\right)V_c - \frac{V_{dc}}{\gamma_{TZ}}\right) = \frac{1}{\gamma_{TZ}}\left[(\gamma_{TZ}+1)V_c - V_{dc}\right] > 0 \qquad (7.30)$$

which is true because $V_c \geq V_{dc}$, as understood from (7.20).

The coefficient of $\left(\dfrac{1}{\gamma_{TZ}}(1-D)-D\right)\left(\left(1+\dfrac{1}{\gamma_{TZ}}\right)V_c - \dfrac{V_{dc}}{\gamma_{TZ}}\right)$ on the right of (7.28), which is the product of (7.29) and (7.30), is thus always positive. Summarizing, the two coefficients on the right of (7.28) have opposite polarities, which in the s-domain, translates to a right-half-plane (RHP) zero associated with the transfer function $\hat{v}_c/\hat{d}$. As with the conventional Z-source inverter, the trans-Z-source inverter therefore exhibits non-minimum-phase response, which means its initial response will always oppose the direction of command change. Tracking will only commence after this initial non-minimum-phase period.

7.1.5 Simulation Results

To show some typical waveforms, the trans-Z-source voltage-fed inverter in Figure 7.4(a) is simulated with the dc source placed at the extreme left. The obtained results are shown in Figure 7.4, where the input current and immediate dc-link voltage to the inverter bridge are seen to be pulsating. These are certainly correct since during shoot-through, the input current will drop to zero because of the blocking of diode D in Figure 7.4(a). At the same

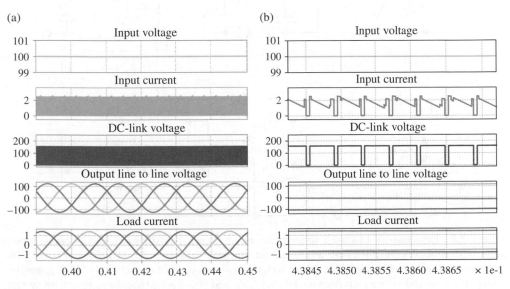

Figure 7.4 Simulation results of T- or Trans-ZSI with shoot-through ratio of 0.14 and modulation ratio of 0.97. (a) Viewed in fundamental period; (b) viewed in control period.

time, the dc-link voltage will drop to zero because of the turning on of at least two switches from the same phase leg of the inverter bridge. Despite these pulsating waveforms at the dc-side of the inverter, the ac line-to-line voltages and currents are still sinusoidal, which is as intended.

7.2 LCCT-ZSI/qZSI

7.2.1 Configuration and Operation of LCCT-ZSI

The LCCT-ZSI is derived from the traditional ZSI, as shown in the schematics in Figure 7.5 [7, 8].

The circuit schematic of the LCCT-ZSI with inverted winding orientation of the transformer and two built-in DC-current blocking capacitors C_1 and C_2 is shown in Figure 7.6(a). The use of the two built-in DC current blocking capacitors is the main benefit of the LCCT-ZSI. The unique property of the LCCT-ZSI is that no energy is stored in the transformer windings. The two DC current blocking capacitors connected in series with the transformer also prevent the transformer core from saturating.

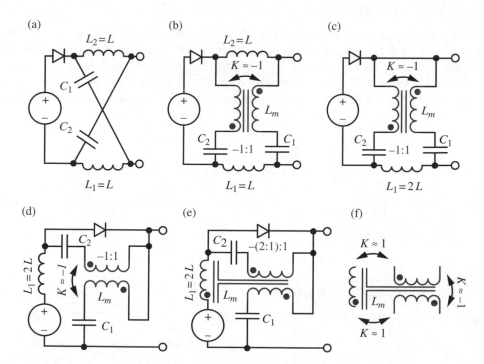

Figure 7.5 Derivation of LCCT-ZS network: (a) basic ZS network, (b) equivalent Z-source circuit with transformer coupling and built-in DC current blocking capacitors, (c) LCCT-ZS network with discontinuous input current, (d) LCCT-ZS network with continuous input current, (e) LCCT-ZS network with transformer and inductor windings on the same core, and (f) integrated transformer-inductor system with ideal couplings (*Source*: Adamowicz 2011 [7] and Adamowicz 2011 [8]. Reproduced with permission of IEEE).

There are three operating states: shoot-through state (duration T_0), non-shoot-through state with $i_D > 0$ (duration T_1), and non-shoot-through state with $i_D = 0$ (duration T_2), as shown in Figure 7.6(b)–(d). For inductor voltages v_{L1}, v_{L2}, v_{L3}, the following is valid [7–8]:

$$v_{L1} = -\left(v_{L2} + v_{L3}\right) \tag{7.31}$$

which enables the use of a common core both for the transformer and the input inductor. Because of the inverted orientation of the transformer windings, the dc-link voltage v_i is given by

$$v_i = V_{C1} - v_{L3} \tag{7.32}$$

For the shoot-through state during the interval T_0, shown in Figure 7.6(b), we have

$$v_{L1} = -\left(1 + \frac{n_2}{n_3}\right) V_{C1}, \quad v_{L3} = V_{C1}, \quad v_{L2} = \frac{n_2}{n_3} V_{C1} \tag{7.33}$$

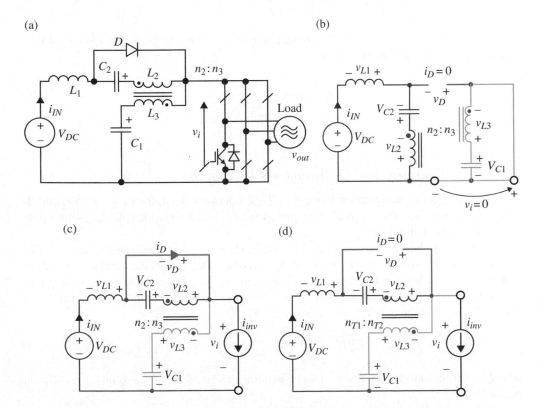

Figure 7.6 LCCT-ZSI in (a) configuration; and the equivalent circuits in three operation states: (b) shoot-through state, (c) non-shoot-through state with $i_D > 0$, and (d) non-shoot-through state with $i_D = 0$ (*Source*: Adamowicz 2011 [8]. Reproduced with permission of IEEE).

For the non-shoot-through state with conducting input diode (first active state), which is during the interval T_1 from the equivalent circuit in Figure 7.6(c), the voltages of the input inductor L_1 and transformer windings L_2 and L_3 can be written as

$$v_{L1} = \left(1 + \frac{n_3}{n_2}\right)V_{C2}, \quad v_{L2} = -V_{C2}, \quad v_{L3} = -\frac{n_3}{n_2}V_{C2} \tag{7.34}$$

During the second active state for an interval T_2 shown in equivalent circuit of Figure 7.6(d), we have

$$v_{L1} = 0, \qquad v_{L2} = 0, \qquad v_{L3} = 0 \tag{7.35}$$

$$v_i = V_{C1} \tag{7.36}$$

In the steady state, the peak dc-link voltage of the LCCT-ZSI can be calculated from

$$V_i = \frac{1}{1 - \left(1 + \dfrac{n_1}{n_2}\right)D}V_{DC} = B \cdot V_{DC} \tag{7.37}$$

The output voltage of the LCCT-ZSI can be controlled by the voltage of capacitor C_1. In the steady state, v_{C1} can be calculated from

$$V_{C1} = \frac{1 - D}{1 - \left(1 + \dfrac{n_2}{n_3}\right)D}V_{DC} \tag{7.38}$$

7.2.2   Configuration and Operation of LCCT-qZSI

Figure 7.7 shows the configuration of the LCCT-qZS network. First, the basic qZS network is shown in Figure 7.7(a), then Figure 7.7(b) shows the LCCT-qZS network, with the T-equivalent circuit shown in Figure 7.7(c).

Figure 7.7(b) shows that the primary winding ($L_2 = L_{T1(dc)}$) of the transformer is a DC winding while the secondary winding is an AC winding. This is possible thanks to the series connection of the built-in DC-current blocking capacitor C_1. The transformer can be represented by an equivalent circuit with three uncoupled inductors, shown in Figure 7.7(c). From Figure 7.7(b) and Figure 7.7(c), we can write [7, 8]

$$L_A = L_M = k\sqrt{L_{T1} \cdot L_{T2}}, \quad L_B = L_{T2} - L_M, \quad L_C = L_{T1} - L_M \tag{7.39}$$

where L_M is mutual inductance of the pair of windings and $0 < k < 1$ is the coupling coefficient defined by

$$k = \frac{L_M}{\sqrt{L_{T1} \cdot L_{T2}}} \tag{7.40}$$

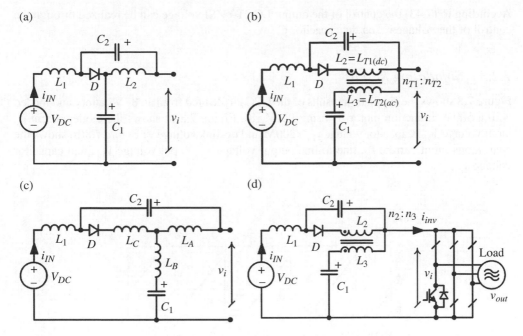

Figure 7.7 The LCCT-qZS network: (a) basic qZS network, (b) LCCT-qZS network, (c) T-equivalent circuit of LCCT-qZS network, and (d) LCCT-qZSI (*Source*: Adamowicz 2011 [7] and Adamowicz 2011 [8]. Reproduced with permission of IEEE).

If the condition ($L_B \to 0, L_C \to 0$) is fulfilled, the impedance networks from Figure 7.7(a) and Figure 7.7(c) are equivalent.

The LCCT-qZS network used for an inverter, named LCCT-qZSI, is shown in Figure 7.7(d). Similar operating states to the LCCT-ZSI can be obtained as in Figure 7.6. The capacitor voltages $v_{CT1} = V_{CT1} = const$ and $v_{CT2} = V_{CT2} = const$ and DC source input voltage $V_{DC} = const$ in the steady state is related by

$$V_{C2} = V_{C1} - V_{DC} \tag{7.41}$$

Moreover, for voltages v_{L1} and v_D of input inductor L_1 and input diode D, and transformer voltages v_{L2} and v_{L3}, we have

$$v_{L1} = v_{L3}, \quad v_{L2} = \frac{n_2}{n_3} v_{L3}, \quad v_{L2} = -v_D + V_{DC} \tag{7.42}$$

The inverter dc-link voltage v_i of LCCT-qZSI is equal to

$$v_i = V_{C1} + v_{L3} \tag{7.43}$$

According to (7.43) the control of the output LCCT-qZSI voltage can be realized through the control of the voltage v_{C1} on the capacitor C_1.

7.2.3 Simulation Results

Figure 7.8 shows the simulation results of the LCCT-qZSI fed from an 87 V battery and loaded with a 500 W induction motor by using PSIM [8]. Figure 7.8(a) shows the diode current i_D, input voltage V_{DC}, capacitor voltage $v_{C1} = 200$ V, and dc-link voltage v_i; Figure 7.8(b) shows the continuous input current i_{IN}, line-to-line output voltage v_{out}, input voltage V_{DC}, and capacitor voltage v_{C1}.

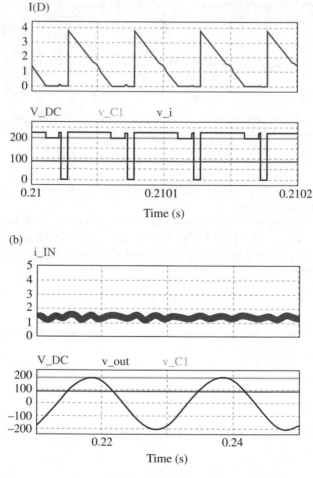

Figure 7.8 Simulation results of the LCCT-qZSI loaded with a 500 W induction motor: (a) diode current i_D, input voltage V_{DC}, capacitor voltage $v_{C1} = 200$ V, and dc-link voltage v_i, (b) continuous input current i_{IN}, line-to-line output voltage v_{out}, input voltage V_{DC} and capacitor voltage v_{C1}.

The input inductor L_1 was built with an inductance of $3\,mH$. The two capacitors are $C_1 = 400\,\mu F$ and $C_2 = 140\,\mu F$, paralleling the dc-link circuit during active mode operation. The switching frequency is 10 kHz. Light load operation of LCCT-qZSI is investigated to examine the theoretical analysis. The reference value of the capacitor voltage is 200 V. Maximum constant boost control strategy is utilized for the inverter bridge. All three modes of operation can be seen in Figure 7.8(a). With the large turns ratio 16:4 of the transformer, shoot-through duty ratio $D = 0.1$ is obtained. Investigated LCCT-qZSI provides continuous input current even in the case of light load, as shown in Figure 7.8(b).

7.3   Conclusion

The two typical transformer-based trans and LCCT-ZSI/qZSI have been presented in this chapter, including configuration, working principles, modeling, and simulation results. They showed improved performance compared to the basic ZSI/qZSI in terms of reduced element counts, compact passive components, and high voltage boosting ability. The unique topology of the LCCT passive input circuit also helped to prevent the transformer core from saturating. Also, by using high switching frequency power devices, such as silicon carbide (SiC) transistors and SiC diodes, high power density is expected for those transformer-based derivations of ZSI/qZSI.

Acknowledgment

Section 7.2 of this chapter was made possible by NPRP-EP grant # [X-033-2–007] from the Qatar National Research Fund (a member of Qatar Foundation). The statements made herein are solely the responsibility of the authors.

References

[1] R. Strzelecki, M. Adamowicz, N. Strzelecka, W. Bury, "New type T-source inverter," in *Proc. CPE'09*, May 2009, pp.191–195.
[2] M. Adamowicz, R. Strzelecki, F. Z. Peng, J. Guzinski, H. Abu-Rub, "New type LCCT-Z-source inverters," in *Proc. European Conference on Power Electronics and Applications (EPE)*, 2011, pp.1–10.
[3] W. Qian, F. Z. Peng, H. Cha, "Trans-Z-source inverters," *IEEE Trans. Power Electron.*, vol.26, no.12, pp.3453–3463, Dec. 2011.
[4] Y. Tang, S. Xie, C. Zhang, Z. Xu, "Improved Z-source inverter with reduced Z-source capacitor voltage stress and soft-start capability," *IEEE Trans. Power Electron.*, vol.24, no.2, pp.409–415, Feb. 2009.
[5] J. Anderson, F. Z. Peng, "A class of quasi-Z-source inverters," in *Proc. IEEE-IAS'08*, Oct. 2008, pp.1–7.
[6] D. Li, P. C. Loh, M. Zhu, F. Gao, F. Blaabjerg, "Cascaded Multi-cell Trans-Z-Source Inverters," *IEEE Trans. Power Electron.*, vol.28, no.2, pp.826–836, Feb. 2013.
[7] M. Adamowicz, J. Guzinski, R. Strzelecki, F. Z. Peng, H. Abu-Rub, "High step-up continuous input current LCCT-Z-source inverters for fuel cells," in *Proc. IEEE Energy Conversion Congress and Exposition (ECCE)*, 2011, pp.2276–2282.
[8] M. Adamowicz, R. Strzelecki, F. Z. Peng, J. Guzinski, H. Abu-Rub, "New type LCCT-Z-source inverters," in *Proc. the 2011–14th European Conference on Power Electronics and Applications (EPE)*, pp.1–10, 2011.

8

Z-Source/Quasi-Z-Source AC-DC Rectifiers

8.1 Topologies of Voltage-Fed Z-Source/Quasi-Z-Source Rectifiers

Active rectifiers are attractive as an interface for ac-dc conversion for use in electric vehicle charging, because of their high efficiency and bidirectional operation [1,2]. The traditional voltage source rectifier (VSR), shown in Figure 8.1(a), and current source rectifier (CSR) lack the buck–boost capability, so a dc-dc converter is generally cascaded to buck–boost required applications for the required voltage. Furthermore, short-circuit for the VSR or open circuit for the CSR of any of the phase legs is arbitrarily forbidden [1]. Those limitations have inspired the Z-source/quasi-Z-source network [3–10]. The voltage-fed Z-source/quasi-Z-source rectifier (ZSR/qZSR), shown in Figures 8.1(b) and (c), respectively, present a more efficient approach than the traditional one, owing to its buck–boost capability and shoot-through availability between phase leg switches, similar to the ZSI/qZSI. By replacing the diode of ZS/qZS network, the ZSR/qZSR are able to bilaterally operate in both inverter and converter modes with unity power factor and low harmonic distortion of the input ac currents [3–5].

Compared to the ZSR and qZSR, the following features are available. The two capacitors C_1 and C_2 in the ZSR have the same high voltage, whereas the voltage of capacitor C_2 in the qZSR is lower than that of C_1. The output capacitor C_2 of the ZSR has to filter larger harmonic amplitude and has to handle the full load current during shoot-through state; the dc-side filter capacitor of the qZSR restricts low amplitude harmonics because one of the qZS inductors performs filtering. Therefore, the current and voltage ratings of the dc-side filter capacitor of the qZSR are much lower than the ZSR. In addition, a common dc rail appears between the dc output and the rectifier in the qZSR, which reduces the EMI and it is easier to assemble.

Impedance Source Power Electronic Converters, First Edition. Yushan Liu, Haitham Abu-Rub, Baoming Ge, Frede Blaabjerg, Omar Ellabban, and Poh Chiang Loh.
© 2016 John Wiley & Sons, Ltd. Published 2016 by John Wiley & Sons, Ltd.

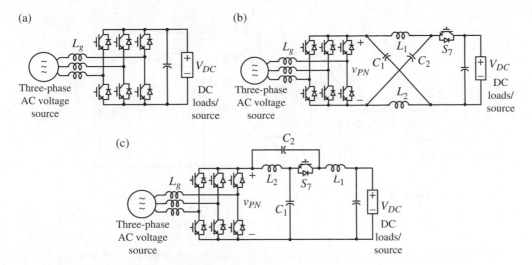

Figure 8.1   Topologies of (a) traditional voltage source rectifier, (b) Z-source rectifier, and (c) quasi-Z-source rectifier.

All the pulsewidth modulation (PWM) methods of ZSI/qZSI, as explained in Chapter 5, could be adopted for the ZSR/qZSR. Furthermore, a triangle-shift strategy is proposed to achieve zero voltage switching of the switch in the impedance network [3]. The dynamic modeling and closed-loop control of the voltage-fed and current-fed ZSR/qZSR have been reported [4,6–9]. A derived ZSR is proposed, aimed at decreasing the capacitor voltages, in [10]. Considering the wider utilization of the voltage-fed rectifiers, the voltage-fed qZSR is discussed in this chapter, including the steady-state working principles, dynamic modeling, control scheme, and simulation results. The analysis is also applicable to the ZSR.

8.2   Operating Principle

Here, the voltage-fed qZSR of Figure 8.1(c) with a resistive load, as shown in Figure 8.2, is used to illustrate the operation of ZSR/qZSR. The circuit variables and directions are as defined in Figure 8.2.

The three-phase ac source has the voltage of

$$\begin{cases} v_{sa}(t) = V_m \sin(\omega t) \\ v_{sb}(t) = V_m \sin(\omega t - 120°) \\ v_{sc}(t) = V_m \sin(\omega t + 120°) \end{cases} \tag{8.1}$$

where V_m is the amplitude of the source voltage.

Similar to the voltage-fed qZSI, in steady state, we have

$$V_{C1} = \frac{1-D}{1-2D}V_o, \quad V_{C2} = \frac{D}{1-2D}V_o, \quad V_o = (1-2D)V_{PN}, \quad V_{PN} = \frac{2V_m}{M}, \quad I_{L1} = I_{L2} = I_o \tag{8.2}$$

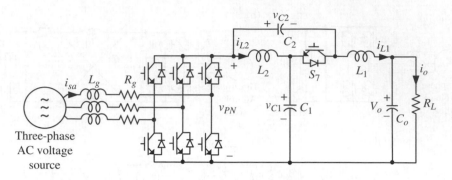

Figure 8.2   Voltage-fed quasi-Z-source rectifier with resistive load.

where V_o and I_o are the load voltage and current of the qZSR, M is the modulation index, D is the shoot-through duty cycle, and V_{PN} is the peak voltage of the rectifier dc-link.

From (8.2), the load voltage of the qZSR is summarized as

$$V_o = \frac{2(1-2D)}{M}V_m \tag{8.3}$$

As we can see from (8.3), the output voltage of qZSR can be regulated by the shoot-through duty cycle D and modulation M for a given ac source voltage. The control strategy will be explained in the following section.

8.3   Dynamic Modeling

The dynamic modeling of the qZSR is addressed in terms of dc-side output with quasi-Z-source network and ac-side input with rectifier bridges. When analyzing the dc side, the rectifier bridge is equivalent to a current source. A similar methodology to the Z-source/quasi-Z-source inverter is applied to model its dynamic model, as follows.

8.3.1   DC-Side Dynamic Model of qZSR

Figure 8.3 shows the equivalent circuit of the qZSR seen from dc side [6–9]. There are shoot-through and non-shoot-through states of the qZSR. In Figure 8.3(a), the rectifier bridge is equivalent to a short-circuit when shoot-through zero vectors are working; Figure 8.3(b) shows the non-shoot-through state, where the rectifier bridge is equivalent to a constant current source.

From Figure 8.3, the dynamic relationship of the shoot-through and non-shoot-through states are, respectively

(a)

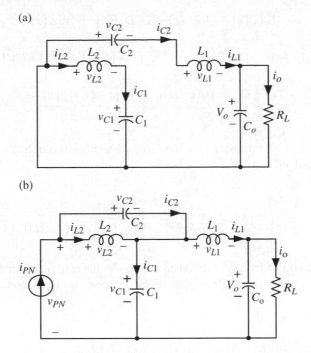

(b)

Figure 8.3 Equivalent circuits of qZSR shown in Figure 8.2: (a) Shoot-through states, (b) non-shoot-through states.

$$
\begin{cases}
L_1 \dfrac{di_{L1}}{dt} = -V_{C2} - V_o \\[2mm]
L_2 \dfrac{di_{L2}}{dt} = -V_{C1} \\[2mm]
C_1 \dfrac{dV_{C1}}{dt} = i_{L2} \\[2mm]
C_2 \dfrac{dV_{C2}}{dt} = i_{L1} \\[2mm]
C_o \dfrac{dV_o}{dt} = i_{L1} - \dfrac{V_o}{R_L}
\end{cases}
\quad
\begin{cases}
L_1 \dfrac{di_{L1}}{dt} = V_{C1} - V_o \\[2mm]
L_2 \dfrac{di_{L2}}{dt} = V_{PN} - V_{C1} \\[2mm]
C_1 \dfrac{dV_{C1}}{dt} = i_{PN} - i_{L1} \\[2mm]
C_2 \dfrac{dV_{C2}}{dt} = i_{PN} - i_{L2} \\[2mm]
C_o \dfrac{dV_o}{dt} = i_{L1} - \dfrac{V_o}{R_L}
\end{cases}
\qquad (8.4)
$$

According to the state-space averaging method and applying the small-signal perturbance to (8.4), the small-signal dynamic equations in the Laplace form are

$$
sC_o \hat{V}_o(s) = \hat{i}_{L1}(s) - \hat{V}_o / R_L \qquad (8.5)
$$

$$
sL_1 \hat{i}_{L1}(s) = (1-D)\hat{v}_{C1}(s) - D\hat{v}_{C2}(s) - \hat{V}_o(s) - V_{11}\hat{d}(s) \qquad (8.6)
$$

$$sL_2\hat{i}_{L2}(s) = -D\hat{v}_{C1}(s) + (1-D)\hat{v}_{C2}(s) - V_{11}\hat{d}(s) \tag{8.7}$$

$$sC_1\hat{v}_{C1}(s) = -(1-D)\hat{i}_{L1}(s) + D\hat{i}_{L2}(s) + (1-D)\hat{I}_{PN}(s) + I_{11}\hat{d}(s) \tag{8.8}$$

$$sC_2\hat{v}_{C2}(s) = D\hat{i}_{L1}(s) - (1-D)\hat{i}_{L2}(s) + (1-D)\hat{I}_{PN}(s) + I_{11}\hat{d}(s) \tag{8.9}$$

where $I_{11} = I_{L1} + I_{L2} - I_{PN}$, $V_{11} = V_{C1} + V_{C2}$.

From (8.5) to (8.9), the transfer functions from the shoot-through duty cycle to output voltage V_o can be solved to give

$$G_{Vod}(s) = \frac{\hat{V}_o(s)}{\hat{d}(s)} = \frac{DR_L V_o}{\left[(1-2D)^2 R_L LC_o + D^2 R_L LC\right]s^2 + (1-2D)^2 Ls + D^2 R_L} \tag{8.10}$$

From (8.10), the controller can be designed to improve the response speed and stabilization of the qZSR's dc output voltage. Bode plots of the compensation are shown in the following section.

8.3.2   AC-Side Dynamic Model of Rectifier Bridge

The qZSR is fed by a three-phase ac source, as shown in (8.1). The ac input to the rectifier dc link has the relationship of

$$\begin{cases} L_g \dfrac{di_{sa}}{dt} = v_{sa} - R_g i_{sa} - d_a v_{PN} \\[2mm] L_g \dfrac{di_{sb}}{dt} = v_{sb} - R_g i_{sb} - d_b v_{PN} \\[2mm] L_g \dfrac{di_{sc}}{dt} = v_{sc} - R_g i_{sc} - d_c v_{PN} \end{cases} \tag{8.11}$$

where d_a, d_b, and d_c are the active duty cycle of rectifier phase-leg bridges, and R_g is the parasitic resistance of the grid-side inductor.

Then the synchronous frame model of the rectifier bridge is

$$\begin{cases} L_g \dfrac{di_{sd}}{dt} = v_{sd} - R_g i_{sd} - \omega L_g i_{sq} - d_d v_{PN} \\[2mm] L_g \dfrac{di_{sq}}{dt} = v_{sq} - R_g i_{sq} + \omega L_g i_{sd} - d_q v_{PN} \end{cases} \tag{8.12}$$

where v_{sd} and v_{sq} are dq-axis components of the three-phase source voltages in the synchronous coordinate, i_{sd} and i_{sq} are the currents in the dq-axis synchronous coordinate, and d_d and d_q are the active duty cycle of the rectifier bridge in synchronous coordinates.

Substituting (8.3) into (8.12), the synchronous frame equations of the qZSR bridge become

$$\begin{cases} L_g \dfrac{di_{sd}}{dt} = v_{sd} - R_g i_{sd} - \omega L_g i_{sq} - d_d \dfrac{V_o}{1-2D} \\[2mm] L_g \dfrac{di_{sq}}{dt} = v_{sq} - R_g i_{sq} + \omega L_g i_{sd} - d_q \dfrac{V_o}{1-2D} \end{cases} \tag{8.13}$$

which shows that the variations of qZSR output voltage depend on the variations of the input current of the rectifier bridge and the shoot-through duty ratio. Therefore, the qZSR could be controlled through the active duty cycle and the shoot-through duty cycle.

Figure 8.4 shows the control block diagram of the qZSR to fulfill the control objectives with the established dc and ac sides modeling of (8.10) and (8.13).

Figure 8.4(a) shows the dc output voltage control by regulating the shoot-through duty cycle D. A feed forward control is applied to quicken the response, with $D_0 = (1 - V_o^*/V_{PN}^*)/2$. Figure 8.4(b) shows the ac current control, where the d-axis current reference i_{sd}^* is set to zero to get unity power factor, either in inverter or rectifier modes. The q-axis current reference i_{sq}^* is determined by the closed-loop compensation of qZS capacitor voltage

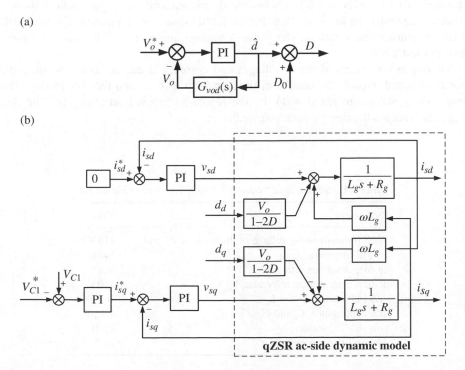

Figure 8.4 Block diagrams of qZSR control scheme: (a) dc output voltage control, (b) ac current control.

$i_{sq}^* = (k_P + k_I/s)(V_{C1} - V_{C1}^*)$. Further, the dq-axis voltage references are then obtained by the inner PI regulators, respectively. Then, through the dq–abc transformation, the voltage references v_{sa}^*, v_{sb}^*, and v_{sc}^*, and the desired shoot-through duty ratio are applied to generate the gate drive signals by the modulation technique of ZSI/qZSI shown in Chapter 4.

8.4 Simulation Results

The control of Figure 8.4 is applied to the qZSR in Figure 8.2. Table 8.1 lists the system specifications. The RMS line-to-line voltage of the three-phase ac grid is 110V, dc output voltage reference $V_o^* = 400$V, dc-link peak voltage reference $V_{PN}^* = 570$V, qZS inductance $L_1 = L_2 = 500\,\mu$H, qZS capacitance $C_1 = C_2 = 470\,\mu$F, grid-side inductance $L_g = 3\,$mH, dc-side filter capacitance $C_o = 2\,$mF. Figure 8.5 shows the compensation of the output voltage control loop of Figure 8.4(a) at the discussed parameters. Figures 8.6 and 8.7 show the simulation results of the qZSR in the rectifier and inverter modes, respectively.

From the Bode plots of the transfer function $G_{Vod}(s)$ and the open-loop transfer function after the compensation in Figure 8.5(a), it can be seen that the low-frequency slope increases after compensation, with the crossover frequency of $G_{Vod}(s)$ reducing to 21 rad/s from 1.09×10^4 rad/s. As a result, the output voltage achieves a fast response and robust stability, as the closed-loop transfer function shown in Figure 8.5(b).

Figure 8.6 shows the phase-A source voltage and current. Note that the current is scaled to ten times the actual value to be comparable with the voltage. A step change load increase is performed at $t = 1$ s. Figure 8.6(b) shows the dc output voltage. Figure 8.6(c) shows the dc-link voltage after the load variation. Figure 8.6(d) shows the transients of source voltage and current during that variation. The phase-A voltage and current in the inverter mode are shown in Figure 8.7.

From simulation results of phase voltages and currents, it can be seen that unity power factor is obtained through the control method, in both inverter and rectifier modes. The dc output voltage stays constant at 400V by the regulation when load changing. The dc-link voltage also reaches the desired peak voltage.

Table 8.1 System specifications for qZSR system in Figure 8.2

Parameters	Values
RMS phase-to-phase voltage of three-phase ac grid	110 V
ac grid frequency, f_0	50 Hz
dc output voltage reference, V_o^*	400 V
dc-link peak voltage reference, V_{PN}^*	570 V
qZS inductance, L_1 and L_2	500 μH
qZS capacitance, C_1 and C_2	470 μF
grid-side inductance, L_g	3 mH
dc-side filter capacitance, C_o	2 mF

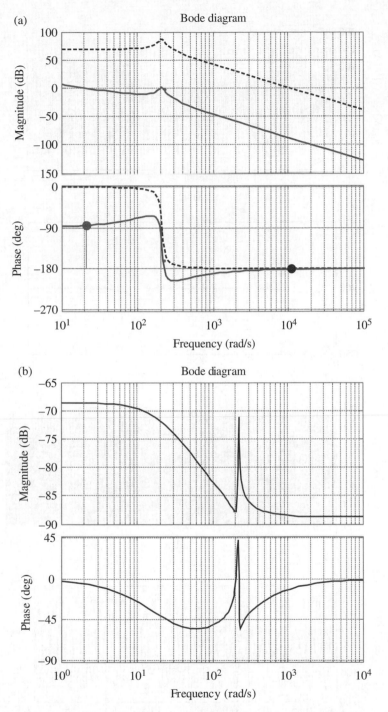

Figure 8.5 Bode plots of (a) $G_{Vod}(s)$ and open-loop transfer function after the compensation by the PI controller, and (b) closed-loop transfer function after the compensation.

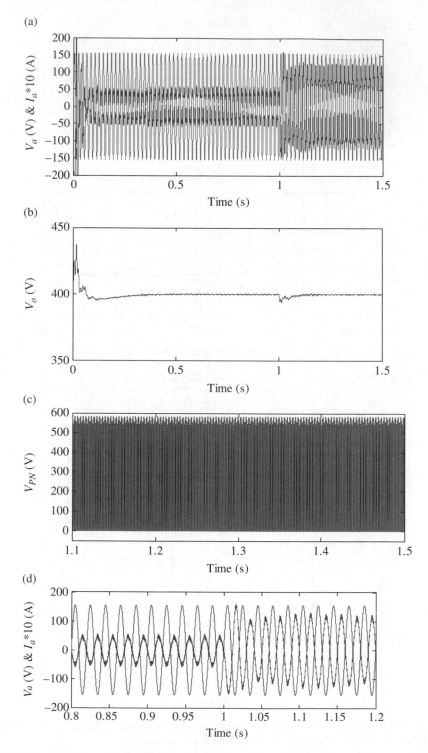

Figure 8.6 Simulation results of qZSR in the rectifier mode with a load increase at $t=1$ s: (a) source voltage and current of phase-A, (b) dc output voltage, (c) dc-link voltage after load changing, (d) transients of source voltage and current during load change.

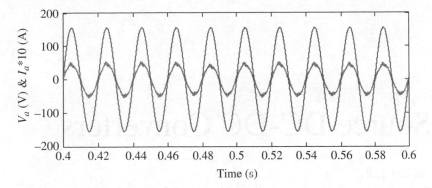

Figure 8.7 Phase-A source voltage and current when qZSR is in the inverter mode.

8.5 Conclusion

This chapter has discussed the voltage-fed Z-source and quasi-Z-source rectifier, which improve the voltage boosting capability of the traditional voltage source rectifier, are immune to short-circuit of the phase legs, and are able to conduct both inverter and converter modes bilaterally. The quasi-Z-source rectifier was further taken as an example to illustrate their steady-state operating principles, dynamic modeling, control scheme, and simulation results, providing the fundamentals for the future development of such rectifiers.

References

[1] B. Wu, *High-Power Converters and ac Drives High-Power Converters and ac Drives*, pp.187–218, Wiley & Sons Ltd, 2005.

[2] A. Keyhani, M. N. Marwali, Min Dai, *Integration of Green and Renewable Energy in Electric Power Systems*, Wiley & Sons Ltd, pp.224–233, 2009.

[3] X. Ding, Z. Qian, Y. Xie, F. Z. Peng, "A novel ZVS Z-source rectifier," in *Twenty-First Annual IEEE Applied Power Electronics Conference and Exposition (APEC)*, pp.1–5, 2006.

[4] X. Ding, Z. Qian, Y. Xie, F. Z. Peng, "Transient modeling and control of the novel ZVS Z-source rectifier," in *Proc. 37th IEEE Power Electronics Specialists Conference (PESC)*, pp.1–5, 2006.

[5] X. Fang, M. Cao, Z. Chen, "Z-source ac-dc-ac converter for mining applications," in *Proc. International Conference on Electrical Machines and Systems (ICEMS)*, pp.44–47, 2010.

[6] G. Lo Calzo, A. Lidozzi, L. Solero, F. Crescimbini, "Three-phase Z-source power supply design and dynamic modeling," in *Proc. IEEE Energy Conversion Congress and Exposition (ECCE)*, pp.1339–1345, 2011.

[7] L. Qin, S. Yang, F. Z. Peng, R. Inoshita, "Three phase current-fed Z-source PWM rectifier," in *Proc. IEEE Energy Conversion Congress and Exposition (ECCE)*, pp.1569–1574, 2009.

[8] S. H. Hosseini, F. Sedaghati, M. Sarhangzadeh, "Improved power quality three phase ac-dc converter," in *Proc. 2010 International Conference on Electrical Machines and Systems (ICEMS)*, pp.148–153, 2010.

[9] C. L. K. Konga, M. N. Gitau, "Three-phase quasi-Z-source rectifier modeling," in *Proc. 2012 Twenty-Seventh Annual IEEE Applied Power Electronics Conference and Exposition (APEC)*, pp.195–199, 2012.

[10] L. Shi, H. Xu, Z. Zhang, Z. Liu, "Design and control of the novel Z-SOURCE PWM rectifier-inverter," in *Proc. 2013 International Conference on Electrical Machines and Systems (ICEMS)*, pp.1830–1835, 2013.

9

Z-Source DC-DC Converters

Poh Chiang Loh

Department of Energy Technology, Aalborg University, Aalborg East, Denmark

When realized as a dc-dc converter, the introduced Z-source impedance network must similarly be short-circuited for a certain defined duty ratio per switching period. The basic operating principles therefore remain unchanged, except that the rear-end converter bridge is changed to one suitable for performing dc-dc conversion. This chapter introduces a few possible dc-dc Z-source topologies for illustration, but will not move deeper into their mathematics since they are almost the same as for dc-ac inversion.

9.1 Topologies

Various isolated and non-isolated dc-dc Z-source converters have been proposed in the literature. They use the same impedance networks as other dc-ac Z-source inverters. For example, a dc-dc converter realized with the quasi-impedance network and a 2- or 3-leg H-bridge switching topology is shown in Figure 9.1 [1, 2]. As expected, this topology is simply assembled with an impedance network added to the front of a traditional H-bridge isolated dc-dc converter. The impedance network can be shorted for voltage boosting by turning S_1 and S_2 on together, S_3 and S_4 on together, or all four switches on together. Its principles are therefore the same as the usual dc-ac Z-source inverter.

The H-bridge configuration shown in Figure 9.1 is obviously only one possibility out of many. If preferred, other dc-dc converter topologies can be assembled with an impedance network for voltage boosting. Some possibilities are shown briefly in Figure 9.2, where only the first two are non-isolated. The third uses the push-pull circuit, and is hence an isolated solution like the H-bridge topology shown in Figure 9.1.

A dc-dc converter with a trans-Z source network has also been investigated and shown in Figure 9.3 [3]. The purpose is to produce a higher boost, while using a smaller shoot-through

Impedance Source Power Electronic Converters, First Edition. Yushan Liu, Haitham Abu-Rub, Baoming Ge, Frede Blaabjerg, Omar Ellabban, and Poh Chiang Loh.
© 2016 John Wiley & Sons, Ltd. Published 2016 by John Wiley & Sons, Ltd.

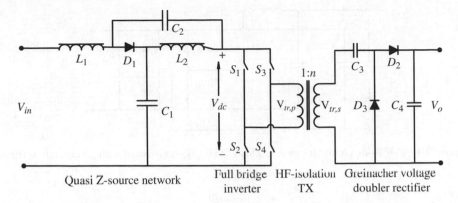

Figure 9.1 A dc-dc converter with intermediate H-bridge switching topology [1, 2]. (*Source*: Siwakoti 2014 [1]. Reproduced with permission of IET).

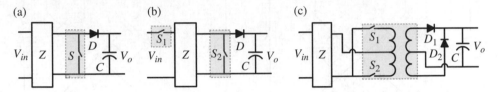

Figure 9.2 Other dc-dc Z-source converters implemented with (a) one switch, (b) two switches, and (c) the push-pull circuitry [1]. (Reproduced by permission of the Institution of Engineering & Technology. Full acknowledgment to the Author, Title, and date of the original work).

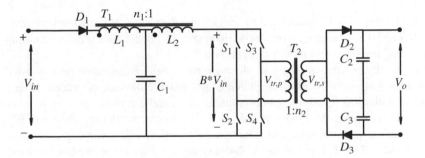

Figure 9.3 Trans-Z-source dc-dc converter implemented with an intermediate H-bridge switching topology (*Source*: Nguyen 2013 [3]. Reproduced with permission of IEEE).

time period for the switches. With its higher boost, the turns ratio of the transformer can be reduced when compared with the dc-dc converter shown in Figure 9.1. This advantage has also been utilized to design multiple converters operating in parallel [4] to achieve a higher power level and premium power quality along with improved system efficiency.

A new dc-dc converter topology called the Z-H converter, inspired by the Z-source inverter, has also been presented in [5] by eliminating the front end diode as shown in Figure 9.4. The concept of the shoot-through duty cycle to control the output voltage is ruled out completely by simple control of the duty cycle of the switches. The converter can achieve two-quadrant

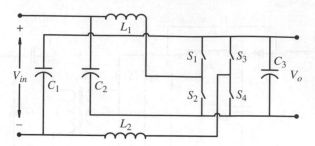

Figure 9.4 Z-H dc-dc converter (*Source*: Zhang 2008 [5]. Reproduced with permission of IEEE).

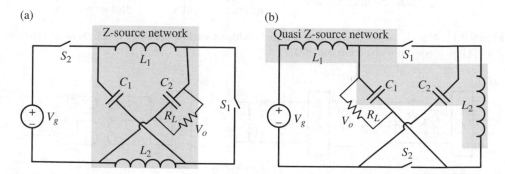

Figure 9.5 Family of four-quadrant dc-dc converters using (a) Z-source and (b) quasi-Z-source networks (*Source*: Cao 2009 [6]. Reproduced with permission of IEEE).

operational modes (II & IV quadrants) by varying the duty cycle of the complementary switch in the 0–0.5 and 0.5–1 ranges, respectively. However, the boost factor remains the same as that of a traditional inverter in all modes of operation.

Another family of four-quadrant dc-dc converters using a Z-source or a quasi-Z-source network with a minimal number of switches and passive devices is presented in [6]. The converter provides four-quadrant operation using four quadrant switches. Two basic converter topologies derived from the Z-source and quasi-Z-source networks are shown in Figure 9.5. The converters have both buck and boost characteristics, while varying their duty cycles between 0 and 1. This feature, along with changing the polarity of the load voltage by just controlling the duty cycles of the switches, makes the converters very simple and more economical for many applications, e.g. dc drives and other renewable energy systems.

9.2 Comparison

It is generally not possible to develop a converter that will suit all applications. Each of the dc-dc converters mentioned above therefore has its unique advantages and disadvantages, which can make them suitable for certain applications but not others. For an overview of their features, Table 9.1 lists some commonly cited ones to show how they compare with each other.

Table 9.1 Summary of features for different boost converters [1] (Reproduced by permission of the Institution of Engineering & Technology. Full acknowledgment to the Author, Title, and date of the original work)

Topologies	Isolation	Voltage gain	No. of switches	Input passive components	Voltage stresses across switches
H-bridge topology in Figure 9.1	Yes	$2n/(1-2D_{st})$ where, $0 \leq D_{st} \leq 0.5$	4	1 diode, 2 inductors, and, 2 capacitors	$V_{in}/(1-2D_{st})$
Two-switch Z-source/ quasi-Z-source in Figure 9.5	No	$(1-2D)/(1-D)$ Buck: $0 \leq D < 0.66$ Boost: $0.66 \leq D_{st} \leq 1$	2	2 inductors and 2 capacitors	$V_{in}/(1-D)$
Push-pull topology in Figure 9.2	Yes	$2n/(3-4D_t)$ where, $0.5 \leq D_{st} \leq 0.75$	2	1 diode, 2 inductors, and 2 capacitors	$2V_{in}/(3-4D_t)$

Note: Considering the turns ratio of the transformer, $n = 1$.

9.3   Example Simulation Model and Results

As an example, the push-pull Z-source converter shown in Figure 9.2 is simulated. The gating sequence for the converter is shown in Figure 9.6(a), where two symmetrical but phase-shifted trains of pulses are used for gating switches S_1 and S_2, respectively. These pulses will not result in flux or volt-second imbalance in the transformer core, but will lead to the desired converter states, as explained below.

Shoot-Through State

The shoot-through equivalent circuit is given in Figure 9.6(b), where it can be seen that switches S_1 and S_2 are conducting, while diodes D_1, D_2, and D_3 are blocking. Capacitor C_1 then charges inductor L_2, while the source and capacitor C_2 charge inductor L_1. Both charging currents flow through the two primary windings equally, before merging at the lower DC terminal. At the output side, capacitor C_3 is effectively disconnected, leaving only capacitor C_4 to power the external load. Voltages across C_1, C_2, and C_4 will therefore drop slightly, while currents through L_1 and L_2 increase almost linearly.

Active State with S_2 ON

By turning on only S_2, the equivalent circuit for representing the proposed converter is shown in Figure 9.6(c). The conducting devices are now D_1 and D_3, in addition to S_2. Conduction of D_1 allows L_1 and L_2 to discharge their earlier stored energy, while conduction of D_3 allows C_3 to charge rapidly through the transformer secondary winding. Conduction of D_3 might, however, end earlier when voltage across C_3 reaches the secondary transformer voltage. Voltage across C_3 can thus be expressed as $V_{C3} = V_{tr,s} = nV_{DC,link}$ based on the notations used in Figure 9.6(c). During this time, D_2 is not conducting, meaning that C_4 is still discharging its energy for powering the load.

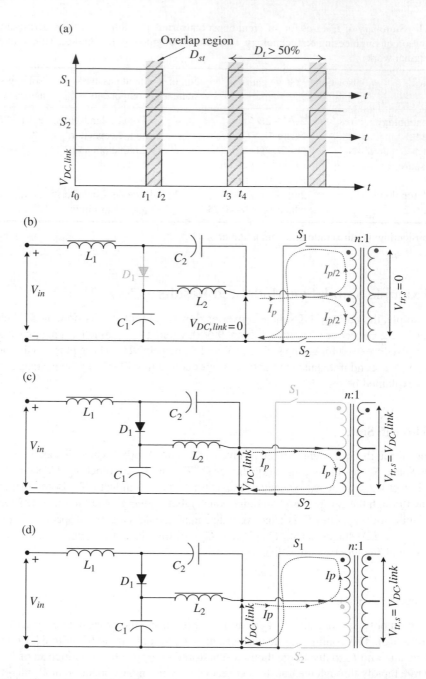

Figure 9.6 Illustration of (a) gating sequence for switches S_1 and S_2, (b) shoot-through equivalent circuit, (c) active equivalent circuit when S_2 is on, and (d) active equivalent circuit when S_1 is on, for the proposed converter [1] (Reproduced by permission of the Institution of Engineering & Technology. Full acknowledgment to the Author, Title, and date of the original work).

Active State with S_1 ON

An active state can also be inserted by turning on only S_1. An equivalent circuit representing the proposed converter can then be drawn, as in Figure 9.6(d), where the conducting devices are indicated as D_1 and D_2, in addition to S_1. Conduction of D_1 again allows L_1 and L_2 to discharge, while conduction of D_2 allows C_4 (and the load) to charge rapidly through the transformer secondary winding and C_3 in series. Conduction of D_2 will, however, end earlier when the voltage across C_4 reaches the series sum of voltages across the transformer secondary winding and C_3. The voltage across C_4, which is also the output voltage, can therefore be expressed as $V_{C4} = V_{tr,s} + V_{C3} = 2 \times n V_{DC,link}$, where the factor of 2 is intentionally introduced by the voltage doubler rectifier formed by D_2, D_3, C_3, and C_4.

Once the states are understood, their normalized times can more meaningfully be expressed in terms of D_t, which for the shoot-through time is determined as $D_{st} = 2(D_t - 0.5) = 2D_t - 1$. This expression, when substituted for the gain expression for a normal Z-source impedance network, leads to the expression in (9.1), where V_{in} and $nV_{DC,link}$ are the input and output voltages of the Z-source network.

$$\frac{V_{DC,link}}{V_{in}} = \frac{1}{1 - 2D_{st}} = \frac{1}{3 - 4D_t} \tag{9.1}$$

From the denominator of (9.1), the maximum value that D_t can assume is determined as 0.75. The boosted voltage is next processed by the middle push-pull stage to form an AC voltage that appears across the secondary winding of the transformer. The gain introduced by the push-pull stage can be written as (9.2), where $n = N_S/N_P$ is the turns ratio of the transformer and $V_{tr,s,max}$ is the magnitude of its secondary winding voltage $V_{tr,s}$.

$$\frac{V_{tr,s,max}}{V_{DC,Link}} = \frac{N_S}{N_P} = n \tag{9.2}$$

Voltage $V_{tr,s}$ is finally fed to a voltage-doubler rectifier which, as mentioned earlier, introduces a gain factor of 2, as in (9.3):

$$\frac{V_o}{V_{tr,s,max}} = 2 \tag{9.3}$$

The gain in (9.3) is obviously independent of the switch duty cycle D_t and shoot-through time D_{st}. This can be an advantage when compared to other rectifiers, which use an LC filter at their outputs for filtering out the average component from the rectified secondary-voltage pulses. For those cases, as the shoot-through duration lengthens or the non-shoot-through active duration narrows to increase the converter gain, the width of the rectified secondary voltage pulses decreases, hence causing their rectifier gain contributions to decrease. This is certainly less effective than the voltage doubler-rectifier, whose gain remains at 2.

Combining (9.1)–(9.3), the overall converter gain is thus determined as (9.4) below, whose value can be adjusted by choosing n and D_t. The turns ratio n is of course fixed during the development stage to give the converter a robust boost, with isolation subject to coupling constraints. The switch duty ratio D_t can, on the other hand, be adjusted dynamically to allow the converter to, for example, track its input voltage variation. This arrangement better distributes gain between the different stages of the converter, while retaining the advantage of isolation that a high-frequency transformer can provide to power circuits. It also helps to reduce instantaneous voltage and current stresses experienced by the switches, which would otherwise be high if the overall gain is concentrated in a single stage [7].

$$\frac{V_o}{V_{in}} = \frac{2n}{3-4D_t} \tag{9.4}$$

As well as the gain expressions, other relevant voltage and current expressions for the components in the converter can be derived:

$$V_{C1} = \frac{2-2D_t}{3-4D_t}V_{in}, V_{C2} = \frac{2D_t-1}{3-4D_t}V_{in} \tag{9.5}$$

$$V_{SW} = 2V_{tr,P1} = 2V_{tr,P2} = \frac{2V_{in}}{3-4D_t} \tag{9.6}$$

$$V_{D1} = \frac{V_{in}}{3-4D_t}, V_{D2} = V_{C4} = V_O = \frac{2nV_{in}}{3-4D_t}, V_{D3} = V_{C3} = \frac{nV_{in}}{3-4D_t} \tag{9.7}$$

where V_{C1}, V_{C2}, V_{C3}, and V_{C4} are voltages across capacitors C_1, C_2, C_3, and C_4, V_{SW} is the voltage across either S_1 or S_2 when it is reverse-biased, and V_{D1}, V_{D2}, and V_{D3} are the voltages across diodes D_1, D_2, and D_3 when they are reverse-biased.

The push-pull converter is simulated with an input voltage of 50 V, while keeping its output voltage at 400 V and under rated load condition. The waveforms obtained are shown in Figure 9.7 with the switch duty ratio set to $D_t = 60\%$ to give a shoot-through time of $D_{st} = 20\%$. This causes the dc-link voltage in the figure to drop to zero for 20% of a switching period. The resulting gain computed from (9.4) is thus 8, which when multiplied with the input voltage, gives the desired output voltage. Other waveforms shown in Figure 9.7 are the transformer voltages and currents in Figure 9.7(c), and switch voltages and currents in Figure 9.7(d). The former shows the effective transfer of ac pulses from the primary to secondary winding of the transformer, while the latter shows no excessive switch current flow, since its magnitude is comparable to the input current shown in Figure 9.7(b).

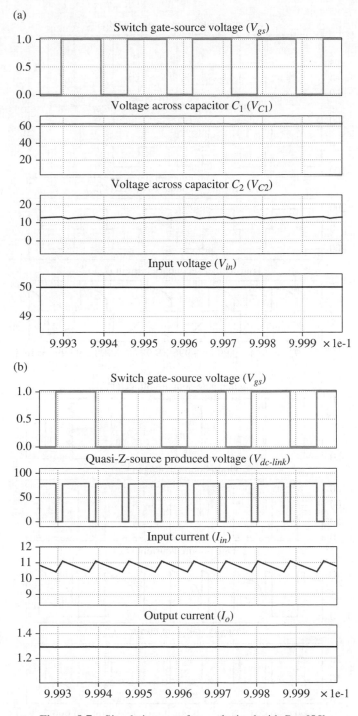

Figure 9.7   Simulation waveforms obtained with $D_t = 60\%$.

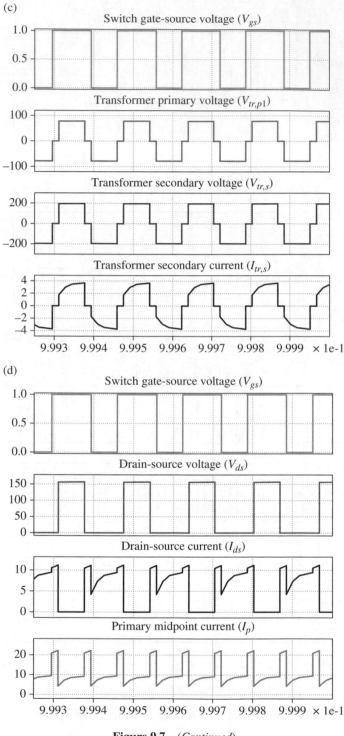

Figure 9.7 (*Continued*)

References

[1] Y. P. Siwakoti, F. Blaabjerg, P. C. Loh, G. E. Town, "High-voltage boost quasi-Z-source isolated dc/dc converter," *IET Power Electron.*, vol.7, no.9, pp.2387–2395, September 2014.

[2] D. Vinnikov, I. Roasto, "Quasi-Z-source-based isolated DC/DC converters for distributed power generation," *IEEE Trans. Ind. Electron.*, vol.58, no.1, pp.192–201, Jan. 2011.

[3] M. K. Nguyen, Q. D. Phan, V. N. Nguyen, Y. C. Lim, J. K. Park, "Trans-Z-source-based isolated DC-DC converters," in *Proc. 2012 IEEE International Symposium on Industrial Electronics (ISIE)*, pp.1–6, May 2013.

[4] H. Lee, H. G. Kim, H. Cha, "Parallel operation of trans-Z-source network full bridge dc/dc converter for wide input voltage range," in *Proc. International Conference on Power Electronics and ECCE Asia (ICPE-ECCE Asia)*, pp.1707–1712, June 2012.

[5] F. Zhang, F. Z. Peng, Z. Qian, "Z-H converter," in *Proc. IEEE Power Electronics Specialists Conference (PESC)*, pp.1004–1007, June 2008.

[6] D. Cao, F. Z. Peng, "A family of Z-source and quasi-Z-source dc-dc converters," in *Proc. The Twenty-Eighth Annual IEEE Applied Power Electronics Conference and Exposition (APEC)*, pp.1097–1101, Feb 2009.

[7] W. Li, X. He, "Review of non-isolated high-step-up dc/dc converters in photovoltaic grid-connected applications," *IEEE Trans. Ind. Electron.*, vol.58, no.4, pp.1239–1250, April 2011.

10

Z-Source Matrix Converter

10.1 Introduction

There are two types of ac-ac conversion systems, the traditional ac-dc-ac converter and the matrix converter (MC). The traditional ac-dc-ac converter consists of a pulse width modulation (PWM) boost rectifier and a PWM inverter with dc-link bus. A bulky dc-link capacitor and heavy input filter inductors may lead to high cost, large size, heavy weight, low reliability, and also high losses [1, 2].

The MC directly connects the ac source to the load without any dc-link capacitor. It has many desirable characteristics: (1) simple and compact power circuit without a dc-link capacitor, (2) output voltages with variable amplitude and variable frequency, (3) sinusoidal input/output currents, and (4) operation with a unity power factor on the input side. The matrix converters can be divided into two categories: the direct matrix converter (DMC) and the indirect matrix converter (IMC), as shown in Figure 10.1. The DMC performs the voltage and current conversion in a one-stage (direct) power conversion, while the IMC features a two-stage (indirect) power conversion. Both of them have the same behavior, but the latter has simpler commutation stress than the former. The DMC and IMC circuit topologies are equivalent in their basic functionality. The difference in the categories results in a difference in loading of the semiconductors and a different commutation scheme. The IMC has a simpler commutation due to its two-stage structure. However, this is achieved at the expense of more series-connected power devices in the high current path, which results in higher semiconductor losses and typically a lower achievable efficiency compared with the DMC. However, the differences between the control performance of DMC and IMC are quite negligible in practice [3].

Despite all these attractive properties, the matrix converter has not yet gained much attention in the industry due to its many unsolved problems. The most critical problem is the reduced voltage transfer ratio, which is defined as the ratio between the output voltage and the input

Impedance Source Power Electronic Converters, First Edition. Yushan Liu, Haitham Abu-Rub, Baoming Ge, Frede Blaabjerg, Omar Ellabban, and Poh Chiang Loh.
© 2016 John Wiley & Sons, Ltd. Published 2016 by John Wiley & Sons, Ltd.

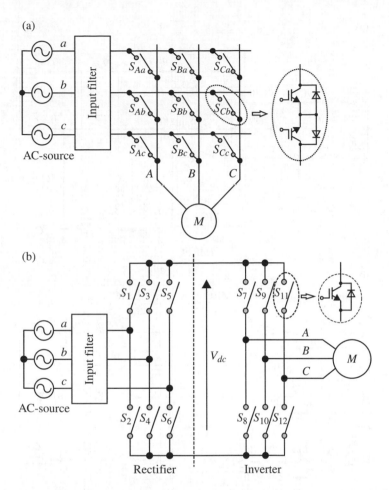

Figure 10.1 Two ac-ac converter topologies: (a) direct matrix converter and (b) indirect matrix converter.

voltage, and has been constrained to 0.866 when using a linear modulation [3]. A lot of research has been done on the over-modulation method to overcome the problem of low voltage transfer ratio. However, the over-modulation can only be achieved at the expense of the quality of both output voltage and input current [4]. Improving the voltage transfer ratio is an important research topic. One easy solution is to connect a transformer between the power supply and the MC. However, a mains transformer is bulky, expensive, and it also affects the system efficiency. Another solution is to use a matrix-reactance frequency converter (MRFC), which consists of an MC and an ac chopper, and has a voltage transfer ratio larger than one. The MRFC converter is categorized into two groups: integrated and cascaded MRFCs, as shown in Figure 10.2. Unfortunately, the IMRFC typology has several disadvantages. First, the control algorithm is complicated due to the required synchronization between the MC and the ac chopper. Second, the voltage transfer ratio strongly depends on the circuit and the load parameters. Finally, the input power factor is lower than other MCs, even for a purely resistive load. The CMRFC

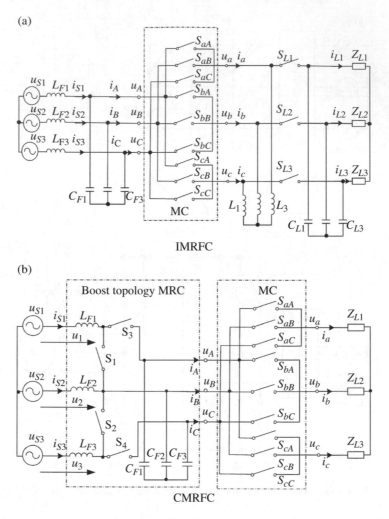

Figure 10.2 Integrated (a) and cascade matrix-reactance frequency (b) converter topologies (*Source*: Ellabban 2015 [5]. Reproduced with permission of IEEE).

topology has fewer passive components compared to the IMRFC topology, but it has limited voltage gain, and more complicated damping control of the input current and disturbed output current [5].

A Z-source (ZS) or quasi-Z-source (QZS) converter/inverter can achieve the high voltage gain in a single-stage power conversion due to its boost ability [6, 7], where the ZS/QZS network is inserted between the dc input source and inverter.

By introducing the ZS/QZS network into different MC topologies, a novel matrix converter called Z-source matrix converter (ZSMC) has been proposed. According to the topology of the MC, they are called a Z-source indirect matrix converter (ZSIMC) and Z-source direct matrix converter (ZSDMC). Therefore, it is possible to overcome the low voltage gain challenge of the traditional MC; in addition, the ZS/QZS network allows the short-circuit, which makes the

ZSMC commutation easier. As a result, the ZSMC provides a low-cost, reliable, and highly efficient structure for buck and boost ac-ac conversion. Moreover, there are two different configurations of the ZSIMC topology, depending on the location of the ZS/QZS network. The ZS/QZS network can be located between the rectifier and the inverter of the indirect matrix converter at the intermediate dc-link; the voltage gain is extended, but it is not an all-silicon solution and will cause larger size and heavier weight than the conventional IMC (it requires large inductors and capacitors in the dc-link bus). The other option, where the ZS/QZS network is inserted between the input ac source and the rectifier of the indirect matrix converter to achieve an all-silicon solution. This configuration can achieve high voltage gain with a small ZS/QZS network, but it requires a large number of switches, so it involves a higher cost.

This chapter will present an updated overview of the different Z source matrix converter topologies including the ZSIMC, with its all-silicon and not all-silicon configurations, and also the ZSDMC. The contents of this chapter are based on the authors' publication of [8].

10.2 Z-Source Indirect Matrix Converter (All-Silicon Solution)

10.2.1 Different Topology Configurations

Figure 10.3 shows the recently proposed ZS/QZSIMC topologies [9–12], which consist of five parts: three-phase AC source, ZS/QZS network, front-end rectifier, back-end inverter, and AC load. The ZS-network includes three inductors (L_a, L_b, L_c), three capacitors (C_a, C_b, C_c), and three bidirectional switches (S_a, S_b, S_c). However, the QZS-network includes six inductors (L_{a1}, L_{a2}, L_{b1}, L_{b2}, L_{c1}, L_{c2}), six capacitors (C_{a1}, C_{a2}, C_{b1}, C_{b2}, C_{c1}, C_{c2}), and three additional bidirectional switches (S_a, S_b, S_c). One gate signal, S_x, can be used to control these three switches because they have the same switching behavior. This unique ZS/QZS network allows the ZS/QZSIMC to work on the buck and boost modes. The ZSIMC topology, Figure 10.3(a), has only three inductors and three capacitors compared to the other QZSIMC topologies which have six inductors and six capacitors. However, the ZSIMC topology has (a) a limited voltage boost ratio (voltage gain can only reach 1.15), (b) inherited phase shift caused by the Z-network, which makes the control inaccurate, (c) discontinuous current in the front of the Z-source network, (d) requires additional input filters to reduce input current THD, and these may increase the whole system volume and cost and reduce efficiency [8]. The discontinuous qZS-IMC QZSIMC topology, Figure 10.3(b), was proposed in [10], by adding three more inductors and three more capacitors compared to the ZS-IMC topology: however, the voltage gain can go to 4–5 times, or even higher, depending on the voltage rating of the switches. Furthermore, there is no phase shift due to the QZ-network. But, it still requires additional input filters. The continuous QZS-IMC topology, Figure 10.3(c), was proposed in [9]. The continuous QZSIMC topology does not require input filters, because the continuous quasi-Z source network is integrated with the LC filter. The three ZS/QZSIMC topologies of Figure 10.3 were compared in detail [13] in terms of voltage gain, current ripple, voltage ripple, inductor current and capacitor voltage stresses, ZS/qZS switch current and voltage stresses, filtering function, input current THD, output voltage THD, and efficiency. Table 10.1 summarizes the evaluation of these three ZS/QZS-MC topologies.

Furthermore, in [11], the common mode voltage (CMV) issue of the QZSIMC with continuous current mode was investigated, and three modulation strategies were developed in order to reduce the CMV for the QZSIMC with the CMV peak value reduction of 42%.

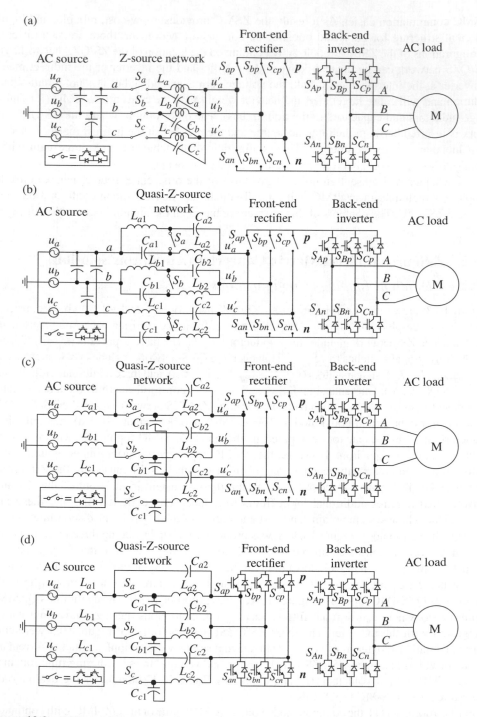

Figure 10.3 Different Z-source matrix converter topologies: (a) ZSIMC with discontinuous current mode (*Source*: Liu 2015 [12]. Reproduced with permission of IEEE), (b) QZSIMC with discontinuous current mode (*Source*: Liu 2014 [10]. Reproduced with permission of IEEE), (c) QZSIMC with continuous current mode (*Source*: Liu 2013 [9]. Reproduced with permission of IEEE), (d) simplified QZSIMC (*Source*: Liu 2014 [13]. Reproduced with permission of IEEE).

Table 10.1 Comparison of three ZS/QZSIMC (all-silicon solution) ac/ac converter topologies (*Source*: Liu 2015 [12]. Reproduced with permission of IEEE)

	ZSIMC	Discont-QZSIM	Cont-QZSIM
Additional input filter	Yes	Yes	No
Voltage gain	Low	High	High
Current ripples	High	Low	Low
Voltage ripples	Low	High	Middle
Inductor current stress	Low	High	Middle
Capacitor voltage stress	Middle	Low	High
Switch current stress	Low	High	High
Switch voltage stress	Low	Low	High
Input current THD	Middle	High	Low
Output voltage THD	High	Middle	Low
Efficiency	High	Low	High

The simplified QZSIMC, proposed in [13] and shown in Figure 10.3(d), is a modification of the topology shown in Figure 10.3(c). This requires fewer power switches (only 12), which reduces the system cost; the 18 switches of Figure 10.3(c) is the same as the back-to-back pulse-width modulation converter without dc-link capacitor. The QZS network integrates the LC filtering function in order to avoid the additional input filter. The simplified QZS-IMC has a performance limitation that input current will have high harmonics if no input current closed-loop compensation is used. Therefore, input current closed-loop control is necessary in order to reduce the harmonics of the input current.

10.2.2   Operating Principle and Equivalent Circuits

Figure 10.4 shows the equivalent circuit of the QZSIMC topology with continuous input current at the shoot-through (ST) and non-shoot-through (NST) states. During the NST state, the switches S_x ($x = a, b, c$) are on ($S = 1$) as shown in Figure 10.4(a), the inductors discharge the capacitors, and each output phase voltage of the QZS network is the sum of two capacitor voltages. On the other hand, during the ST state, switches S_x are off ($S = 0$) as shown in Figure 10.4(b), the input side of front-end rectifier is short-circuited, and the inductors are charged.

During the NST state, from Figure 10.4(a), we have the following voltage and current equations:

$$\begin{bmatrix} u_{Ca2} \\ u_{Cb2} \\ u_{Cc2} \end{bmatrix} = -\begin{bmatrix} u_{La2} \\ u_{Lb2} \\ u_{Lc2} \end{bmatrix} \tag{10.1}$$

$$\begin{bmatrix} u_{La2} \\ u_{Lb2} \\ u_{Lc2} \end{bmatrix} = \begin{bmatrix} u_a \\ u_b \\ u_c \end{bmatrix} - \begin{bmatrix} u_{Ca1} \\ u_{Cb1} \\ u_{Cc1} \end{bmatrix} \tag{10.2}$$

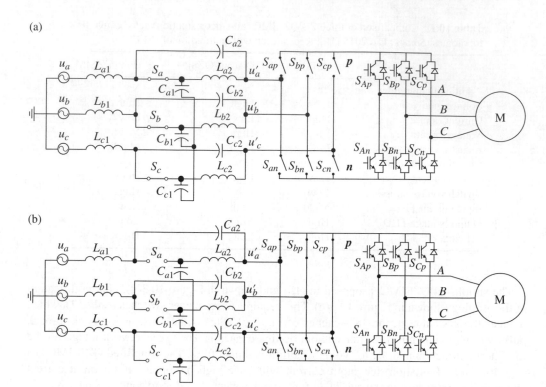

Figure 10.4 Equivalent circuit of the QZS-IMC: (a) non-shoot-through state and (b) shoot-through state.

$$\begin{bmatrix} u_{La2} \\ u_{Lb2} \\ u_{Lc2} \end{bmatrix} = \begin{bmatrix} u_a \\ u_b \\ u_c \end{bmatrix} + \begin{bmatrix} u_{Ca2} \\ u_{Cb2} \\ u_{Cc2} \end{bmatrix} - \begin{bmatrix} u'_a \\ u'_b \\ u'_c \end{bmatrix} \tag{10.3}$$

$$\begin{bmatrix} i_{La1} \\ i_{Lb1} \\ i_{Lc1} \end{bmatrix} - \begin{bmatrix} i_{La2} \\ i_{Lb2} \\ i_{Lc2} \end{bmatrix} = \begin{bmatrix} i_{Ca1} \\ i_{Cb1} \\ i_{Cc1} \end{bmatrix} - \begin{bmatrix} i_{Ca2} \\ i_{Cb2} \\ i_{Cc2} \end{bmatrix} \tag{10.4}$$

And during the ST state, from Figure 10.4(b):

$$\begin{bmatrix} u_a \\ u_b \\ u_c \end{bmatrix} + \begin{bmatrix} u_{Ca2} \\ u_{Cb2} \\ u_{Cc2} \end{bmatrix} = \begin{bmatrix} u_{La1} \\ u_{Lb1} \\ u_{Lc1} \end{bmatrix} \tag{10.5}$$

$$\begin{bmatrix} u_{Ca1} \\ u_{Cb1} \\ u_{Cc1} \end{bmatrix} = \begin{bmatrix} u_{La2} \\ u_{Lb2} \\ u_{Lc2} \end{bmatrix} \tag{10.6}$$

$$\begin{bmatrix} i_{La1} \\ i_{Lb1} \\ i_{Lc1} \end{bmatrix} = -\begin{bmatrix} i_{Ca2} \\ i_{Cb2} \\ i_{Cc2} \end{bmatrix}, \quad \begin{bmatrix} i_{La2} \\ i_{Lb2} \\ i_{Lc2} \end{bmatrix} = -\begin{bmatrix} i_{Ca1} \\ i_{Cb1} \\ i_{Cc1} \end{bmatrix} \tag{10.7}$$

where u denotes the voltage and i the current, and the subscript C_{x1} and C_{x2} are the capacitors 1 and 2 of phase-x; L_{x1} and L_{x2} for the inductors 1 and 2 of phase-x; $x = a$, b, c.

For one switching cycle T_s, if the time interval of ST state is T, the ST duty cycle is defined as $D = T_0/T_s$. The inductor average voltages and the capacitor average currents should be zero over one switching period in steady state. From equations (10.1) to (10.7), it is possible to obtain

$$\begin{bmatrix} i_{La1} \\ i_{Lb1} \\ i_{Lc1} \end{bmatrix} = \begin{bmatrix} i_{La2} \\ i_{Lb2} \\ i_{Lc2} \end{bmatrix} = \begin{bmatrix} i_a \\ i_b \\ i_c \end{bmatrix} \tag{10.8}$$

$$\begin{bmatrix} u_{Ca2} \\ u_{Cb2} \\ u_{Cc2} \end{bmatrix} = \frac{D}{1-D}\begin{bmatrix} u_{Ca1} \\ u_{Cb1} \\ u_{Cc1} \end{bmatrix} \tag{10.9}$$

$$\begin{bmatrix} u_{Ca1} \\ u_{Cb1} \\ u_{Cc1} \end{bmatrix} = \frac{1-D}{1-2D}\begin{bmatrix} u_a \\ u_b \\ u_c \end{bmatrix} \tag{10.10}$$

Combining equations (10.3) and (10.10) we have:

$$\begin{bmatrix} u'_a \\ u'_b \\ u'_c \end{bmatrix} = \frac{1-D}{1-2D}\begin{bmatrix} u_a \\ u_b \\ u_c \end{bmatrix} \tag{10.11}$$

The voltage boost factor B is expressed as

$$B = \frac{u_o}{u_i} = \frac{1}{1-2D} \tag{10.12}$$

where u_i is the amplitude of the input voltage source and u_o is the output voltage amplitude of the QZS-network. The voltage gain G of the proposed QZSIMC will be:

$$G = Bm \tag{10.13}$$

where $m = m_i m_o$ is the modulation index of the IMC, m_i is the modulation index of the front-end rectifier and m_o is the modulation index of the back-end inverter. Figure 10.5 shows the voltage gain versus the modulation index of the QZSIMC. The voltage gain can be greater than 1 by choosing the modulation index and ST duty cycle ($D < 0.5$), which is boost mode. Of course, the buck mode can be achieved through using a lower modulation index.

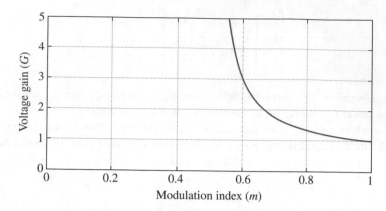

Figure 10.5 Voltage gain versus modulation index of the QZSIMC (*Source*: Liu 2013 [9]. Reproduced with permission of John Wiley & Sons).

10.2.3 Parameter Design of the QZS-Network

The QZS inductor should limit the high-frequency current ripple, and the capacitor should limit high-frequency voltage ripple.

From Figure 10.4(a), the inductor L_{x1} ($x=a$, b, c) voltage is equal to the difference of the input voltage and the capacitor C_{x1} voltage, the inductor L_{x2} voltage is equal to the capacitor C_{x2} voltage, at non-shoot-through state, so we obtain [11]

$$\begin{cases} L_{x1} = \dfrac{\left|u_i - u_{Cx1}\right|(1-D)T_s}{\Delta i_i} \\[4mm] L_{x2} = \dfrac{\left|u_{Cx2}\right|(1-D)T_s}{\Delta i_{L_{x2}}} \end{cases} \tag{10.14}$$

If $\Delta i_i = r_L I_s$ and $\Delta i_{L_{x2}} = r_L I_{L_{x2}}$, we get

$$\begin{cases} L_{x1} \geq \dfrac{\left|u_{i_{max}} - u_{C_{x1max}}\right|(1-D)T_s}{r_L I_s} = \dfrac{\sqrt{2}V_s D(1-D)T_s}{(1-2D)r_L I_s} \\[4mm] L_{x2} \geq \dfrac{\left|u_{C_{x2max}}\right|(1-D)T_s}{r_L I_{L_{x2}}} = \dfrac{\sqrt{2}V_s D(1-D)T_s}{(1-2D)r_L I_{L_{x2}}} \end{cases} \tag{10.15}$$

where $I_s, I_{L_{x2}}$, and V_s are the RMS values of input current, inductor L_{x2} current, and input voltage, respectively; r_L is the designed ripple coefficient of inductor currents.

During the ST state, as shown in Figure 10.4(b), the capacitor C_{x1} current is equal to the inductor L_{x2} current, the capacitor C_{x2} current is equal to the input current, which gives [11]

$$\begin{cases} C_{x1} = \dfrac{i_{L_{x2}}DT_s}{\Delta u_{C_{x1}}} \\[3mm] C_{x2} = \dfrac{i_i DT_s}{\Delta u_{C_{x2}}} \end{cases} \tag{10.16}$$

If $\Delta u_{C_{x1}} = r_C u_{C_{x1}}$ and $\Delta u_{C_{x2}} = r_C u_{C_{x2}}$, we get

$$\begin{cases} C_{x1} \geq \dfrac{i_{L_{x2\max}}DT_s}{\Delta u_{C_{x1}}} = \dfrac{\sqrt{2}I_s D(1-2D)T_s}{(1-D)r_C V_s} \\[4mm] C_{x2} \geq \dfrac{i_{i_{\max}}DT_s}{\Delta u_{C_{x2}}} = \dfrac{\sqrt{2}I_s(1-2D)T_s}{r_C V_s} \end{cases} \tag{10.17}$$

where $u_{C_{x1}}$ and $u_{C_{x2}}$ are the RMS values of the capacitor C_{x1} and capacitor C_{x2} respectively, r_C is the designed ripple coefficient of the capacitor voltages, and I_s is the RMS value of input current.

10.2.4   QZSIMC (All-Silicon Solution) Applications

To demonstrate the application of an all-silicon solution QZSIMC, Figure 10.6 shows a vector control based QZSIMC-fed IM drive. The q-axis component reference i_{qs}^* of the stator current

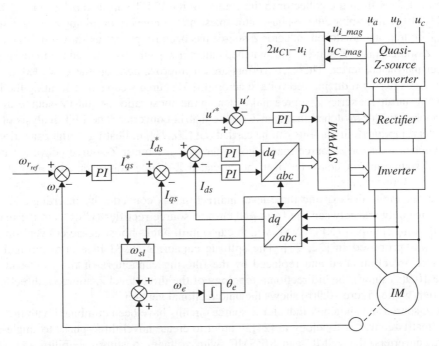

Figure 10.6   Block diagram of a vector control-based QZSIMC-fed IM drive system (*Source*: Liu 2013 [26]. Reproduced with permission of IEEE).

is the output of the speed closed-loop through a PI controller. The d-axis component reference i_{ds}^* of the stator current is constant, which is equal to the excitation current of the IM. The d-axis and q-axis current component closed loops will ensure error-free tracking [8].

10.3   Z-Source Indirect Matrix Converter (Not All-Silicon Solution)

10.3.1   Different Topology Configurations

A second family of ac-ac Z-source converters, based on the indirect matrix converter, was proposed in [14], and is shown in Figure 10.7. Only additional passive components are required, without any increase in the number of semiconductor power switches, which can increase the complexity of control. The main purpose of the proposed topology is to boost the average dc-link voltage, in order to provide higher load voltage when required. In Figure 10.7(a), the switches in the rectification stage are implemented using four-quadrant switches, and the switches in the inversion stage are implemented with an IGBT and an anti-parallel diode. The inductors L_1, L_2, and capacitors C_1 and C_2 make up a symmetrical Z-source network. Furthermore, two simplified ac-ac Z-source converter topologies are presented in Figure 10.7(b) and (c), which are based on the sparse and ultra-sparse matrix converter topologies, respectively. Also, in [15–17], an additional power switch (S_1) is inserted in the dc-link before the Z-source network to ensure the bidirectional power flow, as shown in Figure 10.7(d). While operating in a normal motoring mode, the dc-link current can flow through the integrated anti-parallel diode of the IGBT and feed the load through the Z-source inverter. During operation in the power regenerative braking mode, however, the current can flow from the collector to the emitter of the IGBT and feed back to the grid.

In [18, 19], a three-phase/three-phase ultra-sparse matrix converter utilizing a series Z-source, quasi Z-source, and switched inductor Z-source has been proposed, as shown in Figure 10.8. Series Z-source, quasi Z-source, and switched inductor Z-source is inserted in either rail of the indirect matrix converter. These converters are an improvement on the cascaded Z-source matrix converter by reducing the voltage across the Z-source's capacitor, limiting the inrush current at startup for series Z-source and widening the boost ratio for quasi Z-source and very high boost ratio for switched inductor Z-source matrix converter. The FFT analysis of these converters' input/output currents can be carried out [20, 21], indicating a slight superiority of the switched inductor Z-source matrix converter over the quasi Z-source converter and the series Z-source matrix converter over the cascaded Z-source converter with respect to the quality of input currents.

An extension to the existing three-level indirect matrix converter by inserting a Z-source impedance network between the front-end current-source rectifier (CSR) and the rear-end neutral point clamped (NPC) VSI, which can obtain buck–boost ac-ac conversion capability, was proposed in [22, 23]. The split-dc capacitors used in a conventional NPC converter are eliminated and replaced by the filtering capacitors of the front-end CSR. Figure 10.9(a) shows the bidirectional topology of the three-level Z-source indirect matrix converter, while Figure 10.9(b) shows the unidirectional version.

The Z-source and switched-inductor Z-source circuits have been combined with the super-sparse matrix converter topology in [24] in order to create novel three-phase to single-phase matrix converters, the ZSMC and SIZSMC, with voltage-boosting capability and a unity power factor, as shown in Figure 10.10. These converters are suitable for all applications with unidirectional power flow.

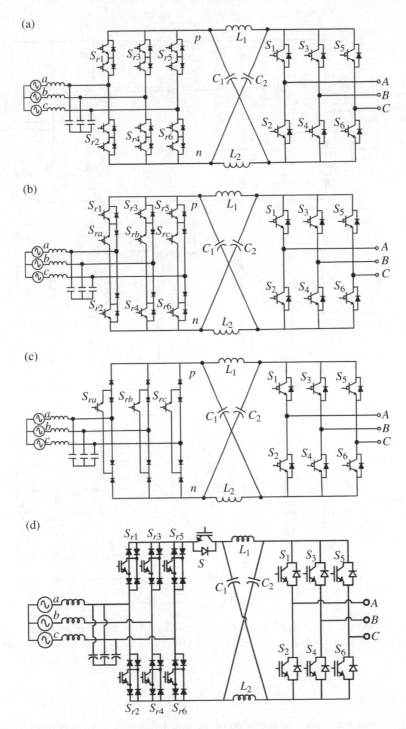

Figure 10.7 Different Z-source indirect matrix converter topologies: (a) Z-source indirect matrix converter [14], (b) Z-source sparse converter, (c) Z-source ultra-sparse converter, (d) bidirectional Z-source sparse converter [15–17]. (*Source*: [15–17]. Reproduced with permission of IEEE.)

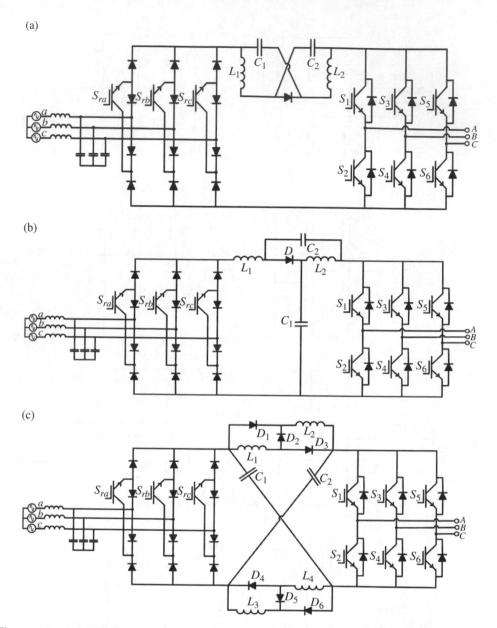

Figure 10.8 (a) Series Z-source matrix converter, (b) quasi-Z-source matrix converter, and (c) switched-inductor Z-source matrix converter topologies [18, 19]. (*Source*: [18, 19]. Reproduced with permission of IEEE.)

10.3.2   *Operating Principle and Equivalent Circuits*

Figure 10.11 shows the modes of operation of the quasi-Z-source matrix converter, shown in Figure 10.7(a), under the NST and ST conditions. Figure 10.11(a) shows the situation when diode *D* is ON. Capacitors C_1 and C_2 are simultaneously charged by the rectified input voltages.

(a)

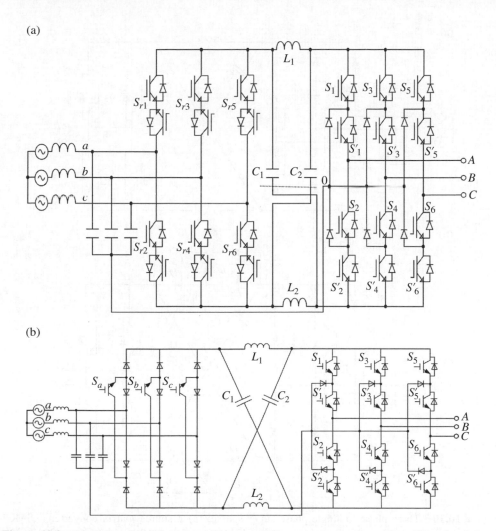

(b)

Figure 10.9 (a) Bidirectional three-level Z-source indirect matrix converter, (b) unidirectional three-level Z-source indirect sparse matrix converter topologies [22, 23]. (*Source*: Effah 2013 [22]. Reproduced with permission of IEEE.)

In Figure 10.11(b), the converter is operating under the ST condition, boosting the output voltage. Now, the diode D is OFF.

During the NST state, from Figure 10.11(a), we get the following voltage and current equations [26]:

$$\begin{cases} C_1 \dfrac{du_{C1}}{dt} = i_{L1} - i_o, & C_2 \dfrac{du_{C2}}{dt} = i_{L2} - i_o \\[2mm] L_1 \dfrac{di_{L1}}{dt} = u_{in} - u_{C1}, & L_2 \dfrac{di_{L1}}{dt} = -u_{C2} \end{cases} \qquad (10.18)$$

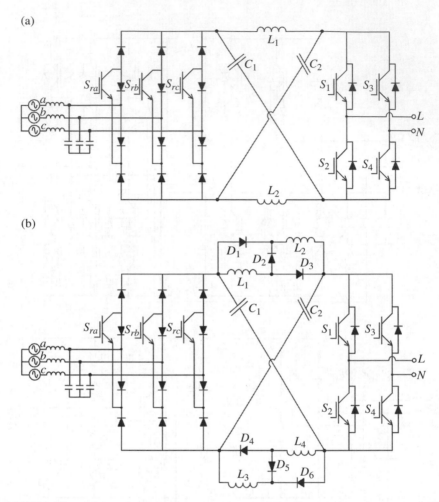

Figure 10.10 Three-phase to single-phase super-sparse: (a) Z-source matrix converter (ZSMC) (b) switched-inductor Z source matrix converter (SIZMC) topologies (*Source*: Karaman 2012 [24]. Reproduced with permission of IEEE).

where i_{L1}, i_{L2}, and i_o are the currents of two inductors and the dc-link bus, respectively; u_{C1}, u_{C2}, and u_{in} are the voltages across the two capacitors and the QZS network input voltage; and u_{in} is the virtual dc-link voltage produced by the rectification state.

During the ST state, from Figure 10.11(b), we get [26]

$$\begin{cases} C_1 \dfrac{du_{C1}}{dt} = -i_{L2}, & C_2 \dfrac{du_{C2}}{dt} = -i_{L1} \\[2mm] L_1 \dfrac{di_{L1}}{dt} = u_{in} + u_{C2}, & L_2 \dfrac{di_{L1}}{dt} = u_{C1} \end{cases} \tag{10.19}$$

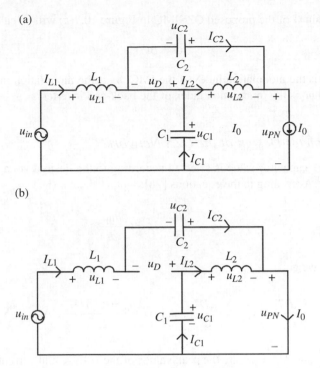

Figure 10.11   Modes of operation of the quasi-Z-source matrix converter: (a) non-shoot-through state, (b) shoot-through state.

For one switching cycle T_s, if the time interval of ST state is T, the ST duty cycle is defined as $D = T_0/T_s$. The inductor average voltages and the capacitor average currents should be zero over one switching period in steady state. From equations (10.18) and (10.19), we have

$$\begin{cases} u_{C1} = \dfrac{1-D}{1-2D} u_{in} \\[3mm] u_{C2} = \dfrac{D}{1-2D} u_{in} \end{cases} \tag{10.20}$$

The output voltage of the QZS network is

$$u_{PN} = u_{C1} + u_{C2} = \frac{1}{1-2D} u_{in} \tag{10.21}$$

The voltage boost factor B is expressed as

$$B = \frac{u_{PN}}{u_{in}} = \frac{1}{1-2D} \tag{10.22}$$

The voltage gain G of the proposed QZSIMC in Figure 10.7(a) will be calculated as

$$G = Bm \tag{10.23}$$

where $m = m_i m_o$ is the modulation index of the IMC, m_i is the modulation index of the front-end rectifier, and m_o is the modulation index of the back-end inverter.

10.3.3 Parameter Design of the QZS Network

The QZS network capacitor value is selected according to the desired voltage ripple and the capacitor current. According to the equations [26]:

$$u_L = L\frac{di_L}{dt}, \quad i_C = C\frac{du_C}{dt} \tag{10.24}$$

From (10.19), we get

$$\Delta u_{C1} = \frac{i_{L2}DT_s}{C_1}, \quad \Delta u_{C2} = \frac{i_{L1}DT_s}{C_2}, \quad \Delta i_{L1} = \frac{(u_{in}+u_{C2})DT_s}{L_1}, \quad \Delta i_{L2} = \frac{u_{C1}DT_s}{L_2} \tag{10.25}$$

where $\Delta u_C = r_C u_C$ and $\Delta i_L = r_L i_L$ are the peak values of the voltage and current ripples, respectively, r_C is the capacitor voltage ripple ratio, and r_L is the inductor current ripple ratio.

Using (10.20)–(10.22) and (10.25), we can get

$$C_1 \geq \frac{PD(1-2D)T_s}{r_C(1-D)u_{in}^2}, \quad C_2 \geq \frac{P(1-2D)T_s}{r_C u_{in}^2}, \quad L_1 = L_2 \geq \frac{u_{in}^2(1-2D)DT_s}{P(1-D)r_L} \tag{10.26}$$

where P is the input power of the system.

10.3.4 ZS/QZSIMC (Not All-Silicon Solution) Applications

The ZSIMC (not all-silicon solution) has been used as part of a variable speed drive system using a three-phase induction motor, as shown in Figure 10.12. A new control method has been proposed to increase the operating voltage range of an indirect matrix converter based motor drive, while at the same time guaranteeing unitary input power factor. The control was separated into two different stages, both depending on the dc-link Z-source capacitor voltage. The rectifier stage is controlled so that the current vectors impose phase and amplitude control, thus guaranteeing near unity power factor. On the other hand, on the inverter side, the inverter stage, along with a closed loop control of the ac three-phase induction motor currents, also enforces specific operation modes, providing the necessary stator voltage, increasing the power flow from the mains to the load [25].

The quasi-Z-source indirect matrix converter (QZSIMC) has been used for induction motor drive systems, and the indirect field oriented control has been used, as shown in Figure 10.13 [26]. The indirect sparse-matrix topology with inductive-capacitive-diode (LCD) networks, Figure 10.14, has been used for gearless wind energy systems [27]. The wind power captured

(a)

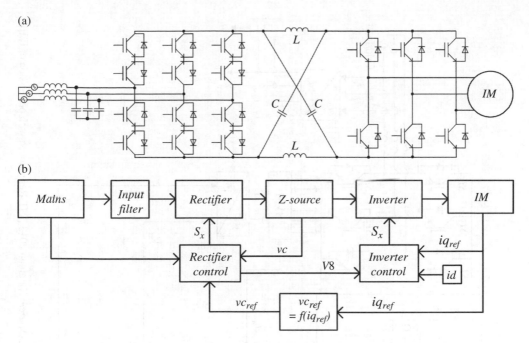

(b)

Figure 10.12   Z-source indirect matrix converter based IM arive system: (a) system topology and (b) control algorithm (*Source*: Sousa 2012 [25]. Reproduced with permission of IEEE).

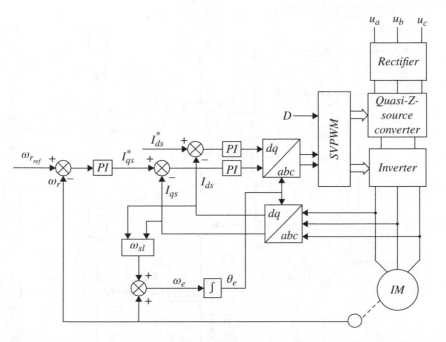

Figure 10.13   Block diagram of the quasi-Z-source indirect matrix converter based induction motor drive (*Source*: Liu 2013 [9]. Reproduced with permission of IEEE).

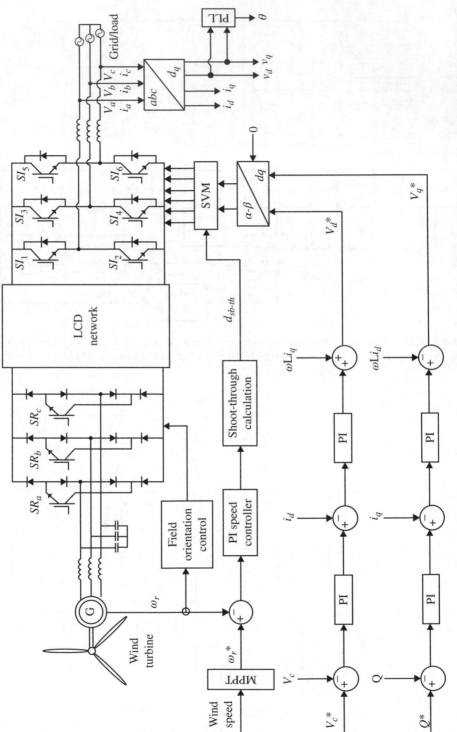

Figure 10.14 Wind-energy system with PMSG, matrix converter, and LCD network (*Source:* Karaman 2013 [27]. Reproduced with permission of IEEE).

by the turbine is converted by the PMSG and transmitted to the grid via a matrix converter. The rectification stage of the converter regulates the power factor and draws sinusoidal currents from the PMSG. The LCD network and the inversion stage step up the voltage on the grid side. The generator-side control strategy is based on the principle of zero d-axis current control. The voltage-oriented control of the rectifier stage allows decoupling of the active and reactive input powers, and the required boost factor is maintained by adjusting the shoot-through duty ratios of the inverter stage.

10.4 Z-Source Direct Matrix Converter

10.4.1 Alternative Topology Configurations

By introducing the Z-source network to the conventional direct matrix converter, which has been proposed as a Z-source direct matrix converter (ZSDMC), Figure 10.15, [28–32], it is possible to overcome the low voltage gain of the traditional direct MC; in addition, the

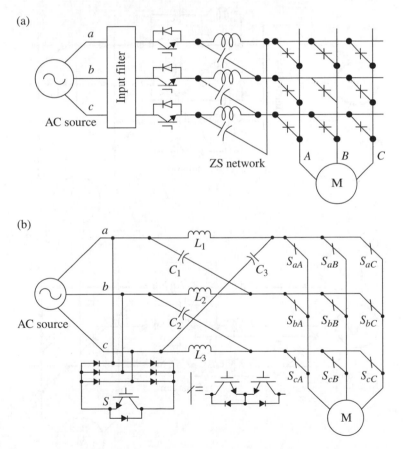

Figure 10.15 Different Z-source direct matrix converter (ZSDMC) topologies (a: *Source*: Ge 2012 [31]. Reproduced with permission of IEEE. b: *Source*: Liu 2012 [32]. Reproduced with permission of IEEE).

Z-source network allows the short-circuit, which makes the ZSDMC commutation easier. The ZSDMC is derived from the traditional DMC by only adding three inductors, capacitors, switches, and diodes. However, the ZSDMC has a limited voltage boost ratio (voltage gain can only reach 1.15), inherited a phase shift caused by the Z-network, which makes the control inaccurate, and also discontinuous current in the front of the Z-source network. However, for the quasi-Z-source direct matrix converter (QZSDMC), as shown in Figure 10.16, the voltage gain can go to 4–5 times or even higher depending on the voltage rating of the switches, no phase shift, which can cause less error in the control, and lower switch voltage and current stress [33]. In addition, the circuit in Figure 10.16(b) has continuous input current [5].

Compared to traditional DMC, the ZSDMC and QZSDMC can both boost the voltage higher than 0.866. The boost ratio depends on the duty cycle of the extra shoot-through state. Also the QZSDMC topology can cause less voltage/current stress on the switch and passive components, less input and output harmonics, and can have a higher power factor than the ZSDMC.

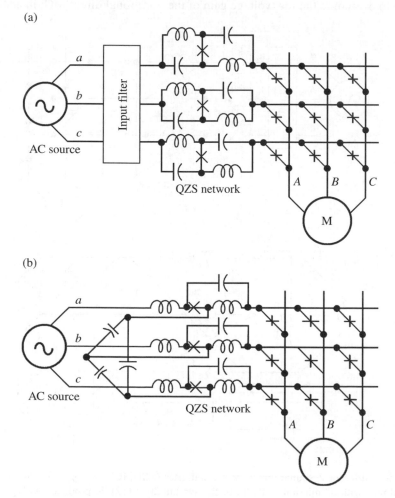

Figure 10.16 Quasi Z-source direct matrix converter topologies with: (a) discontinuous input current and (b) continuous input current (*Source*: Ge 2012 [31]. Reproduced with permission of IEEE).

Moreover, compared to ZSDMC topology, the QZSDMC is a fewer-component, compact, highly efficient and also wide range voltage buck–boost matrix converter [5].

In [34–36], a new family of single-phase AC/AC buck–boost converters based on single-phase matrix converter (SPMC) topology and Z-network topology was proposed, Figure 10.17. The proposed single-phase Z-source matrix converter has the merits that the output voltage can be bucked/boosted and in-phase with the input voltage, and the output voltage can also be bucked/boosted and out-of-phase with the input voltage.

In [36], a single to three-phase Z-source matrix converter (STZMC) was proposed by using the matrix converter theory and Z-source conversion concept, Figure 10.18. The STZMC converters can boost the amplitude of the output voltage at the desired frequency. Furthermore, the limitations of the single to three-phase matrix converter, such as lower input-output voltage transfer ratio and unbalanced output currents, have been avoided.

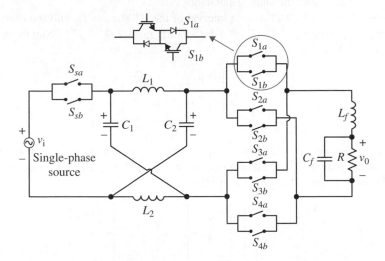

Figure 10.17 Single-phase Z-source matrix converter topology (*Source*: Nguyen 2009 [34]. Reproduced with permission of IEEE).

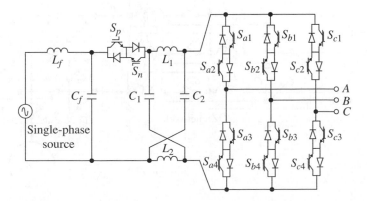

Figure 10.18 Single to three-phase Z-source matrix converter (*Source*: Milan 2012 [37]. Reproduced with permission of IEEE).

10.4.2 Operating Principle and Equivalent Circuits

The main circuit configuration of the QZSDMC is shown in Figure 10.19. It consists of two parts: QZS-network and DMC. The QZS-network includes six inductors (L_{a1}, L_{a2}, L_{b1}, L_{b2}, L_{c1}, L_{c2}), six capacitors (C_{a1}, C_{a2}, C_{b1}, C_{b2}, C_{c1}, C_{c2}), and three additional bidirectional switches (S_a, S_b, S_c). One gate signal can be used to control these three switches because they have the same switching state. Therefore, the drive signal for S_a, S_b, and S_c can be denoted as S_0.

The operation principle of the QZSDMC can be divided into two switching states: shoot-through and non-shoot-through. Figure 10.20 shows the QZSDMC equivalent circuits during these states. During the ST state, Figure 10.20(a), the switch S_0 is off and the output of the QZSDMC is shorted for boost operation. During the NST state, Figure 10.20(b), the switch S_0 is on for normal DMC operation. Owing to the symmetry of the system, inductors of QZS-network (L_{a1}, L_{a2}, L_{b1}, L_{b2}, L_{c1}, L_{c2}) have the same inductance (L), and the capacitors (C_{a1}, C_{a2}, C_{b1}, C_{b2}, C_{c1}, C_{c2}) also have the same capacitance (C).

For one switching cycle, T_s, the time interval of the ST state is T_0, and the time interval of the NST state is T_1, hence, $T_s = T_0 + T_1$, and the ST duty ratio is $D = T_0/T_s$. From Figure 10.20(a), during the ST state, we get

$$\begin{bmatrix} v_{ab} \\ v_{bc} \\ v_{ca} \end{bmatrix} = \begin{bmatrix} v_{La1} \\ v_{Lb1} \\ v_{Lc1} \end{bmatrix} + \begin{bmatrix} v_{Ca1} \\ v_{Cb1} \\ v_{Cc1} \end{bmatrix} - \begin{bmatrix} v_{Cb1} \\ v_{Cc1} \\ v_{Ca1} \end{bmatrix} - \begin{bmatrix} v_{Lb1} \\ v_{Lc1} \\ v_{La1} \end{bmatrix} \qquad (10.27)$$

where v denotes the voltage, and the subscript C_{x1} and C_{x2} are the capacitors 1 and 2 of phase-x; L_{x1} and L_{x2} for the inductors 1 and 2 of phase-x; $x = a$, b, c. During the NST state, its equivalent circuit is shown in Figure 10.20(b), and we have

$$\begin{bmatrix} v_{ab} \\ v_{bc} \\ v_{ca} \end{bmatrix} = \begin{bmatrix} v_{La1} \\ v_{Lb1} \\ v_{Lc1} \end{bmatrix} + \begin{bmatrix} v_{Ca1} \\ v_{Cb1} \\ v_{Cc1} \end{bmatrix} + \begin{bmatrix} v_{a'b'} \\ v_{b'c'} \\ v_{c'a'} \end{bmatrix} - \begin{bmatrix} v_{Cb1} \\ v_{Cc1} \\ v_{Ca1} \end{bmatrix} - \begin{bmatrix} v_{Lb1} \\ v_{Lc1} \\ v_{La1} \end{bmatrix} \qquad (10.28)$$

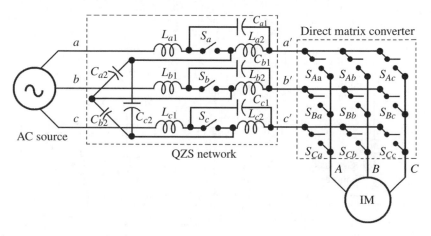

Figure 10.19 Quasi-Z-source direct matrix converter topology (*Source*: Ge 2012 [31]. Reproduced with permission of IEEE).

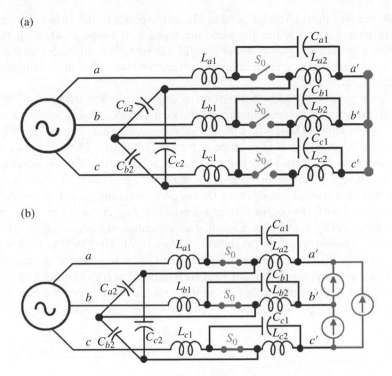

Figure 10.20 Equivalent circuit of the QZSDMC: (a) shoot-through state, (b) non-shoot-through state (*Source*: Ge 2012 [31]. Reproduced with permission of IEEE).

In steady state, the average voltage of the inductors over one switching cycle should be zero, and owing to the symmetric voltages of the three-phase capacitors, we get [31]

$$\begin{bmatrix} v_{a'b'} \\ v_{b'c'} \\ v_{c'a'} \end{bmatrix} = \frac{1}{1-2D} \begin{bmatrix} v_{ab} \\ v_{bc} \\ v_{ca} \end{bmatrix} \tag{10.29}$$

Define B as the boost factor, expressed as

$$B_{QZSDMC} = \frac{v_o}{v_i} = \frac{1}{1-2D} \tag{10.30}$$

where v_i is the amplitude of input voltage source and v_o is the output voltage amplitude of the QZS-network. The voltage gain G of the QZS-network in the one switching cycle is given by:

$$G = BM \tag{10.31}$$

10.4.3 Shoot-Through Boost Control Method

The principle of applying the ST state for the QZSDMC is to replace some of the zero state by the ST state, in order not to affect the output voltage. By using carrier-based pulse width modulation, the zero output voltage state in MC corresponds to the switching state with all three output phases connected to the same input phase. It happens when all three phase output voltages are either higher or lower than the carrier signal. So the ST reference should be either higher than the maximum reference voltage or lower than the minimum reference voltage [33].

All the boost control methods that have been explored for the traditional ZSC, such as simple boost, maximum boost, maximum constant boost, and modified space vector modulation [38], can be applied to the QZSDMC with a modification of the carrier envelope. Figure 10.21 shows a simple boost PWM control strategy for the QZSDMC. The carrier waveform has the same envelope as the three-phase source voltages, v_a, v_b, and v_c. The top envelope consists of the maximum voltage among the three input phase voltages, and the bottom envelope consists of the minimum voltage among them. During each switching period, the modified carrier signal is compared with the output voltage references v_A, v_B, and v_C to produce their PWM switching sequences (S_A, S_B, S_C). The ST pulses are generated by comparing the ST references with the modified carrier waveform, as shown in Figure 10.21. The PWM switching sequences S_A, S_B, S_C should be distributed to nine ac switches in order to generate the expected PWM pulses. For this purpose, six additional logical signals are used, as shown in Figure 10.22, where S_{x1}, S_{y1}, and S_{z1} denote the indicators for their respective phase-a, phase-b, and phase-c of the top voltage envelope. S_{x2}, S_{y2}, and S_{z2} denote the indicators for their respective phase-a,

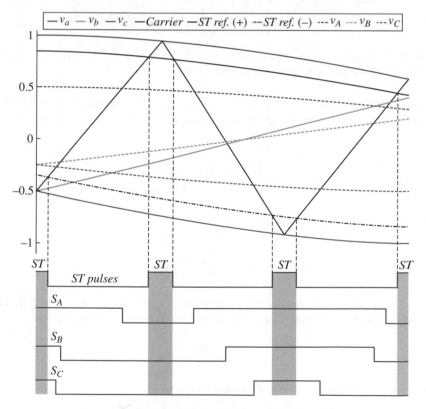

Figure 10.21   QZSDMC switching states generation.

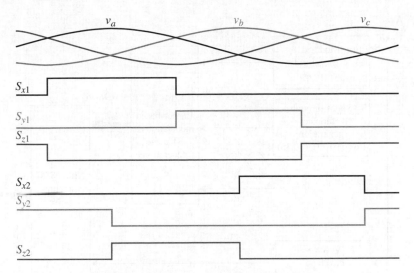

Figure 10.22 Voltage envelope indicators.

phase-b, and phase-c of the bottom voltage envelope. These six voltage envelope indicators are combined with the three PWM switching sequences to generate nine switching signals according to the following logic:

$$
\begin{aligned}
S_{Aa} &= S_{x1}S_A + S_{x2}\overline{S}_A \\
S_{Ab} &= S_{y1}S_A + S_{y2}\overline{S}_A \\
S_{Ac} &= S_{z1}S_A + S_{z2}\overline{S}_A \\
S_{Ba} &= S_{x1}S_B + S_{x2}\overline{S}_B \\
S_{Bb} &= S_{y1}S_B + S_{y2}\overline{S}_B \\
S_{Bc} &= S_{z1}S_B + S_{z2}\overline{S}_B \\
S_{Ca} &= S_{x1}S_C + S_{x2}\overline{S}_C \\
S_{Cb} &= S_{y1}S_C + S_{y2}\overline{S}_C \\
S_{Cc} &= S_{z1}S_C + S_{z2}\overline{S}_C
\end{aligned}
\tag{10.32}
$$

The above logical functions can be used to drive the QZSDMC after inserting the ST states.

A simple boost control is achieved through two ST references, in which both references are related to both envelopes by [31]:

$$
v_{st,r} = \frac{\left(y_{max} - y_{min}\right)n + y_{max} + y_{min}}{2}
\tag{10.33}
$$

where n will determine the ST duty ratio, and its value has a limitation that the resultant minimum value of the top ST reference should be less than 0.5 p.u. and larger than M. Therefore, $1 \geq n \geq (1 + 4M)/3$ for the top ST reference, and its negative value is $(-n)$ for the bottom ST

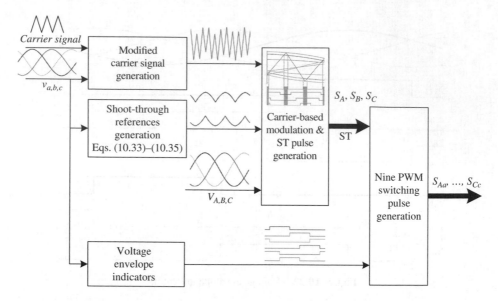

Figure 10.23 Block diagram of QZSDMC PWM generation with ST insertion.

reference. The modulation index should be less than 0.5, given that the output references v_A, v_B, and v_C can be any frequency with any phase angle and with no harmonic injection. Also, y_{max} and y_{min} are the top and bottom envelopes of the source voltages, respectively. For the simple boost control, the shoot-through interval from the top reference can be calculated as

$$T_0 = \frac{1-n}{2} T_c \qquad (10.34)$$

where T_0 and T_c are the ST duration per switching cycle and switching time, respectively, and its ST duty ratio in half carrier cycle is

$$D_h = \frac{1-n}{2} \qquad (10.35)$$

Figure 10.23 shows the complete process for generating the switching signals for the QZSDMC. First, the triangle carrier signal is modulated by the input reference signals v_a, v_b, and v_c to generate the modified carrier signal, which is bounded by the maximum and minimum envelopes of the input reference signals. Second, the ST references are generated from the input reference signals and the desired boost ratio using (10.33)–(10.35). Then, the switching sequences S_A, S_B, S_C and ST pulses are generated by comparing the output voltage references v_A, v_B, and v_C and the ST references with the modified carrier signal. Furthermore, the voltage envelope indicators are generated from the input reference signals v_a, v_b, and v_c. For example, $S_{x1} = 1$ when phase-a voltage is the largest value among the three phase voltages and $S_{x2} = 1$ when phase-a has the minimum voltage among the three phase voltages. Finally, these six voltage envelope indicators are combined with the three PWM switching sequences, S_A, S_B, S_C, and the ST pulses to generate nine switching signals.

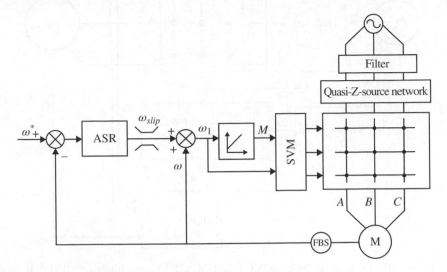

Figure 10.24 Block diagram for a QZSDMC-based ASD system (*Source*: Liu 2012 [32]. Reproduced with permission of IEEE).

10.4.4 Applications of the QZSDMC

In [32], the quasi-Z-source direct matrix converter (QZSDMC) with discontinuous input current was used for feeding IM as part of an adjustable speed drive system. The complete block diagram of the proposed system is shown in Figure 10.24. The speed encoder detects the rotor speed to compare with the reference speed. The speed controller deals with the speed error and adjusts the modulation index of the MC, the ST duty ratio is designed according to the corresponding voltage gain, and the output voltage can be obtained to meet the desired value.

In [5, 39], a four-quadrant adjustable speed drive system based on the QZSDMC feeding a vector controlled induction motor was proposed, which overcomes the reduced voltage transfer ratio limitations of the traditional DMC-based adjustable speed drive (ASD) system, so that the QZSDMC-IM based ASD will increase the application of the DMC in different industry fields. The proposed ASD system can operate at full load with small QZS-network elements, which are suitable for an ASD system. The QZSDMC can achieve buck and boost operation with a reduced number of switches, therefore achieving low cost, high efficiency, and reliability, compared to the traditional DMCs. In addition, there is no requirement of dead time with QZS-network, hence commutation of the QZSDMC is easier than the traditional DMC. The proposed closed-loop speed control system can obtain a voltage gain larger than one and can operate in motoring and regenerating operation modes with perfect reference tracking as verified by MATLAB simulation and dSPACE real-time implementation results. Figure 10.25 shows the control block diagram of an induction motor drive system fed with QZSDMC.

The QZSDMC can be used for interfacing renewable energy sources as it is a direct ac/ac converter with a voltage gain greater than unity [40]. The simulation results presented are attractive enough to verify the proposed system and to justify additional research work to develop a more efficient QZSDMC grid connected system. Figure 10.26 shows the proposed converter with its control algorithm.

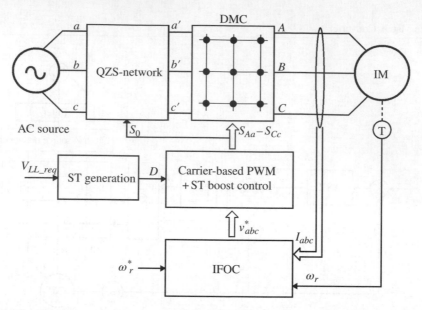

Figure 10.25 Block diagram of the QZSDMC-based IM ASD (*Source*: Ellabban 2015 [5]. Reproduced with permission of IEEE).

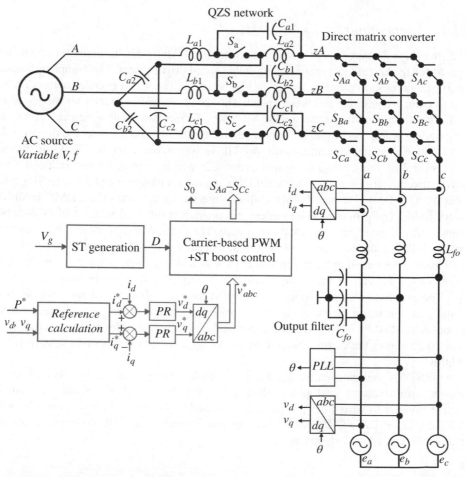

Figure 10.26 Grid-connected QZSDMC system (*Source*: Ellabban 2013 [40]. Reproduced with permission of IEEE).

10.5 Summary

This chapter provided a review of the literature on three different Z-source matrix converter topologies: Z-source indirect matrix converter (all-silicon solution), Z-source indirect matrix converter (not all-silicon solution), and Z-source direct matrix converter. For each topology, the different topology configuration, circuit analysis, parameters design, modulation, and application have been reviewed. The QZSMC topologies overcome the voltage gain limitation of the traditional MC and achieve buck and boost conditions with a reduced number of switches, therefore achieving low cost, high efficiency, and reliability compared to the back-to-back converter. Furthermore, it will lead to more MC industrial applications.

References

[1] J. W. Kolar, T. Friedli, J. Rodriguez, P. W. Wheeler, "Review of three-phase PWM AC–AC converter topologies," *IEEE Trans. Ind. Electron.*, vol.58, no.11, pp.4988–5006, Nov. 2011.

[2] T. Friedli, J. W. Kolar, J. Rodriguez, P. W. Wheeler, "Comparative evaluation of three-phase AC–AC matrix converter and voltage DC-link back-to-back converter systems," *IEEE Trans. Ind. Electron.*, vol.59, no.12, pp.4487–4510, December 2012.

[3] P. W. Wheeler, J. Rodriguez, J. C. Clare, L. Empringham, "Matrix converter: A technology review," *IEEE Trans. Ind. Electron.*, vol.49, no.2, pp.276–288, Apr. 2002.

[4] Y. D. Yoon, S. K. Sul, "Carrier-based modulation technique for matrix converter," *IEEE Trans. Power Electron.*, vol.21, no.6, pp.1691–1703, Nov. 2006.

[5] O. Ellabban, H. Abu-Rub, B. Ge, "A quasi-Z-Source direct matrix converter feeding A vector controlled induction motor drive," *IEEE Journal of Emerging and Selected Topics in Power Electronics*, vol.3, no.2, pp.339–348, June 2015.

[6] F. Z. Peng, "Z-source inverter," *IEEE Trans. Ind. Appl.*, vol.39, no.2, pp.504–510, Mar/Apr. 2003.

[7] J. Anderson, F. Z. Peng, "Four quasi-Z-source inverters," in *Proc. IEEE Power Electronics Specialists Conference (PESC)*, pp.2743–2749, 15–19 June 2008.

[8] O. Ellabban, H. Abu-Rub, S. Bayhan, "Z-source matrix converter: an overview," to appear in *IEEE Trans. Power Electron.*, DOI: 10.1109/TPEL.2015.2471799.

[9] S. Liu, B. Ge, X. Jiang, H. Abu-Rub, F. Z. Peng, "A novel quasi-Z-source indirect matrix converter," *International Journal of Circuit Theory and Applications*, 2013.

[10] S. Liu, B. Ge, X. Jiang, H. Abu-Rub, F. Z. Peng, "Modeling, analysis, and motor drive application of quasi-Z-source indirect matrix converter," *The International Journal for Computation and Mathematics in Electrical and Electronic Engineering (COMPEL)*, vol.33, no.1/2, pp.298–319, 2014.

[11] X. You, B. Ge, S. Liu, N. Nie, X. Jiang, H. Abu-Rub, "Common mode voltage reduction of quasi-Z source indirect matrix converter," *International Journal of Circuit Theory and Applications*, March 2015.

[12] S. Liu, B. Ge, X. Jiang, H. Abu-Rub, F. Z. Peng, "Comparative Evaluation of Three Z-Source/Quasi-Z-Source Indirect Matrix Converters," *IEEE Trans. Ind. Electron.*, vol.62, no.2, pp.692–701, Feb. 2015.

[13] S. Liu, B. Ge, X. Jiang, H. Abu-Rub, F. Z. Peng, "Simplified quasi-Z source indirect matrix converter," *International Journal of Circuit Theory and Applications*, Oct. 2014.

[14] S. Zhang, K. J. Tseng, T. D. Nguyen, "Novel three-phase ac-ac Z-Source converters using matrix converter theory," in *Proc. IEEE Energy Conversion Congress and Exposition (ECCE)*, pp.3063–3070, 20–24 Sept. 2009.

[15] W. Song, Y. Zhong, "A study of Z-source matrix converter with high voltage transfer ratio," in *Proc. IEEE Vehicle Power and Propulsion Conference (VPPC)*, pp.1–6, 3–5 Sept. 2008.

[16] K. Park, K. Lee, "A novel sparse matrix converter with a Z-source network," in *Proc. IEEE 35th Annual Conference of Industrial Electronics (IECON)*, pp.4487–4492, 3–5 Nov. 2009.

[17] K. Park, K. Lee, "A Z-source sparse matrix converter under a voltage sag condition," in *Proc. IEEE Energy Conversion Congress and Exposition (ECCE)*, pp.2893–2898, 12–16 Sept. 2010.

[18] E. Karaman, M. Farasat, A. M. Trzynadlowski, "A 3Φ-3Φ quasi Z-source matrix converter for residential wind energy systems," in *Proc. IEEE Energy Conversion Congress and Exposition (ECCE)*, pp.240–246, 15–20 Sept. 2012.

[19] E. Karaman, F. Niu, A. M. Trzynadlowski, "Three-phase switched-inductor Z-source matrix converter," in *Proc. Twenty-Seventh Annual IEEE Applied Power Electronics Conference and Exposition (APEC)*, pp.1449–1454, 5–9 Feb. 2012.

[20] E. Karaman, M. Farasat, A. M. Trzynadlowski, "Indirect matrix converters as generator–grid interfaces for wind energy systems," *IEEE Journal of Emerging and Selected Topics in Power Electronics*, vol.2, no.4, pp.776–783, Dec. 2014.

[21] E. Karaman, M. Farasat, A. M. Trzynadlowski, "A comparative study of series and cascaded Z-source matrix converters," *IEEE Trans. Ind. Electron.*, vol.61, no.10, pp.5164–5173, Oct. 2014.

[22] F. B. Effah, A. J. Watson, P. W. Wheeler, J. C. Clare, L. De-Lillo, "Space-vector-modulated three-level Z-source hybrid direct ac-ac power converter," in *Proc. 15th European Conference on Power Electronics and Applications (EPE)*, pp.1–10, 2–6 Sept. 2013.

[23] Xiong Liu, Poh Chiang Loh, Fang Zheng Peng, Peng Wang, Feng Gao, "Modulation of three-level Z-source indirect matrix converter," in *Proc. IEEE Energy Conversion Congress and Exposition (ECCE)*, pp.3195–3201, 12–16 Sept. 2010.

[24] E. Karaman, M. Farasat, F. Niu, A. M. Trzynadlowski, "Three-phase to single-phase super-sparse matrix converters," in *Proc. Twenty-Seventh Annual IEEE Applied Power Electronics Conference and Exposition (APEC)*, pp.1061–1066, 5–9 Feb. 2012.

[25] S. Sousa, S. Pinto, F. Silva, J. Maia, "Extended voltage range AC drive using a Z-source indirect matrix converter," in *Proc. International Conference on Electrical Machines (ICEM)*, pp.953–958, 2–5 Sept. 2012.

[26] S. Liu, B. Ge, H. Abu-Rub, X. Jiang, F. Z. Peng, "A novel indirect quasi-Z-source matrix converter applied to induction motor drives," in *Proc. IEEE Energy Conversion Congress and Exposition (ECCE)*, pp.2440–2444, 15–19 Sept. 2013.

[27] E. Karaman, M. Farasat, A. M. Trzynadlowski, "Permanent-magnet synchronous-generator wind-energy systems with boost matrix converters," in *Proc. IEEE Energy Conversion Congress and Exposition (ECCE)*, pp.2977–2981, 15–19 Sept. 2013.

[28] K. Park, S. Jou, and K. Lee, "Z-source matrix converter with unity voltage transfer ratio," in *Proc. Annual Conference of the IEEE Industrial Electronics Society (IECON)*, pp.4523–4528, Nov. 2009.

[29] W. Song, Y. Zhong, "A study of Z-source matrix converter with high voltage transfer ratio," in *Proc. IEEE Vehicle Power and Propulsion Conference (VPPC)*, pp.1–6, 3–5 Sept. 2008.

[30] X. Fang, C. Li, Z. Chen, J. Liu, X. Zhao, "Three-phase voltage-fed Z-source matrix converter," in *Proc. International Conference on Electrical Machines and Systems (ICEMS)*, pp.1–4, 20–23 Aug. 2011.

[31] B. Ge, Q. Lei, W. Qian, F. Z. Peng, "A family of Z-source matrix converters," *IEEE Trans. Ind. Electron.*, vol.59, no.1, p.35–46, Jan. 2012.

[32] S. Liu, B. Ge, H. Abu-Rub, F. Z. Peng, Y. Liu, "Quasi-Z-source matrix converter based induction motor drives," in *Proc. the 38th Annual Conference of the IEEE Industrial Electronics Society (IECON)*, pp.5303–5307, 25–28 October 2012.

[33] Q. Lei, F. Z. Peng, B. Ge, "Pulse-width-amplitude-modulated voltage-fed quasi-Z-source direct matrix converter with maximum constant boost," in *Proc. Twenty-Seventh Annual IEEE Applied Power Electronics Conference and Exposition (APEC)*, pp.641–646, 5–9 Feb. 2012.

[34] M. Nguyen, Y. Jung, Y. Lim, "Single-phase AC/AC buck–boost converter with single-phase matrix topology," in *Proc. 13th European Conference on Power Electronics and Applications (EPE)*, pp.1–7, 8–10 Sept. 2009.

[35] M. Nguyen, Y. Jung, Y. Lim, "Single-phase Z-source buck–boost matrix converter," in *Proc. Twenty-Fourth Annual IEEE Applied Power Electronics Conference and Exposition (APEC)*, pp.846–850, 15–19 Feb. 2009.

[36] X. Fang, M. Cao, C. Li, "Single-phase Z-source matrix converter," in *Proc. 2010 International Conference on Electrical Machines and Systems (ICEMS)*, pp.107–111, 10–13 Oct. 2010.

[37] G. Milan, M. Mohamadian, E. Seifi Najmi, S. M. Dehghan, "A single to three-phase Z-source matrix converter in *Proc. 3rd Power Electronics and Drive Systems Technology (PEDSTC)*, pp.1–6, 15–16 Feb. 2012.

[38] O. Ellabban, J. V. Mierlo, P. Lataire, "Experimental study of the shoot-through boost control methods for the Z-source inverter," *EPE Journal*, vol.21, no.2, pp.18–29, Jun. 2011.

[39] O. Ellabban, H. Abu-Rub, B. Ge, "Field oriented control of an induction motor fed by a quasi-Z-source direct matrix converter," in *Proc. 39th Annual Conference of the IEEE Industrial Electronics Society (IECON)*, pp.4850–4855, 10–13 November 2013.

[40] O. Ellabban, H. Abu-Rub, "Grid connected quasi-Z-source direct matrix converter," in *Proc. 39th Annual Conference of the IEEE Industrial Electronics Society (IECON)*, pp.798–803, 10–13 November 2013.

11

Energy Stored Z-Source/ Quasi-Z-Source Inverters

11.1 Energy Stored Z-Source/Quasi-Z Source Inverters

The ZSI/qZSI with battery used in PV or with fuel cells power systems combines advantages of ZSI/qZSI and energy storage technology [1–8], and presents promising features: (1) it handles the wide PV panel or fuel cell voltage variations in the single-stage power conversion, (2) it injects continuous, stable, and smooth power to the user/grid, (3) it cancels the dead time to improve the output waveforms, and (4) it achieves simple system topology, without an additional dc-dc converter. There are two capacitors in the ZS/qZS network. Reference [1] presented the topology of ZSI paralleling the battery to one of the ZS capacitors, Figure 11.1(a), but no control method was provided. References [2–6] demonstrated maximum power point tracking (MPPT), battery charge, and discharge operation when a battery is paralleled to capacitor C_2 of the quasi-Z-source network, Figure 11.1(b). References [7–9] proposed another type of qZSI with battery system, by paralleling the battery to the capacitor C_1 (rather than C_2).

Due to the non-linear characteristics of the circuit components, such as battery and power switch devices, the energy stored qZSI is a strongly non-linear system and presents an entirely different model and control feature compared to the small-signal model of qZSI presented in Chapter 2. It is extremely important to establish a small-signal dynamic model for the energy stored qZSI, thus to design an effective controller. In this chapter, the qZSI with battery paralleling to C_2 for a PV system is addressed as an example. The dynamic model and the control method, including battery energy management, PV power MPPT, and grid-tie synchronization, are presented. Similar methods can be performed with the ZSI or when battery connecting to C_1. Example simulations are demonstrated in different cases of battery state of charge.

Impedance Source Power Electronic Converters, First Edition. Yushan Liu, Haitham Abu-Rub, Baoming Ge, Frede Blaabjerg, Omar Ellabban, and Poh Chiang Loh.
© 2016 John Wiley & Sons, Ltd. Published 2016 by John Wiley & Sons, Ltd.

(a)   (b)

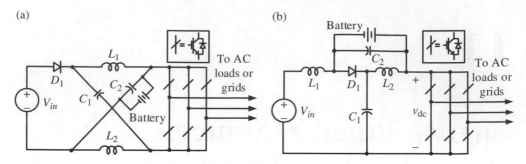

Figure 11.1 ZSI/qZSI with battery: (a) ZSI with battery [1], (b) qZSI with battery (*Source*: Cintron 2011 [2]. Reproduced with permission of IEEE).

11.1.1 Modeling of qZSI with Battery

Figure 11.2(a) shows the topology of the discussed qZSI with battery for grid-connected PV power systems [2–6]. An energy storage battery is paralleled to the capacitor C_2 of a qZS network. Similar to the existing qZSI, the qZSI with battery also has two operating modes in the continuous conduction mode (CCM) – shoot-through state and non-shoot-through state – as the equivalent circuit of Figure 11.2(b) and (c) shows.

In the shoot-through state of Figure 11.2(b), the diode is cut off by the negative voltage, and the PV panel and qZS capacitors charge the inductors. In the non-shoot-through state of Figure 11.2(c), the PV panel and inductors charge the loads and qZS capacitors, and the diode is in forward conduction. Here, r_L is the parasitic resistance of qZS inductor. For the battery, a simple model with the open-circuit voltage V_{OCV} and an internal resistance R_b simulates its basic function [6]. Assume that the battery current direction is positive when discharging.

From Figure 11.2(b), the state-space equations at shoot-through state are written as [6]

$$\begin{bmatrix} L_1 & 0 & 0 & 0 \\ 0 & L_2 & 0 & 0 \\ 0 & 0 & C_1 & 0 \\ 0 & 0 & 0 & C_2 \end{bmatrix} \cdot \begin{bmatrix} \dot{i}_{L1}(t) \\ \dot{i}_{L2}(t) \\ \dot{v}_{C1}(t) \\ \dot{v}_{C2}(t) \end{bmatrix} = \begin{bmatrix} -r_L & 0 & 0 & 1 \\ 0 & -r_L & 1 & 0 \\ 0 & -1 & 0 & 0 \\ -1 & 0 & 0 & 0 \end{bmatrix} \cdot \begin{bmatrix} i_{L1}(t) \\ i_{L2}(t) \\ v_{C1}(t) \\ v_{C2}(t) \end{bmatrix} + \begin{bmatrix} 1 & 0 & 0 \\ 0 & 0 & 0 \\ 0 & 0 & 0 \\ 0 & 0 & 1 \end{bmatrix} \cdot \begin{bmatrix} v_p(t) \\ i_d(t) \\ i_b(t) \end{bmatrix} \quad (11.1)$$

From Figure 11.2(c), the state-space equations at non-shoot-through state are written as

$$\begin{bmatrix} L_1 & 0 & 0 & 0 \\ 0 & L_2 & 0 & 0 \\ 0 & 0 & C_1 & 0 \\ 0 & 0 & 0 & C_2 \end{bmatrix} \cdot \begin{bmatrix} \dot{i}_{L1}(t) \\ \dot{i}_{L2}(t) \\ \dot{v}_{C1}(t) \\ \dot{v}_{C2}(t) \end{bmatrix} = \begin{bmatrix} -r_L & 0 & -1 & 0 \\ 0 & -r_L & 0 & -1 \\ 1 & 0 & 0 & 0 \\ 0 & 1 & 0 & 0 \end{bmatrix} \cdot \begin{bmatrix} i_{L1}(t) \\ i_{L2}(t) \\ v_{C1}(t) \\ v_{C2}(t) \end{bmatrix} + \begin{bmatrix} 1 & 0 & 0 \\ 0 & 0 & 0 \\ 0 & -1 & 0 \\ 0 & -1 & 1 \end{bmatrix} \cdot \begin{bmatrix} v_p(t) \\ i_d(t) \\ i_b(t) \end{bmatrix} \quad (11.2)$$

As the capacitor C_2 voltage is clamped to the battery terminal voltage V_b, we have

$$v_{C2} = V_b = V_{OCV} - R_b i_b(t), \quad \dot{v}_{C2}(t) = -R_b \dot{i}_b(t) \quad (11.3)$$

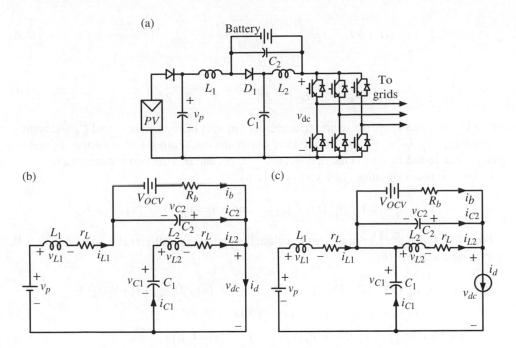

Figure 11.2 Quasi-Z-source inverter with battery: (a) Topology, (b) equivalent circuits in shoot-through state, and (c) non-shoot-through state [6] (Reproduced by permission of the Institution of Engineering & Technology. Full acknowledgment to the Author, Title, and date of the original work).

Then substitute (11.3) into (11.1) and (11.2). Assuming $L_1 = L_2 = L$, $C_1 = C_2 = C$, the state-space average model of qZSI with battery can be deduced as

$$F\dot{x} = \begin{bmatrix} -r_L & 0 & D-1 & -DR_b \\ 0 & -r_L & D & (1-D)R_b \\ 1-D & -D & 0 & 0 \\ D & D-1 & 0 & -1 \end{bmatrix} \cdot x + \begin{bmatrix} 1 & 0 & D \\ 0 & 0 & 1-D \\ 0 & D-1 & 0 \\ 0 & 1-D & 0 \end{bmatrix} \cdot u \qquad (11.4)$$

where $x = [i_{L1}(t) \quad i_{L2}(t) \quad v_{C1}(t) \quad i_b(t)]^T$ is the state vector composed of state variables, $u = [v_p(t) \quad i_d(t) \quad V_{OCV}(t)]^T$ is the input vector composed of input variables, $F = \text{diag}$ $(L \quad L \quad C \quad R_bC)^T$ is the coefficient matrix of the output vector, and D represents the shoot-through duty ratio. Here, the variables and their reference directions are defined in Figure 11.2(b) and (c).

In steady state, the left side of (11.4) is zero, so we have

$$\begin{cases} V_{C1} = \dfrac{1-D}{1-2D} v_p + \dfrac{-r_L(1-D)I_d + r_L DI_b}{1-2D} \\[4mm] V_{C2} = \dfrac{D}{1-2D} v_p - \dfrac{-r_L(1-D)I_d + r_L DI_b}{1-2D} = V_b \end{cases} \qquad (11.5)$$

$$I_{L1} - I_{L2} = I_b, \quad I_d = \frac{(1-2D) \cdot I_{L1} + D \cdot I_b}{1-D}, \quad I_b = \frac{V_b - V_{OCV}}{R_b} \tag{11.6}$$

$$\hat{v}_{dc} = \frac{1}{1-2D} v_p = \frac{V_b}{D}, \quad B = \frac{\hat{v}_{dc}}{v_p} = \frac{1}{1-2D} \tag{11.7}$$

where V_{C1}, V_{C2}, and V_b are the steady-state average voltages of capacitors C_1 and C_2, and battery, respectively, I_{L1}, I_{L2}, I_b, and I_d are the steady-state average currents of inductors L_1 and L_2, battery, and dc-link bus, $\hat{v}_{dc}$ is the dc-link peak voltage, and B is the boost factor of qZSI.

Define the small variations of vectors x and u as

$$\hat{x} = \begin{bmatrix} \hat{i}_{L1}(t) & \hat{i}_{L2}(t) & \hat{v}_{C1}(t) & \hat{i}_b(t) \end{bmatrix}^T \quad \text{and} \quad \hat{u} = \begin{bmatrix} \hat{v}_p(t) & \hat{i}_d(t) & \hat{V}_{OCV}(t) \end{bmatrix}^T$$

From (11.4) and the Laplace transformation, the dynamic small-signal model of qZSI with battery is deduced as [6]

$$\begin{cases} (Ls + r_L)\hat{i}_{L1}(s) = (D-1)\hat{v}_{C1}(s) - DR_b\hat{i}_b(s) + \hat{v}_p(s) + D\hat{V}_{OCV}(s) + V_{11}d(s) \\ Cs\hat{v}_{C1}(s) = (1-2D)\hat{i}_{L1}(s) + D\hat{i}_b(s) + (1-D)\hat{i}_d(s) + I_{11}d(s) \\ (R_bCs + D)\hat{i}_b(s) = (2D-1)\hat{i}_{L1}(s) + (1-D)\hat{i}_d(s) - I_{11}d(s) \end{cases} \tag{11.8}$$

where $I_{11} = I_d - 2I_{L1} + I_b$, $V_{11} = V_{C1} - R_bI_b + V_{OCV}$; $d(s)$ is the small-signal perturbance of shoot-through duty ratio D.

11.1.2   Controller Design

There are three control objectives in the energy-stored qZSI-based grid-tie PV power generation system: (1) PV panel's MPPT, (2) constant grid-injected power immune to PV power fluctuations, and (3) battery power management. Figure 11.3 shows the proposed control strategy to achieve the above purposes [5].

In Figure 11.3, battery current closed-loop control gets the desired shoot-through duty ratio to manage the battery power in real time. To get fast response, from (11.5), a feed forward of shoot-through duty ratio is performed by

$$D_0 = V_b / (v_p^* + 2V_b)$$

where v_p^* is the PV panel voltage reference obtained from the MPPT.

The desired battery power $P_b^* = P_{out}^* - P_{pv}$, where P_{out}^* is the desired grid-injected power according to real PV power and the battery state of charge (SOC). PV panel's MPPT is achieved by regulating the grid-injected power. The inverter's grid-tie control is the same as conventional vector control. The desired shoot-through duty ratio and the inverter output voltage references are combined in the modified Z-source space vector modulation (ZSVM) to fulfill their control goals.

In summary, as shown in Figure 11.3, battery current closed-loop control (or so-called Z-source converter control) and AC grid current control can be controlled independently if the sum of the shoot-through duty ratio and the inverter modulation index is less than one. Moreover, the

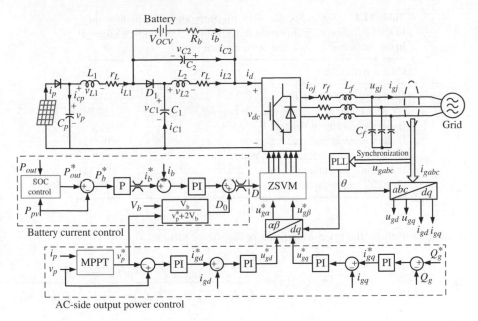

Figure 11.3 Control of energy stored qZSI-based PV power generation system (*Source*: Liu 2013 [5]. Reproduced with permission of IEEE).

Z-source converter control (i.e. shoot-through duty ratio) will control battery power, and AC grid current control will achieve the PV panel's MPPT. Because the PV power is determined for a certain temperature and irradiation, inverter output power is controlled by the battery power through power balance equation $P_{out} = P_b + P_{pv}$. The battery power depends on PV power and battery SOC.

11.1.2.1 Battery Current Closed-Loop Control

From (11.8), the open-loop transfer function from the battery current to the shoot-through duty ratio is given as [6]

$$G_{ibd}(s) = \frac{i_b(s)}{d(s)} = \frac{-sLI_{11} + (2D-1)V_{11} - r_L I_{11}}{R_b L C s^2 + (LD + CR_b r_L)s + \left[D r_L + R_b (1-2D)^2 \right]} \tag{11.9}$$

With the system specifications in Table 11.1, it is found that $G_{ibd}(s)$ demonstrates an RHP (right half plane) zero when the battery current is less than zero. Figure 11.4 demonstrates a compensation process when the battery operating terminal voltage is 70 V (battery open circuit voltage is 72 V) and $I_b < 0$. Figure 11.4(a) shows the proposed battery current compensation through a controller $G_{cpi}(s)$ to weaken the influence from the RHP zero. Figure 11.4(b) shows the pole-zero map of $G_{ibd}(s)$, where the zero is in RHP. The $G_{cpi}(s)$ is performed by a PI regulator

$$G_{cpi}(s) = \left(k_{p\_id} + \frac{k_{i\_id}}{s} \right)$$

in which k_{p_id} is the proportional gain and k_{i_id} is the integral gain.

Table 11.1   System Specifications [6] (Reproduced by permission of the Institution of Engineering & Technology. Full acknowledgment to the Author, Title, and date of the original work).

Circuit parameters	Values
Grid-tie power rating, P_{out}	3 kW
PV panel voltage at maximum power point, V_m	280 V
Battery open circuit voltage, V_{OCV}	72 V
Battery internal resistance, R_b	0.17 Ω
Quasi-Z-source inductance, L_1 and L_2	500 μH
Inductance parasitic resistance, r_L	0.15 Ω
Quasi-Z-source capacitance, C_1 and C_2	470 μF

After compensation, the open-loop transfer function becomes

$$G_{oibd}(s) = \frac{i_b(s)}{i_b^*(s) - i_b(s)} = G_{cpi}(s) \cdot G_{ibd}(s) = \frac{\left(k_{p\_id}s + k_{i\_id}\right)\left[-sLI_{11} + (2D-1)V_{11} - r_L I_{11}\right]}{R_b LC s^3 + (LD + CR_b r_L)s^2 + \left[Dr_L + R_b(1-2D)^2\right]s}$$

(11.10)

and the closed-loop transfer function is

$$G_{cibd}(s) = \frac{d(s)}{i_b^*(s)} = \frac{G_{cpi}(s)}{1 + G_{cpi}(s) \cdot G_{ibd}(s)}$$

(11.11)

The crossover frequency of $G_{oibd}(s)$ to low frequency will improve the stability, but the fast tracking performance may decline; conversely, if the crossover frequency shifts to the corner frequency, the fast tracking response is enhanced. To trade off, the crossover frequency of $G_{oibd}(s)$ is set to one-tenth the corner frequency of $G_{ibd}(s)$. The loop gain of $G_{oibd}(s)$ is equal to 1 at the crossover frequency. Moreover, the zero of the PI regulator is set to the corner frequency of $G_{ibd}(s)$, which is $\omega_n = \sqrt{\dfrac{Dr_L + R_b(1-2D)^2}{R_b LC}} \approx 1.59 \times 10^3$ (rad/s) from (11.9). And also we have

$$\begin{cases} \left| G_{cpi}\left(j\dfrac{\omega_n}{10}\right) \cdot G_{ibd}\left(j\dfrac{\omega_n}{10}\right) \right| = 1 \\ \left| k_{p\_id} \cdot j\omega_n + k_{i\_id} \right| = 0 \end{cases}$$

(11.12)

Therefore, the k_{p_id} and k_{i_id} can be obtained, which are -3.597×10^{-5} and -0.0571, respectively.

Figure 11.4(c) shows the pole-zero map of $G_{oibd}(s)$. It can be seen that all poles and zeros of $G_{oibd}(s)$ are in the LHP (left half plane) after compensation. In addition, the real-axis zero locates at the corner frequency of $G_{ibd}(s)$, ω_n, that is 1.59×10^3 rad/s; the real-axis pole of $G_{oibd}(s)$ that is also the pole of controller $G_{cpi}(s)$ locates at one-tenth of ω_n.

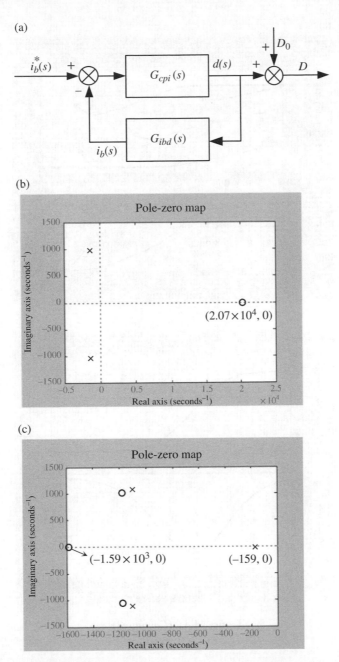

Figure 11.4 Design process of battery current closed-loop controller: (a) block diagram of the proposed closed-loop compensation, (b) pole-zero map of $G_{ibd}(s)$ before compensation, (c) pole-zero map of $G_{oibd}(s)$ after compensation, (d) Bode plots of the compensation process: trace 1 stands for $G_{ibd}(s)$, trace 2 stands for $G_{cpi}(s)$, trace 3 stands for $G_{oibd}(s)$; (e) Bode plot of closed-loop transfer function $G_{cibd}(s)$ [6] (Reproduced by permission of the Institution of Engineering & Technology. Full acknowledgment to the Author, Title, and date of the original work).

(d)

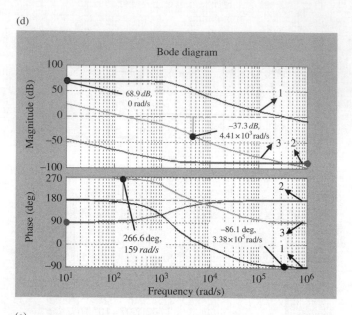

(e)

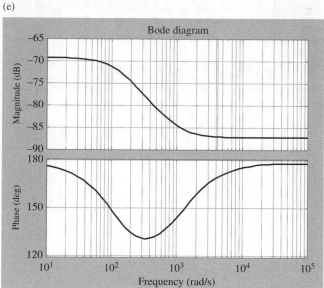

Figure 11.4   (*Continued*)

Figure 11.4(d) shows Bode plots of the compensation process. The amplitude–frequency curve of $G_{oibd}(s)$ decreases at the rate of −20 dB at low frequency, and crosses 0 dB at one-tenth of ω_n; the gain margin of $G_{oibd}(s)$ increases from −68.9 dB to 37.3 dB, and the phase margin is enhanced to 86.6° at 159 rad/s. Thus, the stability and fast response of the battery current closed loop are fulfilled, as the Bode plot of closed-loop transfer function $G_{cibd}(s)$ shows in Figure 11.4(e).

11.1.2.2 Battery Energy Management

The battery current reference is obtained by the battery power reference $P_b{}^*$ through a proportional (P) regulator with the coefficient of $1/V_{OCV}$. The $P_b{}^*$ is based on the system power balance $P_b{}^* = P_{out}{}^* - P_{pv}$. Three power sources are

$$P_{pv} = v_p \cdot i_{L1}, \ P_b = V_b \cdot i_b, \ P_{out} = V_{dc} \cdot I_d = (1-D) \cdot \hat{v}_{dc} \cdot I_d + D \cdot 0 = \frac{1-D}{1-2D} v_p \cdot I_d \quad (11.13)$$

If two of the three powers are controlled, the third one is determined.

First, the PV panel is controlled to capture the maximum power. The battery operating status is based on the inverter output power, and there are three cases: (i) $P_{pv} = P_{out}$, $P_b = 0$, $i_{L1} = i_{L2}$, all of the PV power is injected to the grid. (ii) $P_{pv} < P_{out}$, $P_b > 0$, $i_{L1} > i_{L2}$, all of the PV power is injected to the grid and the battery releases energy to the grid, (iii) $P_{pv} > P_{out}$, $P_b < 0$, $i_{L1} < i_{L2}$, the battery absorbs the redundant power of the PV panel.

However, the practical battery operating status also refers to its SOC. Figure 11.5 shows the battery management scheme, where SOC_{max} is the upper limit of the SOC, SOC_{min} is the lower limit, P_{out} is the actual output power of the inverter, and P^*_{out} is the inverter output power reference [5].

As shown in Figure 11.5, three situations exist during operation:

1. When the SOC is less than SOC_{min}, the battery should no longer discharge, and $P_{out}{}^*$ is equal to the smaller of P_{pv} and P_{out}. If P_{pv} is smaller, $P_{out}{}^* = P_{pv}$, $P_b{}^* = P_{pv} - P_{pv} = 0$. The designed battery current closed-loop will control the battery current to zero, and the battery operates in non-charging or non-discharging state. If P_{out} is smaller, $P_{out}{}^* = P_{out}$, $P_b{}^* = P_{out} - P_{pv} < 0$. The redundant PV power is charged into the battery, and over-discharging is avoided.

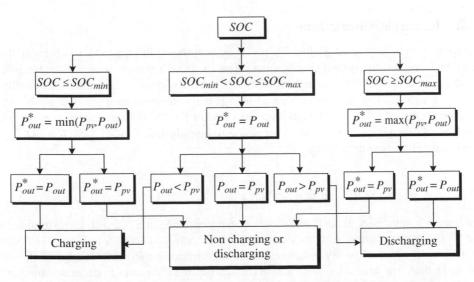

Figure 11.5 Battery SOC control [5] (Reproduced by permission of the Institution of Engineering & Technology. Full acknowledgment to the Author, Title, and date of the original work).

2. When SOC is greater than SOC_{max}, the battery should not be charged any more, and P_{out}^* is equal to the greater of P_{pv} and P_{out}. If P_{pv} is greater, $P_{out}^* = P_{pv}$, $P_b^* = P_{pv} - P_{pv} = 0$, the battery operates in non-charging or non-discharging state. If P_{out} is greater, $P_{out}^* = P_{out}$, $P_b^* = P_{out} - P_{pv} > 0$, the battery discharges. In this way, the battery avoids overcharging.

3. When SOC is between SOC_{min} and SOC_{max}, the battery can charge or discharge freely, and P_{out}^* is equal to P_{out}, then $P_b^* = P_{out} - P_{pv}$. If the PV power is redundant, the battery will charge. Otherwise, the battery will discharge.

11.1.2.3   AC-Side Output Power Control

The AC-side output power controller design is the same as the conventional grid-tie PV system [10], and the output power control is fulfilled through regulating the modulation index. For the system shown in Figure 11.3, the P-Q decoupling control is applied in the grid-tie operation. The active power is adjusted to force the PV panel power to track the maximum power point (MPP). The traditional perturb and observe (P&O) MPPT will provide the PV panel voltage reference v_p^*. The d-axis grid-tie current component reference will be

$$i_{gd}^* = \left(k_P + \frac{k_I}{s} \right) \left(v_p - v_p^* \right).$$

It can be seen that the active power injected into the grid increases when the actual PV voltage is larger than v_p^*. The grid reactive power Q_g can be independently controlled, and unity power factor is achieved at $Q_g = 0$. The d- and q-axis grid current component closed-loops output the related voltage references, respectively. The coordinate transformation from the d-q frame to the α-β frame is used to get the voltage references u_α^* and u_β^*.

11.2   Example Simulations

The qZSI with battery based grid-tie PV power system shown in Figure 11.3 is simulated [6]. Table 11.1 presents the system specifications. The PV panel's power–voltage characteristic in simulation is with 3200 W power and 280 V voltage at MPP under STC (standard test conditions, 1000 W/m², 25 °C). The measured PV panel voltage and current are used to calculate the actual PV power, and the P&O MPPT algorithm searches for the PV voltage reference at the MPP, which is refreshed every 0.05 second. Two cases are investigated: 1) $SOC_{min} < SOC < SOC_{max}$; 2) avoidance of battery overcharging.

11.2.1   Case 1: $SOC_{min} < SOC < SOC_{max}$

The battery's initial SOC is 50%, and the PV panel is working at STC, with 25 °C temperature and 1000 W/m² irradiation. At 0.9 s, the irradiation decreases to 800 W/m². Figure 11.6 shows the simulated results. The PV panel output voltage tracking the MPP voltage is shown in Figure 11.6(a). The irradiation reduction at 0.9 s causes the PV power to decrease, and yet the grid-injected power stays constant, as shown in Figure 11.6(b). The battery is managed to release energy to the grid after 0.9 s, even though the battery is charged to absorb the redundant

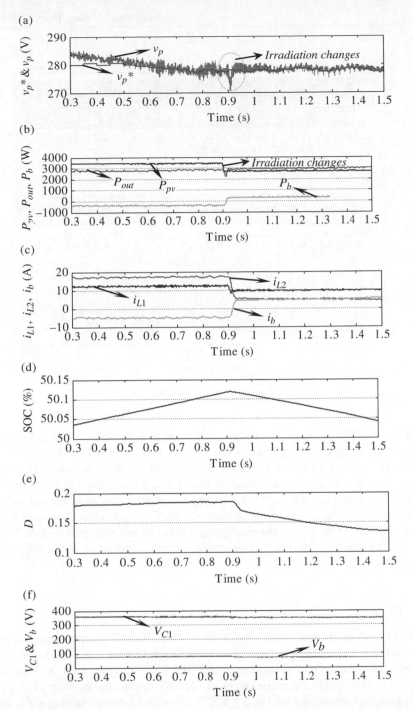

Figure 11.6 Simulated results in $SOC_{min} < SOC < SOC_{max}$: (a) PV panel voltage reference and actual voltage, (b) PV panel power, grid-injected power, and battery power, (c) inductor currents and battery current, (d) battery state of charge, (e) shoot-through duty ratio, (f) capacitor C_1 voltage and battery voltage, (g) dc-link voltage, (h) grid-injected current, (i) zoom-in grid-injected current during the irradiation change [6] (Reproduced by permission of the Institution of Engineering & Technology. Full acknowledgment to the Author, Title, and date of the original work).

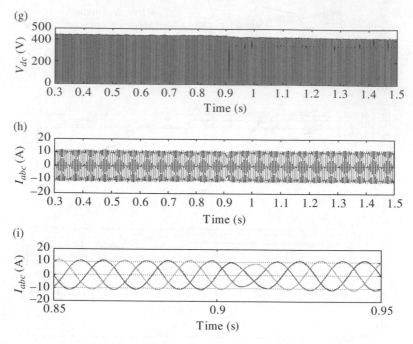

Figure 11.6 (*Continued*)

PV power before 0.9 s. The inductor currents, battery current, and battery SOC also change, as shown in Figure 11.6(c), (d). Before 0.9 s, the inductor L_1 current is less than the inductor L_2 current, the battery charges with the negative current, and the SOC increases. Whereas, after 0.9 s, the inductor L_1 current is greater than the inductor L_2 current, the battery discharges with the positive current, and the SOC decreases. During this process, the shoot-through duty ratio is adjusted to manage the battery, as shown in Figure 11.6(e). The capacitor C_1 voltage, the battery voltage, the dc-link voltage, and the grid-injected currents are shown in Figure 11.6(f)–(i), respectively. The constant grid-injected current verifies the constant grid-injected power, with the constant grid voltage.

11.2.2   Case 2: Avoidance of Battery Overcharging

Assume that the upper limit of battery SOC is 80%. This case will test the system's protection function when the battery SOC is larger than 80%. The initial battery SOC is set to 79.95% because of the limited computer memory space. The PV panel keeps on working at 25 °C, 1000 W/m². The high irradiation causes an increase in PV power, and the redundant PV power is charged into the battery, and the battery SOC keeps rising before reaching 80%, as shown in Figure 11.7(a), (b). When the battery SOC is over 80%, the controller stops the battery charging, the battery SOC becomes constant, and the PV power is fully injected into the grid.

Figure 11.7(c) shows the inductor currents and battery current. The battery current is negative when the SOC is less than 80%, and becomes zero when the SOC is larger than 80%. The larger grid-injected current is presented in Figure 11.7(h), (i) when the battery SOC is

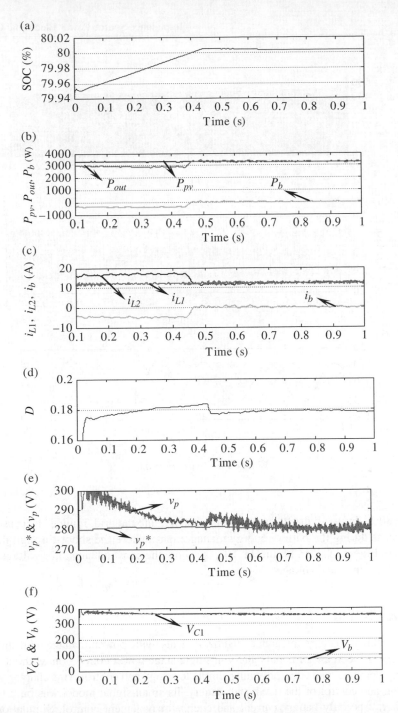

Figure 11.7 Simulated results when the battery SOC reaches the upper limit: (a) battery state of charge, (b) PV panel power, grid-injected power, and battery power, (c) inductor currents and battery current, (d) shoot-through duty ratio, (e) PV panel voltage reference and actual voltage, (f) capacitor C_1 voltage and battery voltage, (g) dc-link voltage, (h) grid-injected current, (i) zoom-in grid-injected current when the SOC has reached 80% [6] (Reproduced by permission of the Institution of Engineering & Technology. Full acknowledgment to the Author, Title, and date of the original work).

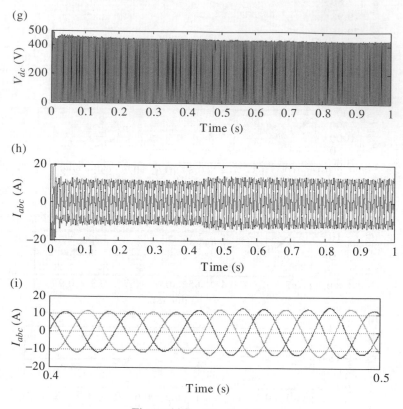

Figure 11.7 (*Continued*)

larger than 80%, due to all the PV power being injected into the grid. Figure 11.7(e) shows the MPPT process during the battery energy management, where the shoot-through duty ratio is adjusted in Figure 11.7(d). Figure 11.7(f), (g) show the constant capacitor voltage, battery voltage, and dc-link peak voltage.

11.3   Conclusion

The energy storage battery integrated ZSI/qZSI deals with power management among the renewable energy sources, grid, and battery in single-stage power conversion, without the extra dc-dc battery charging converter, providing simple topology and control. This chapter presented the modeling and control of the qZSI with battery. Its small-signal model was built to design the controller, especially battery current and energy management control. Simulation results were illustrated for different cases of battery state of charge. The solution is also inheritable to the other derived impedance topologies with battery paralleling to the capacitors.

References

[1] F. Z. Peng, M. Shen, K. Holland, "Application of Z-source inverter for traction drive of fuel cell–battery hybrid electric vehicles," *IEEE Trans. Power Electron.*, vol.22, no.3, pp.1054–1061, May 2007.

[2] J. G. Cintron-Rivera, Y. Li, S. Jiang, F. Z. Peng, "Quasi-Z-source inverter with energy storage for photovoltaic power generation systems," in *Proc. 2011 Twenty-Sixth Annual IEEE Applied Power Electronics Conference and Exposition (APEC)*, pp.401–406, 6–11 March 2011.

[3] H. Abu-Rub, A. Iqbal, Moin Ahmed Sk., F. Z. Peng, Y. Li, B. Ge, "Quasi-Z-source inverter-based photovoltaic generation system with maximum power tracking control using ANFIS," *IEEE Trans. Sustain. Energy*, vol.4, no.1, pp.11–20, Jan. 2013.

[4] D. Sun, B. Ge, D. Bi, F. Z. Peng, "Analysis and control of quasi-Z source inverter with battery for grid-connected PV system," *International Journal of Electrical Power & Energy Systems*, vol.46, pp.234–240, March 2013.

[5] Y. Liu, B. Ge, H. Abu-Rub, F. Z. Peng, "Control system design of battery-assisted quasi-Z-source-inverter for grid-tie photovoltaic power generation," *IEEE Trans. Sustain. Energy*, vol.4, no.4, pp.994–1001, Oct. 2013.

[6] Y. Liu, B. Ge, H. Abu-Rub, F. Z. Peng, "Modeling and controller design of quasi-Z-source inverter with battery based photovoltaic power system," *IET Power Electron.*, vol.7, no.7, pp.1665–1674, July 2014.

[7] B. Ge, H. Abu-Rub, F. Z. Peng, Q. Lei, de Almeida A., Ferreira F., D. Sun, Y. Liu, "An energy stored quasi-Z-source inverter for application to photovoltaic power system," *IEEE Trans. Ind. Electron.*, vol.60, no.10, pp.4468–4481, Oct. 2013.

[8] B. Ge, F. Z. Peng, H. Abu-Rub, F. J. T. E. Ferreira, A. T. de Almeida, "Novel energy stored single-stage photovoltaic power system with constant DC-link peak voltage," *IEEE Trans. Sustain. Energy*, vol.5, no.1, pp.28–36, Jan. 2014.

[9] B. Ge, Q. Lei, F. Z. Peng, D. Sun, Y. Liu, H. Abu-Rub, "An effective PV power generation control system using quasi-Z source inverter with battery," in *Proc. 2011 IEEE Energy Conversion Congress and Exposition (ECCE)*, pp.1044–1050, 17–22 Sept. 2011.

[10] J. Puukko, T. Suntio, "Dynamic properties of a voltage source inverter-based three-phase inverter in photovoltaic application," *IET Renew. Power Gen.*, vol.6, no.6, pp.381–391, November 2012.

12

Z-Source Multilevel Inverters

Poh Chiang Loh[1], Yushan Liu[2], Haitham Abu-Rub[2], and Baoming Ge[3]

[1] *Department of Energy Technology, Aalborg University, Aalborg East, Denmark*
[2] *Electrical and Computer Engineering Program, Texas A&M University at Qatar,*
Qatar Foundation, Doha, Qatar
[3] *Department of Electrical and Computer Engineering, Texas A&M University,*
College Station, USA

There are two hotly researched Z-source/quasi-Z-source (ZS/qZS) multilevel inverters: the ZS/qZS neutral-point-clamped (NPC) inverter and ZS/qZS cascaded multilevel inverter (CMI). The ZS/qZS concept, integrating with those traditional multilevel inverters, inherits the merits of both topologies with improved inverter reliability due to short-circuit immunity. This chapter presents the configurations, operating principles, and modulation schemes of the two ZS/qZS multilevel inverters.

12.1 Z-Source NPC Inverter

12.1.1 Configuration

Three-level NPC inverters, having many inherent advantages, are commonly used as the preferred topology for medium voltage ac drives [1]. They have also been recently explored for low voltage renewable grid-interfacing applications [2, 3]. Despite their generally favorable output performance, NPC inverters are constrained by their ability to perform only voltage-buck operation if no additional dc-dc boost stages are added to their front-ends. Overcoming this limitation, a buck–boost Z-source NPC inverter is proposed in [4], where two additional X-shaped impedance networks are added between two isolated dc sources and traditional NPC circuitry, as illustrated in Figure 12.1. The added impedance networks are responsible for balanced inductive voltage boosting upon shooting through any of the inverter phase-legs without causing damage to their semiconductor switches. This protection from sudden current surge is provided by the inductors found within the Z-source impedance networks.

Impedance Source Power Electronic Converters, First Edition. Yushan Liu, Haitham Abu-Rub, Baoming Ge, Frede Blaabjerg, Omar Ellabban, and Poh Chiang Loh.
© 2016 John Wiley & Sons, Ltd. Published 2016 by John Wiley & Sons, Ltd.

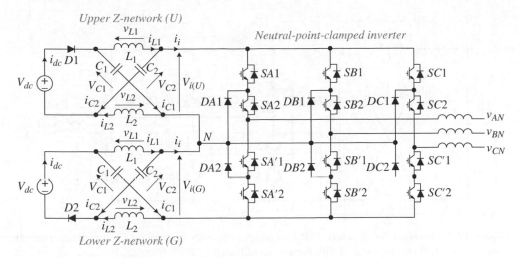

Figure 12.1 Topology of Z-source NPC inverter with two LC impedance networks (*Source*: Loh 2007 [4]. Reproduced with permission of IEEE).

Although theoretically feasible and with optimized switching pattern noted at its output, the inverter in Figure 12.1 is not an economical solution. The reason is linked to its two isolated dc sources and the large number of passive LC elements, which can significantly increase the cost, size, and weight of the inverter. Therefore, instead of two networks per inverter, alternative topologies that use only one impedance network will likely be more attractive in terms of economics. However, from past developmental trends where fewer components are used, performance degradations are usually expected, implying a trade-off, which must be seriously considered before deciding on the final topology.

At first sight, the design of a Z-source NPC inverter will experience the same trade-off, but through detailed operational analysis, it can gradually be shown that this trade-off will not occur. In other words, the improved Z-source NPC inverter will produce the same performance as its two-network precedent, while using one fewer network. This unique complementary, rather than supplementary, behavior is attributed to two new operating modes introduced to the inverter modulation control, explained as follows, with reference to the improved Z-source NPC inverter shown in Figure 12.2.

12.1.2 Operating Principles

Figure 12.2 shows a Z-source NPC inverter with only a single Z-source impedance network [4]. The network consists of a split-inductor (L_1 and L_2) and two capacitors (C_1 and C_2), connected between the input dc source and a traditional NPC inverter circuitry. The input source can be a split-dc source formed by two series-connected capacitors, rather than two isolated sources. On the other hand, the rear-end NPC circuitry allows the inverter to assume three distinct voltage levels per phase leg, whose expressions and corresponding gating signals are shown in Table 12.1.

Compared with the traditional NPC inverter, Table 12.1 includes four non-traditional states. Like the two-level Z-source inverter [5, 6], these non-traditional states are for boosting

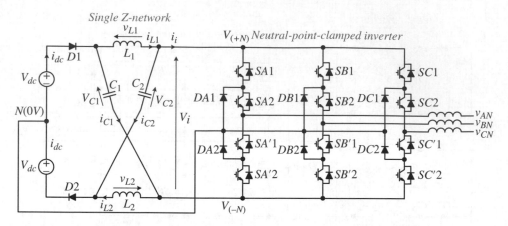

Figure 12.2 Topology of Z-source NPC inverter using only a single LC impedance network (*Source*: Loh 2007 [4]. Reproduced with permission of IEEE).

Table 12.1 Switching States of Z-source NPC Inverter (ST ≡ shoot-through) (*Source*: Loh 2007 [4]. Reproduced with permission of IEEE)

State Type	ON Switches	ON Diodes	V_A
Non-ST	$SA1, SA2$	$D1, D2$	$+v_i/2$
Non-ST	$SA2, SA'1$	$D1, D2,$	0
		$\{DA1 \text{ or } DA2\}$	
Non-ST	$SA'1, SA'2$	$D1, D2$	$-v_i/2$
Full-ST (not preferred)	$SA1, SA2, SA'1, SA'2$	—	0
Full-ST (preferred)	$SA1, SA2, SA'1, SC2, SC'1, SC'2$	$DA2, DC1$	0
Upper-ST	$SA1, SA2, SA'1$	$DA2, D1$	See (12.16)
Lower-ST	$SA2, SA'1, SA'2$	$DA1, D2$	See (12.17)

voltages carried by the Z-source NPC inverter. In particular, the two full shoot-through states help to short circuit the dc-link fully, and can be used together with other non-shoot-through active and null states in a typical modulation state sequence for a Z-source NPC inverter. For distinctly representing the non-shoot-through and shoot-through states, Figure 12.3 shows their simplified equivalent circuits for analysis, where in Figure 12.3(a) the inverter circuitry and external load have been represented by a simplified current source in the non-shoot-through state. Using this equivalent representation with input diodes $D1$ and $D2$ conducting, the inductive voltage of the single symmetrical Z-source network ($\Rightarrow v_{L1} = v_{L2} = v_L$ and $V_{C1} = V_{C2} = V_C$) and the three distinct dc-link voltage levels ($v_{(+N)}$, V_N, $v_{(-N)}$) are respectively expressed as

$$v_L = 2V_{dc} - V_C \tag{12.1}$$

$$v_{(+N)} = +v_i/2; \quad V_N = 0; \quad v_{(-N)} = -v_i/2; \quad v_i = 2(V_C - V_{dc}) \tag{12.2}$$

Alternatively, the full dc-link shoot-through state can be entered by, for example, turning ON all switches $\{SA1, SA'1, SA2, SA'2\}$ from phase A simultaneously. The simplified circuit

(a)

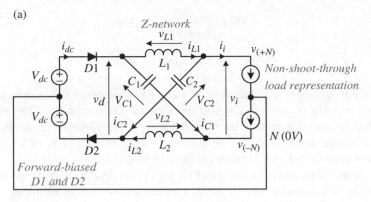

(b)

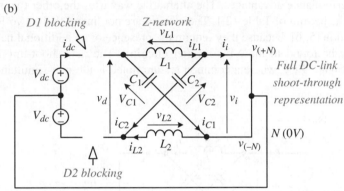

Figure 12.3 Simplified representations of Z-source NPC inverter of Figure 12.1 with a single impedance network when in (a) non-shoot-through and (b) full dc-link shoot-through states (*Source*: Loh 2007 [4]. Reproduced with permission of IEEE).

representation is shown in Figure 12.3(b) with input diodes $D1$ and $D2$ blocking. The relevant mathematical expressions can be rewritten as

$$v_L = V_C \tag{12.3}$$

$$v_{(+N)} = V_N = v_{(-N)} = 0 \text{ since } v_i = 0 \tag{12.4}$$

Averaging capacitor voltage over a switching cycle T then gives:

$$V_C = \frac{2V_{dc}\left(1 - T_0/T\right)}{1 - 2T_0/T} \tag{12.5}$$

where T_0 represents the shoot-through duration. Using (12.5), the peak inverter dc-link voltage $\hat{v}_i$, peak ac output voltage $\hat{v}_x$ ($x=A$, B or C), and the three distinct voltage levels when in a non-shoot-through state are derived as

$$\hat{v}_i = V_C - v_L = 2V_C - 2V_{dc} = \frac{2V_{dc}}{1 - 2T_0/T} \tag{12.6}$$

$$\hat{v}_x = M\hat{v}_i/2 = \frac{MV_{dc}}{1-2T_0/T} = MBV_{dc} \qquad (12.7)$$

$$v_{(+N)} = +\hat{v}_i/2; \quad V_N = 0; \quad v_{(-N)} = -\hat{v}_i/2 \qquad (12.8)$$

where M is the modulation ratio and $B = 1/(1-2T_0/T)$ is the boost factor, which preferably should be set to unity ($T_0/T = 0$) for voltage-buck operation and $B > 1$ for voltage-boost operation. In particular, for voltage-buck operation, $\hat{v}_i = 2V_{dc}$ since $B = 1$, implying that the three distinct voltage levels that the inverter can assume are $+V_{dc}$, $0\,V$, and $-V_{dc}$ which, in principle, are similar to those of a traditional three-level inverter.

Although the non-shoot-through and shoot-through states described above are sufficient for voltage boosting when used correctly, there is an alternative way of boosting voltage with some gained performance advantages. The alternative way uses the other two shoot-through states listed at the bottom of Table 12.1. These states are not linked to the two-level Z-source inverter reported in [5, 6], because they require the presence of an additional neutral-point N, which can only be found in the NPC inverter circuitry. The new shoot-through states are represented in Figure 12.4, where a common feature noted is the non-simultaneous shorting

(a)

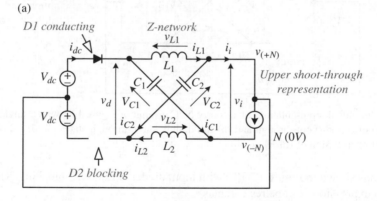

(b)

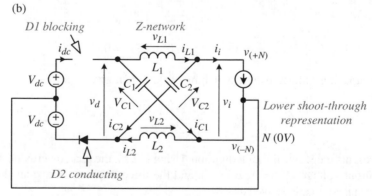

Figure 12.4   Simplified representations of Z-source NPC inverter with a single impedance network as shown in Figure 12.2 when in (a) upper-shoot-through and (b) lower-shoot-through states (*Source*: Loh 2007 [4]. Reproduced with permission of IEEE).

of either the upper or lower dc source through single Z-source impedance network. Moreover, unlike the full dc-link shoot-through state, where both input diodes $D1$ and $D2$ are reverse-biased, the new operating states result in only $D2$ blocking for the upper-shoot-through and only $D1$ blocking for the lower-shoot-through. Deduced from these equivalent circuits, (12.9) and (12.10) can be written to represent the upper shoot-through state, while (12.11) and (12.12) represent the lower shoot-through state.

$$v_L = V_{dc} \tag{12.9}$$

$$v_{(+N)} = V_N = 0\text{V}; \quad v_{(-N)} = V_{dc} - V_C \tag{12.10}$$

$$v_L = V_{dc} \tag{12.11}$$

$$v_{(+N)} = -V_{dc} + V_C; \quad v_{(-N)} = V_N = 0 \tag{12.12}$$

If only the non-shoot-through state in Figure 12.3(a) and the shoot-through operating modes in Figure 12.4 are used for modulating the inverter, state-space averaging performed on v_L in (12.1), (12.9), and (12.11) gives rise to

$$V_C = \frac{2V_{dc}\left(1 - T_0'/(2T)\right)}{1 - T_0'/T} \tag{12.13}$$

where T_0' represents the common duration set for both upper and lower-shoot-through states needed for achieving balanced voltage-boosting. Substituting (12.13) into (12.2), (12.10), and (12.12), the peak dc-link voltage $\hat{v}_i$ and dc-rail potentials in the respective operating states are summarized as

$$\hat{v}_i = \frac{2V_{dc}}{1 - T_0'/T} \tag{12.14}$$

Non-shoot-through

$$v_{(+N)} = +\hat{v}_i/2; \quad V_N = 0; \quad v_{(-N)} = -\hat{v}_i/2 \tag{12.15}$$

Upper-shoot-through

$$v_{(+N)} = V_N = 0; \quad v_{(-N)} = -V_{dc}/\left(1 - T_0'/T\right) = -\hat{v}_i/2 \tag{12.16}$$

Lower-shoot-through

$$v_{(+N)} = V_{dc}/\left(1 - T_0'/T\right) = \hat{v}_i/2; \quad v_{(-N)} = V_N = 0 \tag{12.17}$$

Comparing (12.14) and (12.5), it is noted that the same peak dc-link voltage can be obtained by setting $T_0' = 2T_0$, where the maximum values for T_0' and T_0 are T and $0.5T$ respectively. Also, by comparing (12.15) and (12.17), an interesting feature clearly seen is that the lower dc-rail voltage $v_{(-N)}$ remains unaltered when transmitting from a non-shoot-through to an upper-shoot-through state. Similarly, transmitting from a non-shoot-through to a lower-shoot-through state will not affect the upper dc-rail voltage $v_{(+N)}$. Therefore, for the two phases connected to either the neutral point or the dc-rail not shorted by the third phase, their potentials remain unaffected, which is an important feature explored for the Z-source NPC inverter with only a single impedance network.

12.1.3 Modulation Scheme

While inserting full dc-link or a balanced combination of upper- and lower-shoot-through states to the inverter state sequence, it is important to ensure that the correct normalized volt-sec average is preserved at all instances, regardless of the reference phasor position on the three-level vector diagram shown in Figure 12.5. With this criterion in view, two possible modulation techniques are proposed in the following subsections with their advantages and disadvantages discussed in detail.

12.1.3.1 Alternative Phase Opposition Disposition Scheme

As known from past literature [7–9], a method for controlling the traditional NPC inverter is to use two vertically disposed, 180° phase-shifted triangular carriers for comparison with a set of normalized three-phase sinusoidal references (labeled as V_x, $x = a$, b, or c), to produce two independent gating signals per phase (e.g. $SA1$ and $SA2$ in Figure 12.2(a) with their logical NOT signals used for driving $SA'1$ and $SA'2$). This style of carrier arrangement is commonly referred to as the alternative phase opposition disposition (APOD) scheme with a typical state

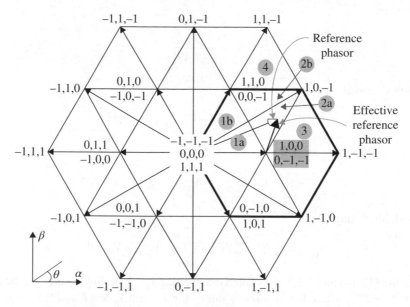

Figure 12.5 Space vector representation of a three-level inverter.

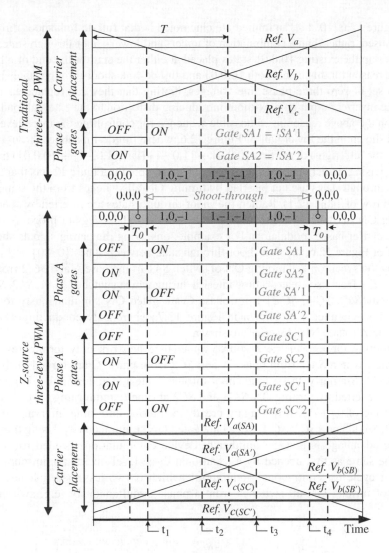

Figure 12.6 APOD modulation of traditional and three-level Z-source inverters (*Source*: Loh 2003 [10] and Loh 2007 [11]. Reproduced with permission of IEEE).

sequence produced by it shown in the upper half of Figure 12.6. (Note that in that figure, a triplen offset, defined as $v_{off} = -0.5(\max(V_x) + \min(V_x))$, is intentionally added to the sinusoidal references to gain an important switching advantage, whose details are described later.)

Certainly, the illustrated traditional sequence in Figure 12.6 can be used for controlling the three-level Z-source inverter, since it uses NPC circuitry at its rear-end. However, that sequence will only give rise to voltage-buck operation with the voltage-boosting capability of the Z-source network left unexplored. To initiate this voltage-boosting ability, shoot-through states must appropriately be inserted into the inverter state sequence without altering its normalized volt-sec average appearing across the externally connected load. Based on this insertion principle and by referring to the traditional APOD state sequence shown in the upper half of Figure 12.6 [10, 11], an immediate clue noted is that full dc-link shoot-through states should be used, rather than a combination of upper- and lower-shoot-through states, and they should overlap the existing {0,0,0} states placed at either the start or the end of a half carrier cycle. The reason for this is that both {0,0,0} and full dc-link shoot-through states indistinctly produce a set of zero three-phase line voltages, hinting that they can replace each other for altering the inverter output gain, without introducing distortion to the external load.

Based on the above reasoning, a possible sequence for controlling the three-level Z-source inverter is shown in the lower half of Figure 12.6, where full dc-link shoot-through states are inserted to the left (right) of the first {0,0,0} → {1,0,−1} (last {1,0,−1} → {0,0,0}) transition of the falling (rising) upper carrier edge. Another feature noted in Figure 12.6 is that the full dc-link shoot-through states are not inserted by turning ON all switches from the same phase leg (see fourth row of Table 12.1). Rather, the shoot-through states are inserted by synchronizing the turning ON of switches from two naturally selected phase-legs (a process auto-built-in within the carrier-based modulator). For example, observing the gating signals shown in the lower half of Figure 12.6, the first shoot-through state between states {0,0,0} and {1,0,−1} is initiated by the synchronized turning ON of switch $SA1$ from phase A and $SC'2$ from Phase C at time $(t_1 - T_0)$. Doing so creates a time interval during which switches {$SA1, SA2, SA'1$} from phase A and {$SC2, SC'1, SC'2$} from phase C are gated ON simultaneously to create the "bolded" short-circuited path shown in Figure 12.7, where only conducting switches and diodes of the Z-source NPC inverter are shown.

This shooting through of the Z-source impedance network only terminates at time t_1, during which switch $SA'1$ from phase A and $SC2$ from phase C are turned OFF to initiate state {1,0,−1}. Similarly, the second shoot-through state added during the rising upper carrier edge is inserted by turning ON $SA'1$ and $SC2$ at time t_4 and turning OFF $SA1$ and $SC'2$ at a later time of $(t_4 + T_0)$, which again create an intermediate time interval during which {$SA1, SA2, SA'1$} and {$SC2, SC'1, SC'2$} are gated ON simultaneously. Using this approach, the visible advantage is that the number of device commutations needed for a switching cycle is the same as that needed by a traditional three-level inverter (minimum of six per half carrier cycle for continuous modulation), which is two less than that needed by the approach of turning ON all switches from a phase simultaneously (e.g. switching phase

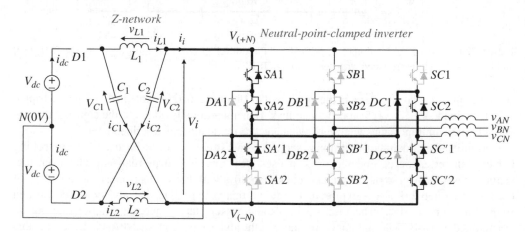

Figure 12.7   Intermediate shoot-through path of a Z-source NPC inverter.

A from the positive dc rail → shoot-through → 0 V would require {$SA1$, $SA2$, $SA'1$, $SA'2$} to transit from {ON, ON, OFF, OFF} → {ON, ON, ON, ON} → {OFF, ON, ON, OFF}).

Given that the four gating signals per phase leg are now not explicitly arranged as complementary pairs in all instances, they need to be controlled using two references per phase (instead of one) if the same APOD carrier set is used for comparison. Therefore, in total, three additional references are needed and their expressions can easily be formulated by first linearly converting the horizontal shoot-through time of T_0 to a normalized vertical offset of T_0/T (assuming a carrier peak of unity). The obtained vertical offset should then be added to the uppermost waveform of the original three-phase reference set – represented by $\max(V_x)$ – where in the upper half of Figure 12.6, it corresponds to V_a. The resulting two references used for controlling phase A are then written as $V_{a(SA)} = V_a + T_0/T$ and $V_{a(SA')} = V_a$, shown in the lower half of Figure 12.6. Because both references are shown intersecting with the upper disposed carrier, their generated gating signals are used for driving $SA1$ and $SA'1$, also shown in the same figure. In particular, the reference $V_{a(SA)}$ is used for turning $SA1$ ON at an earlier intersection time of $(t_1 - T_0)$, and $V_{a(SA')}$ is used for turning $SA'1$ OFF at a later intersection time of t_1. This in turn creates an interval during which three switches $SA1$, $SA2$, and $SA'1$ from phase A are ON.

Continuing to use the same approach, but now with T_0/T subtracted from the lowermost waveform V_c shown in the upper half of Figure 12.6, the derived references for controlling phase C are written as $V_{c(SC)} = V_c$ and $V_{c(SC')} = V_c - T_0/T$. Since both references cross the lower carrier, the produced gating signals are for switching $SC2$ and $SC'2$ respectively. Indeed, an interval during which $SC'1$, $SC2$, and $SC'2$ are ON, is easily created by using $V_{c(SC')}$ to turn $SC'2$ ON at a time advance of T_0, before using $V_{c(SC)}$ to turn $SC2$ OFF at a later instant. Moving forward to create a full dc-link short-circuit, a second offset $v_{off} = -0.5(\max(V_x) + \min(V_x))$ is added to the reference set to always center its maximal and minimal envelopes vertically within the full carrier band. This naturally aligns the intersecting points of the uppermost and lowermost references to always create a short-circuit interval, during which three switches from phase A and another three from phase C are turned ON simultaneously for voltage boosting.

Generalizing the above reference generation process to all sextants of the vector diagram, the generic rules for producing the six references are to add the vertical offset of T_0/T to the "maximum" reference V_{max}, and subtract the same offset from the "minimum" reference V_{min}, while keeping the "middle" reference V_{mid} unchanged. In equation form, these rules are summarized from (12.18) to (12.20), with their implementation giving rise to the voltage transfer gains expressed from (12.6) to (12.8), and the sequence shown in Figure 12.6.

$$\begin{cases} V_{max(SX)} = V_{max} + V_{off} + T_0/T \\ V_{max(SX')} = V_{max} + V_{off} \end{cases} \tag{12.18}$$

$$\begin{cases} V_{mid(SX)} = V_{mid} + V_{off} \\ V_{mid(SX')} = V_{mid} + V_{off} \end{cases} \tag{12.19}$$

$$\begin{cases} V_{min(SX)} = V_{min} + V_{off} \\ V_{min(SX')} = V_{min} + V_{off} - T_0/T \end{cases} \tag{12.20}$$

$$X = A, B, \text{ or } C$$

12.1.3.2 Phase Disposition Scheme

An alternative shoot-through scheme that can be developed is based on the traditional phase disposition (PD) carrier-based approach, which is known to produce a lower harmonic content [7–9]. For that approach, two vertically disposed in-phase carriers are used instead, for comparison with the three-phase references to produce the typical state sequence shown in the upper half of Figure 12.8, assuming that the reference phasor is in triangle 3 on the vector diagram in Figure 12.5. An implication immediately noted from Figure 12.8 is that the {0,0,0} state is no longer produced, hinting that the full dc-link shoot-through state might not be a favorable choice for insertion if the normalized volt-sec average is to be preserved. Therefore, for the PD case, inserting a combination of upper- and lower-shoot-through states might be the only option left with its analysis presented as follows [10, 11].

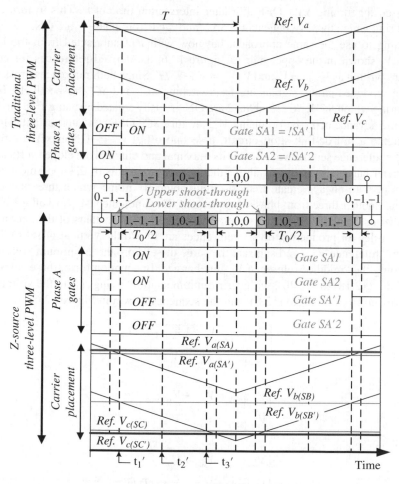

Figure 12.8 PD modulation of traditional and three-level Z-source inverters (*Source*: Loh 2003 [10] and Loh 2007 [11]. Reproduced with permission of IEEE).

Beginning with the traditional state sequence shown in the upper half of Figure 12.8, an observation noted is the switching of phase A from $0V$ to the positive dc rail potential labeled as $v_{(+N)}$ to give the first state transition $\{0,-1,-1\} \rightarrow \{1,-1,-1\}$. During this transition, phases B and C remain clamped to the negative dc rail $v_{(-N)}$, and according to (12.16), they will stay unaltered even if an upper-shoot-through state is intentionally inserted by switching $\{SA1, SA2, SA'1, SA'2\}$ from $\{0,1,1,0\} \rightarrow \{1,1,1,0\} \rightarrow \{1,1,0,0\}$ (bold signals correspond to the inserted upper-shoot-through). The observed three-phase pole voltages during the upper-shoot-through interval therefore resemble those of state $\{0,-1,-1\}$, implying that they can supplement each other for voltage-boosting without modifying the produced volt-sec average.

Applying the same analysis and moving on to the second transition ($\{1,-1,-1\} \rightarrow \{1,0,-1\}$), where phase B is triggered to switch from the negative dc rail to the neutral $0V$ potential, no shoot-through state should be inserted since its inclusion will cause either phase A or C to collapse to zero unintentionally, introducing volt-sec error in turn. Moving forward again to the third transition ($\{1,0,-1\} \rightarrow \{1,0,0\}$) initiated by raising the potential of phase C from $v_{(-N)}$ to $0V$, inserting a lower-shoot-through state appears feasible since the gating of $\{SC1, SC2, SC'1, SC'2\}$ from $\{0,0,1,1\} \rightarrow \{0,1,1,1\} \rightarrow \{0,1,1,0\}$ (bold signals represent the inserted lower-shoot-through) will not affect phases A and B, which remain clamped to $v_{(+N)}$ and $0V$ respectively, according to (12.17). The resulting pole voltages therefore resemble those of $\{1,0,0\}$, implying that both states are interchangeable for voltage gain tuning without introducing volt-sec distortion.

Based on the above reasoning, a modified state sequence that can be used for controlling the Z-source NPC inverter is shown in the lower half of Figure 12.8, where an upper- and a lower-shoot-through interval are inserted immediately to the left of the first transition and to the right of the third transition per half carrier cycle respectively. The reason for having adjacent placements is to again make use of the time advance/delay technique for inserting shoot-through states without increasing the inverter commutation count. Specifically, for the example shown in Figure 12.8, the first shoot-through state is inserted by using a modified reference $V_{a(SA)} = V_a + 0.5T_0'/T$ for turning $SA1$ ON at an earlier time of $(t_1' - 0.5T_0')$, and another reference $V_{a(SA')} = V_a$ for turning $SA'1$ OFF at a later time of t_1' (maximum envelope). Since $SA2$ is already ON within the illustrated carrier cycle, an interval with $SA1 = SA2 = SA'1 = DA2 = ON$ is therefore created for shorting the upper dc source as intended, with the commutation count maintained at two (also reflected by the gating signals shown in Figure 12.8). Similarly, by using $V_{c(SC)} = V_c$ to turn ON $SC2$ at t_3' and $V_{c(SC')} = V_c - 0.5T_0'/T$ to turn OFF $SC'2$ at $t_3' + 0.5T_0'$ (minimum envelope), an interval, during which $SC2, SC'1, SC'2$ and $DC1$ are ON, is created for shorting the lower dc source with the commutation count kept unchanged.

The same analysis, when applied to the second distinct triangle on the vector diagram for $0° \leq \theta \leq 30°$, will again validate the technique of adding $0.5T_0'/T$ to the maximum envelope and subtracting the same offset from the minimum envelope to always guarantee the correct insertion of shoot-through states. (Triangles 1a and 1b are not considered here since they are used mainly for low modulation condition, during which B is set to unity – no shoot-through state inserted – to avoid unnecessary stressing of the semiconductor devices, while producing a low output [4]). Therefore, generalizing across the full fundamental cycle, the set of six modified references needed for PD shoot-through insertion is again given by (12.18)–(12.20), except for the replacement of T_0/T by $0.5T_0'/T$ in those expressions.

Note that although the triplen offset V_{off} is now not needed for aligning the commutations of phases A and C, as discussed in Subsection 12.1.3.1 for the APOD scheme, its inclusion is still needed for the PD scheme to maintain (for example) equal {0,−1,−1} and {1,0,0} intervals at the start and end of each half carrier cycle for inserting balanced upper- and lower-shoot-through states. Conceptually, balanced insertion is needed for undistorted voltage boosting, and is automatically guaranteed in all sextants on the vector diagram, as long as V_{off} is added to the set of modified references. (For knowledge sharing, it is highlighted here that balanced shoot-through insertion at both edges of a switching cycle is also encouraged in [6] for the reduction of network current ripple flowing through a two-level Z-source inverter. This reason is equally valid for the three-level Z-source inverter, but avoiding unbalanced voltage boosting is a more prominent concern here.)

12.2 Z-Source/Quasi-Z-Source Cascade Multilevel Inverter

12.2.1 Configuration

The CMI is a promising candidate for the next generation of PV power converters because it has features such as separate PV sources, transformerless, low harmonics, small filters, and low switching frequency. Nevertheless, an inherent voltage imbalance exists in a conventional CMI-based PV system because its H-bridge inverter (HBI) modules lack the boost function to deal with the PV voltage's wide range of variation. Some research works to mitigate this disadvantage mainly deal with modifying modulation methods [12, 13], and developing various control schemes such as continuous-time linear proportional and integral control [14–16], discrete-time linear control [17], and model predictive control [18, 19]. However, an overrating design is still necessary in order to fit a wide PV voltage range.

Recently, an extra dc-dc boost converter was added between the HBI module and the PV panel to achieve each module's constant dc-link voltage, even if the PV panel voltage changes [20–22]. However, each module fulfills two-stage power conversion, and many extra dc-dc converters make the whole PV system complex, bulky, costly, and of low efficiency.

The Z-source/quasi-Z-source cascaded multilevel inverter (ZS/qZS-CMI) combines advantages of ZS/qZSI and CMI [23–30] for applications in PV power systems. Figure 12.9 shows their topologies. Each module achieves voltage step-up/down function during dc to ac single-stage power conversion, and independent dc-link voltage control. Therefore, the qZS-CMI has the balanced dc-link peak voltage when implementing the distributed maximum power point tracking (MPPT); also, the system presents high reliability due to allowing shoot-through states, and low cost due to saving one-third of the modules when compared to conventional CMI based PV systems [24].

Independent dc-link peak voltage controls of all modules, distributed MPPT, grid-tie power control, and pulse-width modulation (PWM) are keys for implementing the qZS-CMI based grid-tie PV system. Reference [23] focuses on efficiency evaluation of GaN switches-based qZS-CMI for PV power system; reference [24] demonstrates the qZS-CMI's advantages over the conventional CMI in terms of reliability, cost, and efficiency when applied to MW-scale PV power systems; references [25–28] present control methods for single-phase/three-phase qZS-CMI based PV power system, including independent dc-link peak voltage control, distributed MPPT, and grid-tie power control; the qZS-CMI's phase shift PWM methods were presented in [26, 29, 30].

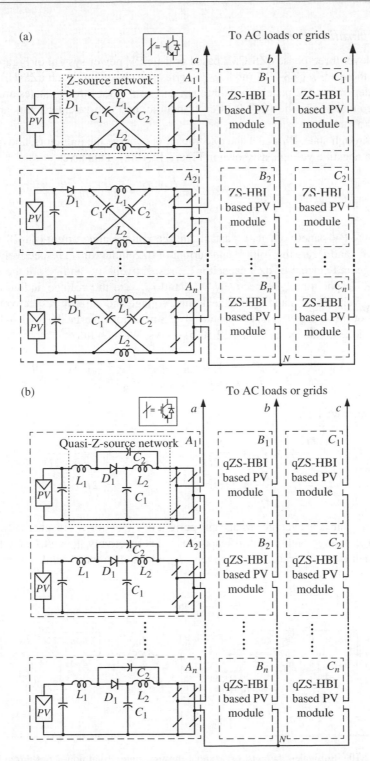

Figure 12.9 Topologies of (a) ZS-CMI and (b) qZS-CMI based PV power system.

12.2.2   *Operating Principles*

Using the *n*-layer three-phase qZS-CMI based grid-tie PV power system of Figure 12.9(b) as an example, there are *n* qZS-HBI modules in series per phase, and each qZS-HBI module is fed by an independent PV panel. Shoot-through and non-shoot-through states are fulfilled in continuous conduction mode (CCM) of each module, and equivalent circuits are shown in Figure 12.10(a) and (b).

At shoot-through state, PV panel and qZS capacitors charge the inductors, and qZS diode is cut off due to negative voltage, as shown in Figure 12.10(a). This gives

$$v_{L1} = v_{PV} + v_{C2} = v_{C1}, v_{L2} = v_{C1}, v_{DC} = 0, v_{diode} = v_{C1} + v_{C2}, i_{diode} = 0,$$
$$i_{C1} = -i_{L2}, i_{C2} = -i_{L1}, i_{DC} = i_{L1} + i_{L2}$$

(12.21)

where v_{PV} is the PV panel voltage, v_{L1} and i_{L1} are the voltage and current of qZS inductor L_1, respectively, v_{L2} and i_{L2} are the voltage and current of qZS inductor L_2, respectively, v_{C1} and i_{C1} are the voltage and current of qZS capacitor C_1, respectively, v_{C2} and i_{C2} are the voltage and current of qZS capacitor C_2, respectively, v_{diode} and i_{diode} are the voltage and current of qZS diode, respectively, and v_{DC} and i_{DC} are the dc-link voltage and current, respectively.

At non-shoot-through state, the PV panel and qZS inductors charge the load and capacitors, and the qZS diode conducts, as Figure 12.10(b) shows. We then have

$$v_{L1} = v_{PV} - v_{C1}, v_{L2} = -v_{C2}, v_{DC} = \hat{v}_{DC} = v_{C1} + v_{C2}, v_{diode} = 0,$$
$$i_{diode} = i_{L1} + i_{L2} - i_{DC}, i_{C1} = i_{L1} - i_{DC}, i_{C2} = i_{L2} - i_{DC}$$

(12.22)

In steady state, we have

$$V_{C1} = \frac{1-D}{1-2D} V_{PV}, V_{C2} = \frac{D}{1-2D} V_{PV}, \hat{v}_{DC} = V_{C1} + V_{C2} = \frac{1}{1-2D} V_{PV},$$
$$I_{L1} = I_{L2} = I_L = \frac{P}{V_{PV}}, I_{C1} = I_{C2} = I_L - I_{DC}, I_{diode} = 2I_L - I_{DC}$$

(12.23)

where V_{C1} and I_{C1} are the average voltage and current of capacitor C_1, respectively, V_{C2} and I_{C2} are the average voltage and current of capacitor C_2, respectively, I_{L1} and I_{L2} are the average currents of inductors L_1 and L_2, respectively, I_{DC} is the average dc-link current, D is the

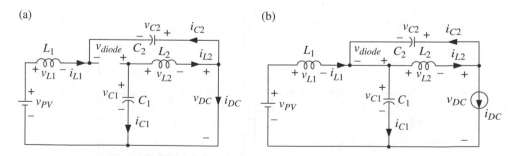

Figure 12.10   Equivalent circuits: (a) at shoot-through state, (b) at non-shoot-through state.

shoot-through duty cycle of one qZS-HBI module, $D = T_0/T_s$, T_0 is the shoot-through time interval and T_s is the switching cycle.

Total output voltage of qZS-CMI per phase is

$$v_{Hx} = v_{Hx1} + v_{Hx2} + \cdots + v_{Hxk} + \cdots + v_{Hxn} \tag{12.24}$$

where v_{Hxk} is the kth qZS-HBI module's output voltage in phase x; $x = a$, b, and c; $k = 1, 2, \ldots, n$.

12.2.3 Modulation Scheme

Three modulation methods – phase-shifted sinusoidal PWM (PS-SPWM), phase-shifted pulse-width-amplitude modulation (PS-PWAM), and modular multilevel space vector modulation (MM-SVM) – were proposed for the qZS-CMI based PV power system in [25, 29, 30], respectively.

12.2.3.1 PS-SPWM

The PS-SPWM shown in Figure 12.11 is a basic method derived from the conventional CMI [25]. The PS-PWAM was proposed to reduce the switching number and loss. The MM-SVM was for the three-phase qZS-CMI to simplify the modulation implementation and enhance the voltage utilization ratio.

As Figure 12.11 shows, each qZS-HBI module of the qZS-CMI is controlled by the unipolar PWM with the shoot-through references $1 - D_n$ and $D_n - 1$; the m_n are the modulation signals of qZS-HBI modules; the carriers of upper and lower switches for one module are at 180° phase shift. When the carrier is higher than the top shoot-through reference $1 - D_n$, or lower than the bottom shoot-through reference $D_n - 1$, a shoot-through state is produced. The step-like multilevel voltage waveform is achieved by a π/n phase shift between the adjacent cascaded modules. When compared with the traditional CMI, there are additional shoot-through switching actions in each module of the qZS-CMI, which results in extra switching losses.

12.2.3.2 PS-PWAM

Figure 12.12 shows the PS-PWAM for the ZS/qZS-CMI [29]. As shown in Figure 12.12, the phase-shifted PAM is realized by using the amplitude-varied carriers for a seven-level qZS-CMI, where $u_{\{A,B,C\}L}$ and $u_{\{A,B,C\}R}$ are the three-phase modulation signals for left-bridge and right-bridge legs, respectively. Carrier$_{A\{1, 2, 3\}}$ are three carriers for three modules of phase A, and the carrier amplitudes are the upper and lower envelopes consisting of u_{AL}, u_{BL}, and u_{CL}, also the subscript $A\{1, 2, 3\}$ defines the module – for example, $A1$ represents the module 1 of phase A.

The simple boost control can be implemented for the PAM if the shoot-through references V_{PA1} and V_{NA1} (using module A_1 as an example) are defined as λy_{max} and λy_{min}, respectively, where λ is with $0.5 < \lambda < 1$, and

$$y_{max} = \max\left\{u_{AL}, u_{BL}, u_{CL}\right\}, \quad y_{min} = \min\left\{u_{AL}, u_{BL}, u_{CL}\right\}$$

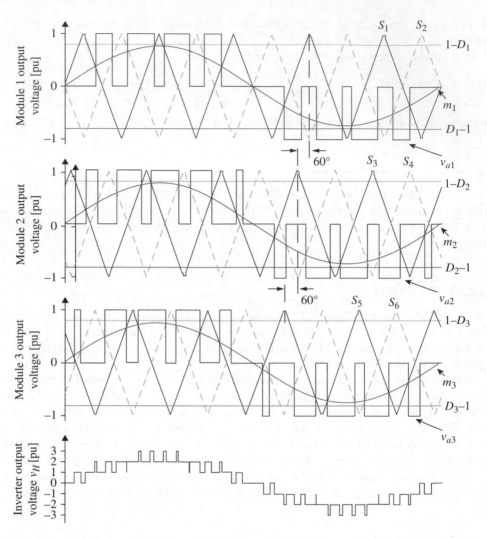

Figure 12.11 Sketch map of the PS-SPWM for the qZS-CMI (*Source*: Sun 2012 [25]. Reproduced with permission of IEEE).

Thus, the shoot-through duty ratio is deduced as

$$D = \frac{\overline{T}_{sh}}{T_s} = 1 - \lambda \qquad (12.25)$$

Figure 12.12 shows the switch patterns $S_{a1\{1,2,3,4\}}$ of module A1 in the PAM of qZS-CMI. From $\pi/6$ to $5\pi/6$, the modulation signal u_{AL} is the maximal in $u_{\{A,B,C\}L}$, the upper switch S_{a11} of left bridge leg keeps on-state without any switching action; during this interval, the lower switch S_{a12} is turned on when the carrier is lower than V_{NA1}, so the shoot through occurs. From $7\pi/6$ to $11\pi/6$, the modulation signal u_{AL} is the minimum in $u_{\{A,B,C\}L}$, the lower switch S_{a12} of

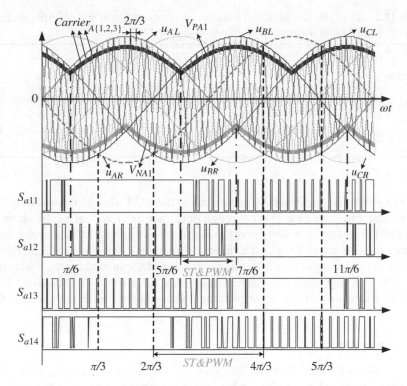

Figure 12.12   PS-PWAM of qZS-CMI (*Source*: Liu 2014 [29]. Reproduced with permission of IEEE).

left bridge leg keeps on-state, and the upper switch S_{a11} achieves the shoot-through action by comparison of carrier and reference V_{PA1}. During the intervals of $5\pi/6$ to $7\pi/6$, 0 to $\pi/6$, and $11\pi/6$ to $12\pi/6$, both of the upper and lower switches act to fulfill the shoot-through state and active states, through comparing the carrier with the shoot-though references and modulation signals. For the two switches of the right bridge leg, the modulation signal u_{AR} has 180° phase shift to that of the left bridge leg. As shown in Figure 12.12, from $\pi/3$ to $2\pi/3$, the lower switch S_{a14} of right bridge leg keeps on-state, and the upper switch S_{a13} achieves the shoot-through action by comparison of carrier and reference V_{PA1}; from $4\pi/3$ to $5\pi/3$, the upper switch S_{a13} of right bridge leg keeps on-state without any switching action, and the lower switch S_{a14} acts the shoot through when the carrier is lower than V_{NA1}; during 0 to $\pi/3$, $2\pi/3$ to $4\pi/3$, and $5\pi/3$ to $6\pi/3$, both of the upper and lower switches act to fulfill the shoot-through state and active states, through comparing the carrier with the shoot-through references and modulation signals. Three carriers of three modules in phase A present $2\pi/3$ phase shift to each other to generate the stepwise voltage waveform. Each layer module has the same carrier; for example, Carrier$_{A1}$ is for the modules A_1, B_1, and C_1. The shoot-through references may be different for all modules, which depend on the PV panel voltage of each module; u_{BL} and u_{BR} are the modulation signals of left and right bridge legs in phase B; u_{CL} and u_{CR} are those for phase C.

From Figure 12.12, each switch presents three types of behavior during one fundamental period: no switching, only shoot-through switching (represented by ST), and both ST and traditional active states (represented by ST&PWM), as shown in Table 12.2. It is obvious that

Table 12.2 Switching actions of four switches when using PAM in one qZS-HBI module (*Source*: Liu 2014 [29]. Reproduced with permission of IEEE)

Actions Switches	No switching	ST	ST & PWM
S_{a11}	$\pi/6 \sim 5\pi/6$	$7\pi/6 \sim 11\pi/6$	$-\pi/6 \sim \pi/6;\ 5\pi/6 \sim 7\pi/6$
S_{a12}	$7\pi/6 \sim 11\pi/6$	$\pi/6 \sim 5\pi/6$	$-\pi/6 \sim \pi/6;\ 5\pi/6 \sim 7\pi/6$
S_{a13}	$4\pi/3 \sim 5\pi/3$	$\pi/3 \sim 2\pi/3$	$-\pi/3 \sim \pi/3;\ 2\pi/3 \sim 4\pi/3$
S_{a14}	$\pi/3 \sim 2\pi/3$	$4\pi/3 \sim 5\pi/3$	$-\pi/3 \sim \pi/3;\ 2\pi/3 \sim 4\pi/3$

each module's switching actions greatly decrease in the PAM when compared to the SPWM. For a particular case when the shoot-through duty ratio is zero, that is, no shoot-through actions, each left-bridge leg's switch only performs the switching actions in the interval of $2\pi/3$ for the whole period of 2π, which achieves 2/3 switching reduction; for each right-bridge leg's switch, the switching action is fulfilled only in the $4\pi/3$ interval for the whole period of 2π, with 1/3 switching reduction. It will contribute to power loss reduction. Therefore, significant loss reduction will be achieved if the PAM is applied to the conventional CMI.

12.2.3.3 MM-SVM

Figure 12.13 shows the MM-SVM of ZS/qZS-CMI [30]. For the three-phase ZS/qZS-CMI, each layer has six half-H-bridge legs, where three left half-H-bridge legs from phases *a*, *b*, and *c* can be considered as a three-phase inverter; three right half-H-bridge legs from phases *a*, *b*, and *c* are considered as another three-phase inverter. Therefore each layer includes two three-phase inverters, and *n* layers with $2n$ three-phase inverters. Conventional SVM is applied to each of these three-phase inverters and as a result a new multilevel SVM for three-phase qZS-CMI. There are two key points: (1) the modulation signals of two three-phase inverters in the same layer have 180° phase shift in order to ensure that each module has three-level output voltages; (2) there is a $2\pi/(nK)$ phase shift between the reference vectors of the two adjacent layers, in which K is the number of reference voltages per control cycle, to produce stepwise multilevel voltages.

The first layer's three modules are used as an example to demonstrate the above idea. With SVM, two three-phase inverters have two groups of switching times: t_{a1L}, t_{b1L}, and t_{c1L} for three left half-H-bridge legs (one three-phase inverter), t_{a1R}, t_{b1R}, and t_{c1R} for three right half-H-bridge legs (another three-phase inverter). These switching times are distributed to three modules in the first layer: t_{a1L} and t_{a1R} for the module A_1 of phase *a*, t_{b1L} and t_{b1R} for the module B_1 of phase *b*, t_{c1L} and t_{c1R} for the module C_1 of phase *c*.

Figure 12.13 shows the switching pattern of the module A_1, where S_{a1L1}, S_{a1L2}, S_{a1R1}, and S_{a1R2} are the switching control signals of four switches in the module A_1 shown in Figure 12.9, and $t_{min} = \min(t_{a1L}, t_{a1R})$, $t_{max} = \max(t_{a1L}, t_{a1R})$. When the shoot-through time interval (T_{sh}) per control cycle is equally divided into four parts, we define $t_{min-} = (t_{min} - T_{sh}/4)$ and $t_{max+} = (t_{max} + T_{sh}/4)$, and meantime the active vector has the same time interval T_v. For the modules B_1 and C_1, there are the same method, but $t_{min} = \min(t_{b1L}, t_{b1R})$ and $t_{max} = \max(t_{b1L}, t_{b1R})$ for the module B_1, $t_{min} = \min(t_{c1L}, t_{c1R})$ and $t_{max} = \max(t_{c1L}, t_{c1R})$ for the module C_1.

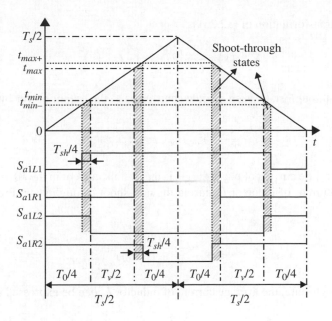

Figure 12.13 Switching pattern of module A_1 in three-phase qZS-CMI (*Source*: Liu 2013 [30]. Reproduced with permission of IEEE).

Compared to the PS-SPWM and PS-PWAM, the MM-SVM does not need the extra two references to produce the shoot-through duty cycle, so its implementation is easy to reduce the hardware resource burden.

12.2.4   System-Level Modeling and Control

The system-level modeling and control of a qZS-CMI is demonstrated for a three-phase PV power system, which is based on the authors' publication of [28]. Control objectives are to achieve: (1) the distributed MPPT for all PV panels, (2) the balanced dc-link peak voltages for all modules, (3) unity power factor in grid-injected power.

In the following derivation, the suffix x denotes the phase a, b, and c in Figure 12.9(b); and $k = 1, 2, \ldots, n$ denotes the kth module in phase x.

12.2.4.1   Modeling of Each Module

As shown in Figure 12.10, the PV voltage of the kth module in phase x meets

$$i_{L1xk} = i_{PVxk} - C_p \frac{dv_{PVxk}}{dt} \tag{12.26}$$

where i_{PVxk} and i_{L1xk} are the currents of PV panel and inductor L_1 of the kth module in phase x.

The Laplace transformation of (12.26) is

$$V_{PVxk}(s) = \frac{1}{C_p s}\left[I_{PVxk}(s) - I_{L1xk}(s)\right] \tag{12.27}$$

At non-shoot-through state, each module contributes to the output power by

$$\frac{\hat{i}_{gx}\hat{v}_{Hxk}}{2} = \hat{v}_{DCxk}\overline{i}_{DCxk} = v_{PVxk}\overline{i}_{L1xk\_nsh} \tag{12.28}$$

where i_{gx} is the grid-tie current of phase x; i_{DCxk} is the dc-link current in the kth module of phase x; i_{L1xk_nsh} is the current of inductor L_1 during the non-shoot-through state. From (12.23) and (12.28), there is

$$\overline{i}_{L1xk\_nsh} = \frac{\hat{i}_{gx}\hat{v}_{Hxk}}{2\hat{v}_{DCxk}\left(1 - 2D_{xk}\right)} \tag{12.29}$$

At shoot-through state, the average current of inductor L_1 can be expressed as

$$\overline{i}_{L1xk\_sh} = \overline{i}_{Pvxk} = i_{PVxk} \tag{12.30}$$

From (12.29) and (12.30), the inductor L_1 has average current

$$\overline{i}_{L1xk} = D_{xk}\overline{i}_{L1xk\_sh} + \left(1 - D_{xk}\right)\overline{i}_{L1xk\_nsh} = D_{xk}i_{PVxk} + \frac{\hat{i}_{gx}\left(1 - D_{xk}\right)\hat{v}_{Hxk}}{2\hat{v}_{DCxk}\left(1 - 2D_{xk}\right)} \tag{12.31}$$

The transfer function of each H-bridge module is

$$G_{invxk}(s) = \frac{V_{Hxk}(s)}{V_{mxk}(s)} = \hat{v}_{DCxk} \tag{12.32}$$

where V_{mxk} is the kth module's modulation signal of phase x.

Figure 12.14 shows the module model of (12.27), (12.31), and (12.32).

12.2.4.2   Modeling of qZS-CMI Based PV System

Figure 12.9(b) is redrawn here with grid connection, shown in Figure 12.15.

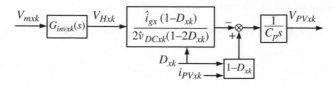

Figure 12.14 Block diagram of the kth module model in phase x (*Source*: Liu 2014 [28]. Reproduced with permission of IEEE).

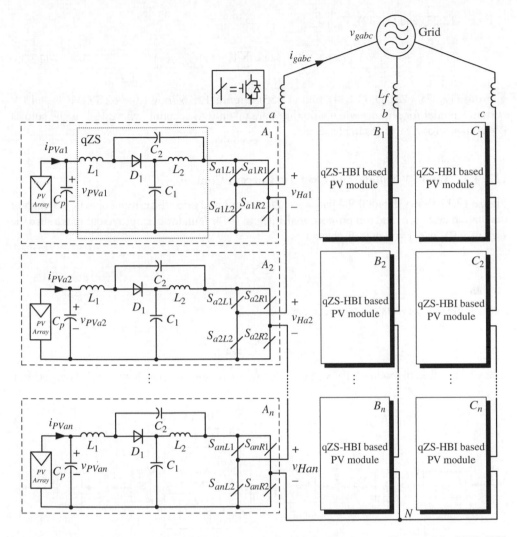

Figure 12.15 Topology of *n*-layer qZS-CMI based grid-tie PV system (*Source*: Liu 2014 [28]. Reproduced with permission of IEEE).

From Figure 12.15, the qZS-CMI based grid-tie PV system has the dynamic equation

$$v_{Hx} = v_{gx} + L_f \frac{di_{gx}}{dt} + r_f i_{gx} \tag{12.33}$$

where L_f and r_f are the inductance and parasitic resistance of the filter and v_{gx} is the grid voltage of phase *x*. Then, the transfer function is

$$G_{fx}(s) = \frac{I_{gx}(s)}{V_{Hx}(s) - V_{gx}(s)} = \frac{1}{L_f s + r_f} \tag{12.34}$$

From (12.23) and (12.32), V_{Hx} is

$$V_{Hx}(s) = \sum_{k=1}^{n} V_{Hxk}(s) = \sum_{k=1}^{n} V_{mxk}(s) G_{invxk}(s) \tag{12.35}$$

From (12.27), (12.31), (12.34), and (12.35), Figure 12.16 shows the qZS-CMI based PV system's model for phase x when n modules are taken into account, where V_{PVtx} is the sum of PV voltages for n modules in phase x.

12.2.4.3   Total PV Voltage and Grid-Tie Current Control

Figure 12.17 shows the total PV panel voltage control and grid-tie current control. For each of the three phases a, b, and c, a proportional-integral (PI) regulator is employed to make sure the sum of n PV panel voltages of phase x

$$v_{PVtx} = \sum_{k=1}^{n} v_{PVxk}$$

to track the sum of n PV voltage references

$$v_{PVtx}^{*} = \sum_{k=1}^{n} v_{PVxk}^{*}$$

where v_{PVxk}^{*} is the kth module's PV voltage reference in phase x when using the MPPT algorithm.

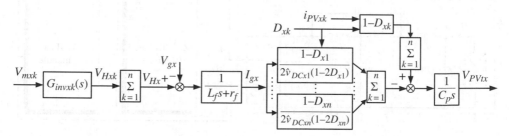

Figure 12.16   Block diagram of qZS-CMI based PV system's model for phase x (*Source*: Liu 2014 [28]. Reproduced with permission of IEEE).

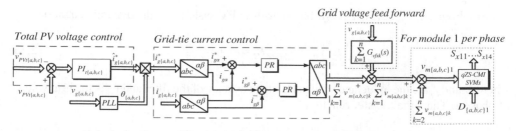

Figure 12.17   The total PV voltage and grid-tie current control of three-phase qZS-CMI based PV system (*Source*: Liu 2014 [28]. Reproduced with permission of IEEE).

As shown in Figure 12.17, the grid-tie current peak value references are the outputs of three-phase total PV voltage controllers. The phase-lock loop (PLL) senses the phase angle of grid voltage to ensure that the grid current references are in phase with the grid voltages. The grid-tie current closed-loop control is fulfilled in the two-phase stationary α-β frame through using three-phase/two-phase transformation. The practical grid-tie current will track the sinusoidal references with zero-error when using the proportional and resonant (PR) regulator

$$G_{PRj}(s) = k_{Pj} + \frac{k_{Rj}\omega_0}{s^2 + \omega_0^2} \tag{12.36}$$

where $j = \alpha, \beta$. The resonant term introduces an infinite gain at grid frequency ω_0 by k_{Rj}, whereas the system dynamics in terms of bandwidth, phase, and gain margin are determined by the proportional term k_{Pj}, which can be designed in the same way as for a PI controller. Therefore, considering the pulsation of grid frequency, a compromise of k_{Pj} should be taken into account between frequency, bandwidth, and stability.

A grid voltage feed-forward control for each module will reduce the PI regulator's burden, achieve fast dynamic response, and minimize the grid voltage's impact on grid-tie currents. Then, the kth module in phase x has the voltage modulation signal

$$V_{mxk} = V'_{mxk} + V_{gx}(s)G_{vfxk}(s) \tag{12.37}$$

with

$$G_{vfxk}(s) = \frac{1}{nG_{invxk}(s)} \tag{12.38}$$

where V'_{mxk} is the regulated modulation signal from the kth module's PV voltage control in phase x.

From (12.35), (12.37), and (12.38), the phase x of qZS-CMI has the voltage V_{Hx} as

$$V_{Hx}(s) = \sum_{k=1}^{n} V'_{mxk}G_{invxk}(s) + V_{gx}(s) \tag{12.39}$$

According to (12.34) and (12.39), the grid current in phase x is

$$I_{gx}(s) = \frac{1}{L_f s + r_f} \sum_{k=1}^{n} V'_{mxk}G_{invxk}(s) \tag{12.40}$$

which shows that feed-forward control eliminates the effect of grid voltage on the grid-tie current.

As shown in Figure 12.17, the output of grid-tie current loop per phase is the sum of modulation signals from all cascaded qZS H-bridge modules in that phase. The first module's modulation signal is obtained by total modulation signals and other $n - 1$ module modulation signals (v_{mx2} to v_{mxn}):

$$V_{mx1} = \sum_{k=1}^{n} V'_{mxk} + V_{gx}(s)\sum_{k=1}^{n} G_{vfxk}(s) - \sum_{k=2}^{n} V_{mxk} \tag{12.41}$$

12.2.4.4 PV Voltage Controls of Other Modules

Figure 12.18 shows the PV panel voltage control for modules 2 to n in phase x.

As shown in Figure 12.18, each PV panel voltage is controlled by an independent PI regulator to track the reference v^*_{PVxk}. The MPPT algorithm outputs the desired v^*_{PVxk}.

From the module model of Figure 12.15, we have

$$G_{voxk}(s) = \frac{V_{PVxk}(s)}{V_{mxk}(s)} = -\frac{\hat{i}_{gx}(1-D_{xk})}{2\hat{v}_{DCxk}(1-2D_{xk})C_p s} \quad G_{invxk}(s) = -\frac{\hat{i}_{gx}(1-D_{xk})}{2C_p s(1-2D_{xk})} \quad (12.42)$$

For the PI regulator of the kth module in phase x

$$G_{PIxk}(s) = k_{Pxk} + \frac{k_{Ixk}}{s}$$

the open-loop transfer function of PV voltage control is

$$G_{vcomxk}(s) = G_{voxk}(s)G_{PIxk}(s) = -\frac{\hat{i}_{gx}(1-D_{xk})(k_{Pxk}s + k_{Ixk})}{2C_p s^2(1-2Dx_k)} \quad (12.43)$$

and the closed-loop transfer function will be

$$G_{vcxk}(s) = \frac{V_{PVxk}(s)}{V^*_{PVxk}(s)} = \frac{-G_{vcomxk}(s)}{1 - G_{vcomxk}(s)} \quad (12.44)$$

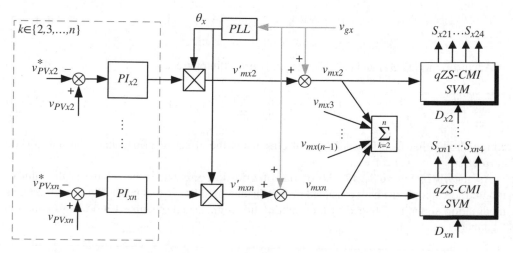

Figure 12.18 PV panel voltage controls for modules 2 to n in phase x (*Source*: Liu 2014 [28]. Reproduced with permission of IEEE).

12.2.4.5 Balanced DC-Link Voltage Control

Figure 12.19 shows the dc-link peak voltage control for the kth module in phase x, where the inductor current closed loop with proportional (P) regulator is employed to improve the dynamic response; the PI regulator of dc-link peak voltage closed loop will ensure zero-error tracking. The actual dc-link peak voltage is calculated by capacitor C_1 voltage and shoot-through duty cycle. The dc-link peak voltages of all modules in three phases are set to the same desired value, the balanced dc-link voltage is achieved.

The obtained modulation signals v_{mxk} and the desired shoot-through duty ratios D_{xk} ($x = a$, b, c; $k = 1, 2, \ldots, n$) are used to operate the qZS-CMI switches by the modulation scheme in Subsection 12.2.3 to achieve the aforementioned control objectives.

12.2.5 Simulation Results

Simulation results are presented here for the qZS-CMI-based grid-tie PV system in Figure 12.15, with the addressed system-level control method in Subsection 12.2.4. Table 12.3 lists the system specifications.

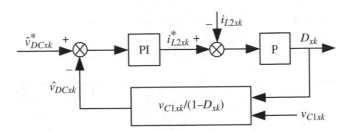

Figure 12.19 Independent control of dc-link peak voltage of the kth module in phase x (*Source*: Liu 2014 [28]. Reproduced with permission of IEEE).

Table 12.3 System specifications (*Source*: Liu 2014 [28]. Reproduced with permission of IEEE)

Parameters	Values
Rated power of each module, P_m	600 W
PV source's voltage range, v_{PV}	60~120 V
Desired dc-link peak voltage of each module	136 V
qZS inductance, L_1 and L_2	1.8 mH
qZS capacitance, C_1 and C_2	3300 μF
PV panel terminal capacitance, C_p	1100 μF
Grid-side filter inductance, L_f	1 mH
Line-to-line voltage of three-phase grid	400 V
Grid frequency	50 Hz

12.2.5.1 Grid-Tie Control and Voltage Balance Control

In this case, the three-phase qZS-CMI based PV system is connected to the grid. The different PV voltages between modules are used to investigate the qZS-CMI's voltage balance control capability. At the beginning, all PV panels work at 38 °C and 1000 W/m²; at 1 s, PV panel temperature of module A_3 changes to 25 °C in simulation (the assumption of temperature change is just for test purposes, the practical temperature cannot change immediately), which leads to the PV panel's maximum power point change from (105 V, 525 W) to (120 V, 600 W) according to the P-V characteristics.

As shown in Figure 12.20(a), (b), different temperatures result in different PV voltages at the maximum power points for modules A_1 and A_3; lower temperature causes the higher PV

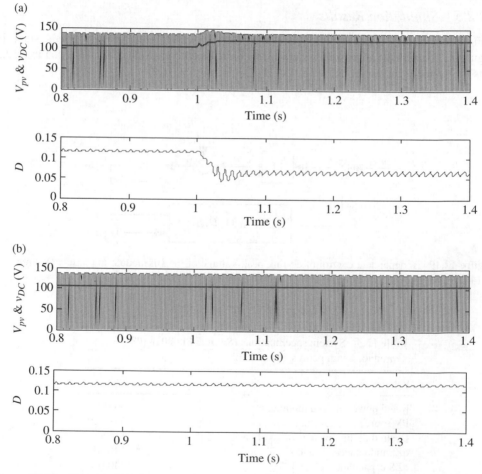

Figure 12.20 Simulated results when PV panel of module A_3 has temperature change. (a) PV voltage, dc-link voltage, and shoot-through duty cycle of the module A_3, (b) PV voltage, dc-link voltage, and shoot-through duty cycle of module A_1, (c) modulation signals of three modules in phase a, (d) grid-tie current in phase with grid voltage, (e) three-phase grid-tie currents and qZS-CMI's seven-level three-phase voltages (*Source*: Liu 2014 [28]. Reproduced with permission of IEEE).

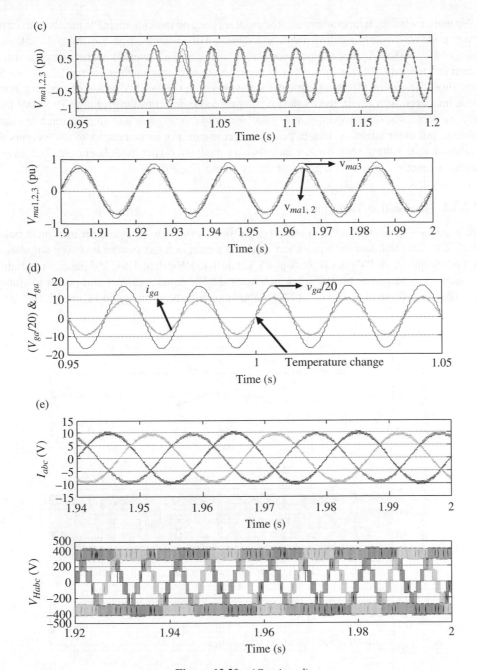

Figure 12.20 (*Continued*)

voltage (120 V) of module A₃, original voltage is 105 V. To ensure the constant dc-link peak voltage (136 V), the shoot-through duty cycle of module A₃ is adjusted to a smaller value (*D*=0.058) from the original *D*=0.114. The balanced voltage is shown in seven-level output voltage with equal stair step voltage, as shown in Figure 12.20(e).

No matter what the temperatures are, the modules keep on tracking their PV panels' maximum power points by modifying the modulation signals. For example, with the same dc-link peak voltage, the module A_3 has a larger modulation signal at the lower PV panel temperature, as shown in Figure 12.20(c), which leads to higher module output voltage for module A_3. At the same time, the grid-tie phase current will increase because the total PV power of modules in this phase increases. As a result, the module A_3 outputs more power to the grid after PV panel temperature decreases; other modules have smaller modulation signals and smaller module output voltages, but other modules' output powers are constant. For instance, module A_1 outputs the smaller module voltage with smaller modulation signal and higher module current. The grid-tie current is exactly in phase with grid voltage, as Figure 12.20(d) shows.

12.2.5.2 Irradiation Variations

The irradiation variation of a PV panel will cause the PV current to change at the maximum power point. This case will test the irradiation variation's effect on the system by using simulation. At the beginning, all PV panels work at 25 °C and 1000 W/m²; at 1.5 s, PV panel's irradiation of module A_3 changes to 800 W/m², which leads the PV panel's current and power to change from (5 A, 600 W) to (4 A, 480 W) at MPP. Simulation results are shown in Figure 12.21.

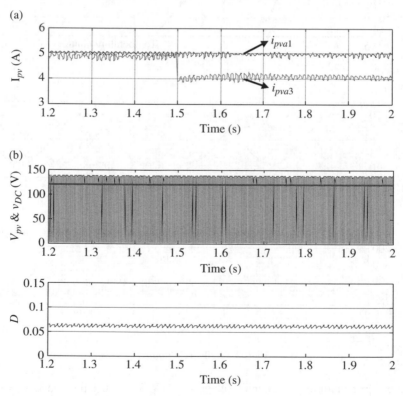

Figure 12.21 Simulated results when PV panel's irradiation of module A_3 decreases: (a) PV currents of modules A_1 and A_3, (b) PV voltage, dc-link voltage, and shoot-through duty cycle of module A_3, (c) modulation signals of three modules in phase a, (d) grid-tie current in phase with grid voltage, and (e) qZS-CMI's seven-level three-phase voltages and currents (*Source*: Liu 2014 [28]. Reproduced with permission of IEEE).

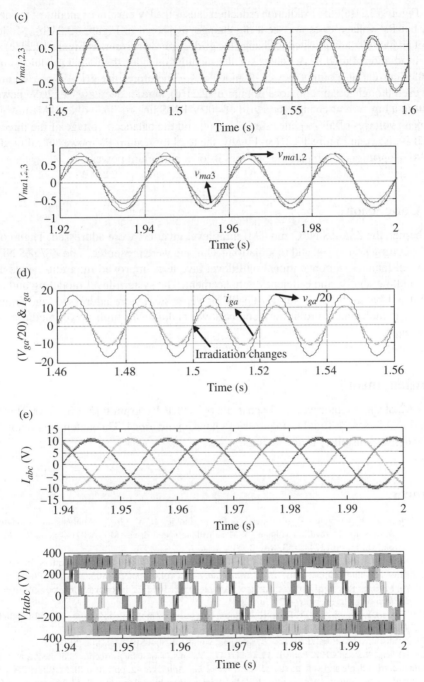

Figure 12.21 (*Continued*)

From Figure 12.21(a), the irradiation reduction causes the PV current of module A_3 to decrease at MPP, but PV voltage does not see a big change, as shown in Figure 12.21(b). Modulation signals of all modules are adjusted to change the grid-tie phase current, as Figure 12.21(c) shows. A lower irradiation of module A_3 causes smaller modulation index than other modules, implying the smaller current and output power of module A_3. Therefore, the grid-tie phase current is reduced even though it is not obvious in Figure 12.21(d), since a decrease of 120 W power will not result in a big current change for a grid of 400 V RMS line-to-line voltage. Meanwhile, the dc-link peak voltages of all modules stay constant, and the balanced voltage of the three-phase qZS-CMI is shown in Figure 12.21(e). Finally, the total maximum PV power of all modules is injected to the grid, and each module provides its own maximum power.

12.3   Conclusion

In this chapter, the ZS/qZS NPC and CMI multilevel inverters were addressed. Their configurations, operating principles, and modulation technique were presented. The ZS/qZS NPC and CMI showed features of conventional multilevel inverters, improved reliability, wide dc-link voltage handling ability, and efficiency enhancement. The system-level modeling and control of a qZS-CMI-based three-phase grid-tie PV power system were addressed, and simulation results were demonstrated, providing a competitive candidate for high-power medium-voltage power systems.

Acknowledgment

Section 12.2 of this chapter was made possible by NPRP-EP grant # [X-033-2–007] from the Qatar National Research Fund (a member of Qatar Foundation). The statements made herein are solely the responsibility of the authors.

References

[1] D. A. Rendusara, E. Cengelci, P. N. Enjeti, V. R. Stefanovic, J. W. Gray, "Analysis of common mode voltage – 'neutral shift' in medium voltage PWM adjustable speed drive (MV-ASD) systems," *IEEE Trans. Power Electron.*, vol.15, pp.1124–1133, Nov. 2000.

[2] R. Teichmann, S. Bernet, "A comparison of three-level converters versus two-level converters for low-voltage drives, traction and utility applications," *IEEE Trans. Ind. Appl.*, vol.41, pp.855–865, May/Jun. 2005.

[3] B. Welchko, M. B. de Rossiter Correa, T. A. Lipo, "A three-level MOSFET inverter for low-power drives," *IEEE Trans. Ind. Electron.*, vol.51, pp.669–674, Jun. 2004.

[4] P. C. Loh, F. Gao, F. Blaabjerg, S. Y. Feng, K. N. Soon, "Pulse-width-modulated Z-source neutral-point-clamped inverter," *IEEE Trans. Ind. Appl.*, vol.43, pp.1295–1308, Sep./Oct. 2007.

[5] F. Z. Peng, "Z-source inverter," *IEEE Trans. Ind. Appl.*, vol.39, pp.504–510, Mar/Apr. 2003.

[6] M. Shen, J. Wang, A. Joseph, F. Z. Peng, L. M. Tolbert, D. J. Adams, "Constant boost control of the Z-source inverter to minimize current ripple and voltage stress," *IEEE Trans. Ind. Appl.*, vol.42, pp.770–778, May/Jun. 2006.

[7] B. P. McGrath, D. G. Holmes, "Multi-carrier PWM strategies for multilevel inverters," *IEEE Trans. Ind. Electron.*, vol.49, pp.858–867, Aug. 2002.

[8] B. P. McGrath, D. G. Holmes, T. A. Lipo, "Optimized space vector switching sequences for multilevel inverters," *IEEE Trans. Power Electron.*, vol.18, pp.1293–1301, Nov. 2003.

[9] D. G. Holmes, T. A. Lipo, *Pulse Width Modulation for Power Converters: Principles and Practice*, Wiley & Sons Ltd, 2003.

[10] P. C. Loh, D. G. Holmes, Y. Fukuta, T. A. Lipo, "Reduced common mode modulation strategies for cascaded multilevel inverters," *IEEE Trans. Ind. Appl.*, vol.39, pp.1386–1395, Sep/Oct. 2003.

[11] P. C. Loh, F. Blaabjerg, C. P. Wong, "Comparative evaluation of pulse-width modulation strategies for Z-source neutral-point-clamped inverter," *IEEE Trans. Power Electron.*, vol.22, no.3, pp.1005–1013, May 2007.

[12] H. Sepahvand, J. S. Liao, M. Ferdowsi, K. A. Corzine, "Capacitor voltage regulation in single-DC-source cascaded H-bridge multilevel converters using phase-shift modulation," *IEEE Trans. Ind. Electron.*, vol.60, no.9, pp.3619–3626, 2013.

[13] J. Napoles, A. J. Watson, J. J. Padilla, J. I. Leon, L. G. Franquelo, P. W. Wheeler, M. A. Aguirre, "Selective harmonic mitigation technique for cascaded H-bridge converters with non-equal DC link voltages," *IEEE Trans. Ind. Electron.*, vol.60, no.5, pp.1963–1971, 2013.

[14] S. Vazquez, J. I. Leon, J. M. Carrasco, L. G. Franquelo, E. Galvan, M. Reyes, J. A. Sanchez, E. Dominguez, "Analysis of the power balance in the cells of a multilevel cascaded H-bridge converter," *IEEE Trans. Ind. Electron.*, vol.57, no.7, pp.2287–2296, 2010.

[15] A. R. Beig, A. Dekka, "Experimental verification of multilevel inverter-based standalone power supply for low-voltage and low-power applications," *IET Power Electron.*, vol.5, no.6, pp.635–643, 2012.

[16] A. Dell'Aquila, M. Liserre, V. G. Monopoli, P. Rotondo, "Overview of PI-based solutions for the control of DC buses of a single-phase H-bridge multilevel active rectifier," *IEEE Trans. Ind. Appl.*, vol.44, no.3, pp.857–866, 2008.

[17] J. Chavarria, D. Biel, F. Guinjoan, C. Meza, J. J. Negroni, "Energy-balance control of PV cascaded multilevel grid-connected inverters under level-shifted and phase-shifted PWMs," *IEEE Trans. Ind. Electron.*, vol.60, no.1, pp.98–111, 2013.

[18] P. Cortes, A. Wilson, S. Kouro, J. Rodriguez, H. Abu-Rub, "Model predictive control of multilevel cascaded H-bridge inverters," *IEEE Trans. Ind. Electron.*, vol.57, no.8, pp.2691–2699, 2010.

[19] C. D. Townsend, T. J. Summers, J. Vodden, A. J. Watson, R. E. Betz, J. C. Clare, "Optimization of switching losses and capacitor voltage ripple using model predictive control of a cascaded H-bridge multilevel STATCOM," *IEEE Trans. Power Electron.*, vol.28, no.7, pp.3077–3087, 2013.

[20] S. Rivera, S. Kouro, B. Wu, J. I. Leon, J. Rodriguez, L. G. Franquelo, "Cascaded H-bridge multilevel converter multi-string topology for large scale photovoltaic systems," in *Proc. IEEE International Symposium on Industrial Electronics (ISIE)*, pp.1837–1844, 2011.

[21] S. Kouro, C. Fuentes, M. Perez, J. Rodriguez, "Single dc-link cascaded H-bridge multilevel multi-string photovoltaic energy conversion system with inherent balanced operation," in *Proc. 38th Annual Conference on IEEE Industrial Electronics Society (IECON)*, pp.4998–5005, 2012.

[22] Z. Wang, S. T. Fan, Y. Zheng, M. Cheng, "Design and analysis of a CHB converter based PV-battery hybrid system for better electromagnetic compatibility," *IEEE Trans. Magn.*, vol.48, no.11, pp.4530–4533, 2012.

[23] Y. Zhou, L. Liu, H. Li, "A high-performance photovoltaic module-integrated converter (MIC) based on cascaded quasi-Z-source inverters (qZSI) using EGANFETS," *IEEE Trans. Power Electron.*, vol.28, no.6, pp.2727–2738, 2013.

[24] Y. Xue, B. Ge, F. Z. Peng, "Reliability, Efficiency, and Cost comparisons of MW-scale photovoltaic inverters," in *Proc. IEEE Energy Conversion Congress and Exposition (ECCE)*, 2012, pp.1627–1634.

[25] D. Sun, B. Ge, F. Z. Peng, H. Abu-Rub, D. Bi, Y. Liu, "A new grid-connected PV system based on cascaded H-bridge quasi-Z source inverter," in *Proc. IEEE International Symposium on Industrial Electronics (ISIE)*, 2012, pp.951–956.

[26] D. Sun, B. Ge, X. Yan, D. Bi, H. Zhang, Y. Liu, H. Abu-Rub, L. Ben-Brahim, F. Z. Peng, "Modeling, impedance design, and efficiency analysis of quasi-Z source module in cascade multilevel photovoltaic power system," *IEEE Trans. Ind. Electron.*, vol.61, no.11, pp.6108–6117, 2014.

[27] Y. Liu, B. Ge, H. Abu-Rub, F. Z. Peng, "An effective control method for quasi-Z-source cascade multilevel inverter based grid-tie single-phase photovoltaic power system," *IEEE Trans. Ind. Informat.*, vol.10, no.1, pp.399–407, Feb. 2014.

[28] Y. Liu, B. Ge, H. Abu-Rub, F. Z. Peng, "An effective control method for quasi-Z-source cascade multilevel inverter based three-phase grid-tie photovoltaic power system," *IEEE Trans. Ind. Electron.*, vol.61, no.12, pp.6794–6802, Dec. 2014.

[29] Y. Liu, B. Ge, H. Abu-Rub, F. Z. Peng, "Phase-shifted pulse-width-amplitude modulation for quasi-Z-source cascade multilevel inverter based photovoltaic power system," *IET Power Electron.*, vol.7, no.6, pp.1444–1456, June. 2014.

[30] Y. Liu, B. Ge, H. Abu-Rub, F. Z. Peng, "A modular multilevel space vector modulation for photovoltaic quasi-Z-source cascade multilevel inverters," in *Proc. Twenty-Eighth Annual IEEE Applied Power Electronics Conference and Exposition (APEC)*, pp.714–718, 2013.

13

Design of Z-Source and Quasi-Z-Source Inverters

13.1   Z-Source Network Parameters

Inductance and capacitance are the two main parameters to be designed in the Z-source/quasi-Z-source inverter. This chapter focuses on the network parameters, loss calculations, voltage and current stress, coupled inductor of Z-source/quasi-Z source inverters and also efficiency, cost, and volume, in comparison to the conventional inverter.

For three-phase qZSI, the inductor and capacitor should be designed to limit the switching frequency current and voltage ripple. However, for single-phase qZSI, the double line-frequency ripple will be the main concern for design of the inductor and capacitor, because larger inductance and capacitance are required to limit the switching frequency ripple.

13.1.1   Inductance and Capacitance of Three-Phase qZSI

As shown in Figure 13.1, the inductor in the quasi-Z source network limits the current ripple during the boost mode. For the shoot-through duration, the inductor current increases linearly. The inductance can be calculated by

$$L_1 = L_2 = \frac{V_L \Delta T}{\Delta I} \tag{13.1}$$

where ΔT is the shoot-through time, ΔI is the current variation, and V_L is the voltage across the inductor. Here, it is assumed that the two inductors have the same inductance. When coupled inductors are used to minimize the size and weight, the flux is doubled for each inductor.

Two capacitors are in series in the quasi-Z source network at the non-shoot-through state. These two capacitors absorb the current ripple and limit the voltage ripple on the inverter

Impedance Source Power Electronic Converters, First Edition. Yushan Liu, Haitham Abu-Rub, Baoming Ge, Frede Blaabjerg, Omar Ellabban, and Poh Chiang Loh.
© 2016 John Wiley & Sons, Ltd. Published 2016 by John Wiley & Sons, Ltd.

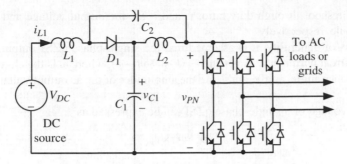

Figure 13.1 Three-phase quasi-Z source inverter.

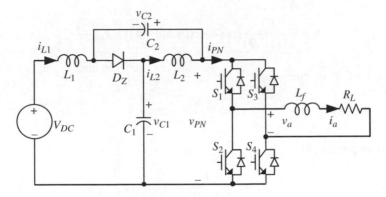

Figure 13.2 Single-phase quasi-Z source inverter.

bridge. For the same capacitance of two capacitors, the total dc-link capacitance will be maximum. The capacitance can be calculated as

$$C_1 = C_2 = \frac{2I_c \Delta T}{\Delta(V_{C1} + V_{C2})} \tag{13.2}$$

13.1.2 Inductance and Capacitance of Single-Phase qZSI

Figure 13.2 shows a single-phase quasi-Z source inverter. It has different concerns when designing the inductance and capacitance owing to the double line-frequency power flowing through the dc side.

There are also two operating states for the single-phase quasi-Z source inverter: non-shoot-through and shoot-through states. At non-shoot-through state, the power is transmitted from dc side to ac side, but at shoot-through state, there is no power transmission because the dc-link voltage is zero. The power balance to be met is [1]

$$v_{PN} \times i_{PN} \times (1-D) + 0 \times D = v_a \times i_a \tag{13.3}$$

where v_{PN} represents the dc-link voltage envelope which consists of a dc value plus a 2ω residual ripple, i_{PN} represents the current entering the H-bridge averaged on the switching period, D is the shoot-through duty ratio, v_a and i_a are the output voltage and current of the H-bridge module, respectively.

Without loss of generality, it is assumed that the single-phase qZSI outputs fundamental voltage and current with $v_a = V_a \cdot \sin\omega t$ and $i_a = I_a \cdot \sin(\omega t - \phi)$, where ω is the angular frequency, ϕ is the impedance angle, and V_a and I_a are the amplitudes of the ac output voltage and current, respectively.

The output voltage of a single-phase qZSI can be expressed as

$$v_a = m \times v_{PN} \tag{13.4}$$

where $m = M \cdot \sin\omega t$, M is the modulation index.

From (13.3) and (13.4), i_{PN} is deduced as

$$i_{PN} = \frac{MI_a}{2(1-D)}\left(\cos\phi - \cos(2\omega t - \phi)\right) = I_{PN} + \tilde{i}_{PN} \tag{13.5}$$

which consists of the dc component

$$I_{PN} = \frac{MI_a}{2(1-D)}\cos\phi$$

and the 2ω component [1]

$$\tilde{i}_{PN} = -\frac{MI_a}{2(1-D)}\cos(2\omega t - \phi)$$

Both I_{PN} and $\tilde{i}_{PN}$ affect the state variables i_{L1}, i_{L2}, v_{C1}, and v_{C2}. As a result, they also consist of dc and 2ω components. In particular, we have

$$i_{L1} = I_{L1} + \tilde{i}_{L1}, i_{L2} = I_{L2} + \tilde{i}_{L2}, \quad v_{C1} = V_{C1} + \tilde{v}_{C1}, v_{C2} = V_{C2} + \tilde{v}_{C2}.$$

13.1.2.1 Equivalent Model of DC Component

The dc component has steady-state model as

$$\begin{cases} V_{C1} = \dfrac{1-D}{1-2D}V_{DC} \\[2mm] V_{C2} = \dfrac{D}{1-2D}V_{DC} \\[2mm] I_{L1} = \dfrac{MI_a}{2(1-2D)}\cos\phi \\[2mm] I_{L2} = \dfrac{MI_a}{2(1-2D)}\cos\phi \end{cases} \tag{13.6}$$

13.1.2.2   Equivalent Model of 2ω Component

In steady state, it can be assumed that the dc source voltage V_{DC} is constant. For the non-shoot-through state, the dynamic equations are [1]

$$
\begin{cases}
L_1 \dfrac{d\tilde{i}_{L1}}{dt} = -\tilde{v}_{C1} \\[2mm]
L_2 \dfrac{d\tilde{i}_{L2}}{dt} = -\tilde{v}_{C2} \\[2mm]
C_1 \dfrac{d\tilde{v}_{C1}}{dt} = \tilde{i}_{L1} - \tilde{i}_{PN} \\[2mm]
C_2 \dfrac{d\tilde{v}_{C2}}{dt} = \tilde{i}_{L2} - \tilde{i}_{PN}
\end{cases}
\tag{13.7}
$$

For shoot-through state, they are

$$
\begin{cases}
L_1 \dfrac{d\tilde{i}_{L1}}{dt} = \tilde{v}_{C2} \\[2mm]
L_2 \dfrac{d\tilde{i}_{L2}}{dt} = \tilde{v}_{C1} \\[2mm]
C_1 \dfrac{d\tilde{v}_{C1}}{dt} = -\tilde{i}_{L2} \\[2mm]
C_2 \dfrac{d\tilde{v}_{C2}}{dt} = -\tilde{i}_{L1}
\end{cases}
\tag{13.8}
$$

For the 2ω components, the average inductor currents and average capacitor voltages over one switch cycle are not zero. With $L_1 = L_2 = L$, $C_1 = C_2 = C$, from (13.7) and (13.8):

$$
\begin{cases}
L \dfrac{d<\tilde{i}_{L1}>_T}{dt} = (1-D)\cdot(-\tilde{v}_{C1}) + D\cdot(\tilde{v}_{C2}) \\[2mm]
L \dfrac{d<\tilde{i}_{L2}>_T}{dt} = (1-D)\cdot(-\tilde{v}_{C2}) + D\cdot(\tilde{v}_{C1}) \\[2mm]
C \dfrac{d<\tilde{v}_{C1}>_T}{dt} = (1-D)\cdot(\tilde{i}_{L1} - \tilde{i}_{PN}) + D\cdot(-\tilde{i}_{L2}) \\[2mm]
C \dfrac{d<\tilde{v}_{C2}>_T}{dt} = (1-D)\cdot(\tilde{i}_{L2} - \tilde{i}_{PN}) + D\cdot(-\tilde{i}_{L1})
\end{cases}
\tag{13.9}
$$

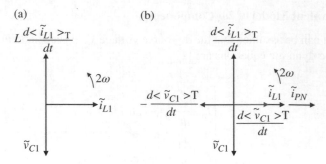

Figure 13.3 Phasor diagram of 2ω voltage and current components: (a) capacitor voltage and inductor current, (b) capacitor voltage and dc-link current (*Source*: Sun 2014 [1]. Reproduced with permission of IEEE).

From (13.9), $\tilde{i}_{L1}$ and $\tilde{i}_{L2}$ can be replaced by each other, and $\tilde{v}_{C1}$ and $\tilde{v}_{C2}$ can be replaced by each other. Therefore, $\tilde{i}_{L1} = \tilde{i}_{L2}$ and $\tilde{v}_{C1} = \tilde{v}_{C2}$, and the fourth-order model in (13.9) is simplified to

$$\begin{cases} L\dfrac{d<\tilde{i}_{L1}>_{\mathrm{T}}}{dt} = (2D-1)\tilde{v}_{C1} \\[3mm] C\dfrac{d<\tilde{v}_{C1}>_{\mathrm{T}}}{dt} = (1-2D)\tilde{i}_{L1} - (1-D)\tilde{i}_{PN} \end{cases} \tag{13.10}$$

From the first equation of (13.10), Figure 13.3(a) can be obtained in the 2ω frame, where $D < 0.5$. From the second equation of (13.10), we find

$$\tilde{i}_{PN} = \frac{1}{1-D}\left[(1-2D)\tilde{i}_{L1} - C\frac{d<\tilde{v}_{C1}>_{\mathrm{T}}}{dt}\right] \tag{13.11}$$

From (13.11), Figure 13.3(b) can be obtained, where $\tilde{i}_{L1}$ has the same phase as $\tilde{i}_{PN}$, and it can be defined as

$$\tilde{i}_{L1} = \hat{i}_{L1}\cos(2\omega t - \phi), \tilde{i}_{L2} = \hat{i}_{L2}\cos(2\omega t - \phi), \tilde{v}_{C1} = \hat{v}_{C1}\sin(2\omega t - \phi), \tilde{v}_{C2} = \hat{v}_{C2}\sin(2\omega t - \phi).$$

Using these equations and (13.9) gives

$$\begin{cases} \tilde{i}_{L1} = \tilde{i}_{L2} = \dfrac{(1-2D)}{4LC\omega^2 - (1-2D)^2}\cdot\dfrac{MI_a}{2}\cdot\cos(2\omega t - \phi) \\[4mm] \tilde{v}_{C1} = \tilde{v}_{C2} = \dfrac{2\omega L}{4LC\omega^2 - (1-2D)^2}\cdot\dfrac{MI_a}{2}\cdot\sin(2\omega t - \phi) \end{cases} \tag{13.12}$$

13.1.2.3   Complete Model of qZSI Module

Adding dc components in (13.6) and 2ω components in (13.12) together, in steady state, the complete model is

$$
\begin{cases}
i_{L1} = I_{L1} + \tilde{i}_{L1} = \dfrac{1}{(1-2D)} \cdot \dfrac{MI_a}{2} \cdot \cos\phi + \dfrac{(1-2D)}{4LC\omega^2 - (1-2D)^2} \cdot \dfrac{MI_a}{2} \cdot \cos(2\omega t - \phi) \\[2ex]
i_{L2} = I_{L2} + \tilde{i}_{L2} = \dfrac{1}{(1-2D)} \cdot \dfrac{MI_a}{2} \cdot \cos\phi + \dfrac{(1-2D)}{4LC\omega^2 - (1-2D)^2} \cdot \dfrac{MI_a}{2} \cdot \cos(2\omega t - \phi) \\[2ex]
v_{C1} = V_{C1} + \tilde{v}_{C1} = \dfrac{1-D}{1-2D} \cdot V_{DC} + \dfrac{2\omega L}{4LC\omega^2 - (1-2D)^2} \cdot \dfrac{MI_a}{2} \cdot \sin(2\omega t - \phi) \\[2ex]
v_{C2} = V_{C2} + \tilde{v}_{C2} = \dfrac{D}{1-2D} \cdot V_{DC} + \dfrac{2\omega L}{4LC\omega^2 - (1-2D)^2} \cdot \dfrac{MI_a}{2} \cdot \sin(2\omega t - \phi)
\end{cases}
\tag{13.13}
$$

The impedance design of the qZS network aims to limit the 2ω components of the dc-link voltage and inductor currents within the engineering tolerance ranges.

From (13.13), the dc-link voltage envelope is deduced as [1]

$$
v_{PN} = \frac{V_{DC}}{1-2D} + \frac{2\omega LMI_a}{4LC\omega^2 - (1-2D)^2} \sin(2\omega t - \phi)
\tag{13.14}
$$

where the dc component is

$$
V_{PN} = V_{DC} / (1-2D),
\tag{13.15}
$$

and the 2ω voltage component's amplitude is

$$
\hat{v}_{PN} = 2\hat{v}_{C1} = 2\hat{v}_{C2} = \frac{2\omega LMI_a}{4LC\omega^2 - (1-2D)^2}.
\tag{13.16}
$$

The dc-link voltage envelope's peak-to-peak ripple ratio is defined by

$$
a = \frac{2\hat{v}_{PN}}{V_{PN}},
\tag{13.17}
$$

then it will be

$$
a = \frac{4\omega LMI_a (1-2D)}{\left[4LC\omega^2 - (1-2D)^2\right]V_{DC}}
\tag{13.18}
$$

For the 2ω current component in (13.12), the amplitude is

$$
\hat{i}_{L1} = \frac{(1-2D)MI_a}{2\left[4LC\omega^2 - (1-2D)^2\right]}
$$

when defining the inductor 2ω current ripple ratio as

$$b = \frac{\hat{i}_{L1}}{I_{L1}}, \tag{13.19}$$

the defined 2ω current ripple ratio is

$$b = \frac{(1-2D)^2}{\left[4LC\omega^2 - (1-2D)^2\right]\cos\phi} \tag{13.20}$$

Figure 13.4(a), (b) show the ripple ratios a and b versus the inductance and capacitance of the qZS impedance network, respectively, where $D = 0.25$, $M = 0.7$, $\omega = 314$ rad/s, $\cos\phi = 0.9986$, $I_a = 5$ A, and $V_{DC} = 25$ V. As shown in Figure 13.4(a), the dc-link voltage envelope

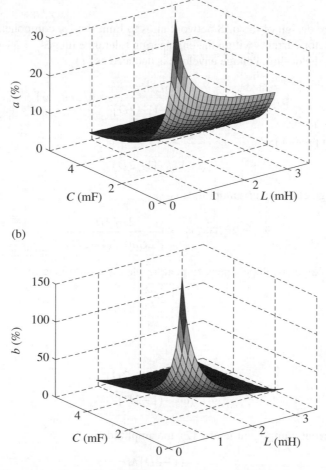

Figure 13.4 The 2ω ripple ratio of (a) dc-link voltage envelope, and (b) inductor current (*Source*: Sun 2014 [1]. Reproduced with permission of IEEE).

ripple significantly decreases with the capacitance increasing, but slowly decreases with the inductance increasing. However, the inductor current ripple significantly decreases when the inductance and capacitance increases, as shown in Figure 13.4(b).

From (13.18), (13.20), and Figure 13.4, the minimum capacitance and inductance should meet

$$C \geq \frac{(1-2D)(1+b^* \cos\phi) MI_a}{a^* \omega V_{DC}},$$

$$L \geq \frac{a^* V_{DC}(1-2D)}{4\omega b^* MI_a \cos\phi} \tag{13.21}$$

to limit the voltage and current ripples within the required ratios b^* and a^*, respectively.

13.2 Loss Calculation Method

The single-phase qZSI of Figure 13.2 is used as an example to demonstrate the loss calculations for the ZSI/qZSI. The DC source is a PV panel. Generally, the loss includes one caused by the traditional states and one caused by the shoot-through states. The total loss consists of those of the H-bridge device, qZS diode, inductor, and capacitor of qZS network.

The switch states of qZSI can be presented in the positive half fundamental cycle, as shown in Figure 13.5, because there are the same states in the negative half cycle.

When comparing the traditional H-bridge module operating states, shoot-through states 1 and 2 are added in the single-phase qZSI. As a result, there are five operating states in one switching cycle T_s: traditional zero states 1 and 2, shoot-through states 1 and 2, and active state in Figure 13.5, and the equivalent circuit of each operating state is shown in Figure 13.6.

As shown in Figure 13.6, the devices of qZSI module have different on-state currents during one switching cycle, which affects the device power losses. The conduction currents and time intervals for all devices are summarized in Table 13.1.

13.2.1 H-bridge Device Power Loss

1. Conduction loss – The conduction loss includes the switch loss in forward conduction and the conduction loss of the free-wheeling diode. In traditional states we have

$$P_{con\_S\_tr} = \frac{2}{\pi} \int_0^{\pi} \left[R_{DS(on)} \cdot i_a^2 \cdot \frac{1+m}{2} \right] d\omega t + \frac{2}{\pi} \int_0^{\pi} \left[u_{D0} \cdot i_a + R_D \cdot i_a^2 \right] \cdot \left(1 - \frac{1+m}{2} \right) d\omega t \tag{13.22}$$

where $R_{DS(on)}$ is the drain-source on-state resistance of MOSFET, and u_{D0} and R_D are the on-state zero-current voltage and the on-state resistance of the anti-parallel diode, respectively. They are available in the datasheets.

The conduction loss in shoot-through states is [1]

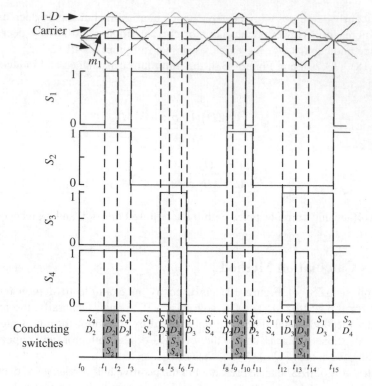

Figure 13.5 Switching states of single-phase qZSI (*Source*: Sun 2014 [1]. Reproduced with permission of IEEE).

$$P_{con\_S\_th} = 4 \cdot \left[R_{DS(on)} \cdot (2i_{L1})^2 \cdot \frac{D}{2} \right] \tag{13.23}$$

2. Switching loss – The switching loss of the MOSFET includes the turn on/off loss, and the reverse recovery loss of the anti-parallel diode.

The switching loss in the traditional states is

$$P_{sw\_S\_tr} = \frac{2}{\pi} \int_0^\pi \left[V_{PN} \cdot i_a \cdot \frac{tri + tfu}{2} \cdot f_s \right] d\omega t$$
$$+ 2 \cdot Q_{rr} \cdot V_{PN} \cdot f_s + \frac{2}{\pi} \int_0^\pi \left[V_{PN} \cdot i_a \cdot \frac{tru + tfi}{2} \cdot f_s \right] d\omega t \tag{13.24}$$

where *tri* is typically the current rise time, *tfu* is the voltage fall time, *tru* is the voltage rise time, *tfi* is the current fall time, Q_{rr} is the reverse recovery charge. They are available in the datasheets.

The switching loss in shoot-through states is

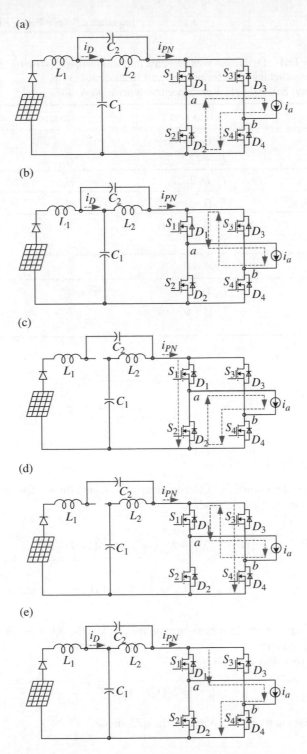

Figure 13.6 Equivalent circuits in one switching cycle: (a) traditional zero state 1, $S_{1234} = \{0\ 1\ 0\ 1\}$, (b) traditional zero state 2, $S_{1234} = \{1\ 0\ 1\ 0\}$, (c) shoot-through state 1, $S_{1234} = \{1\ 1\ 0\ 1\}$, (d) shoot-through state 2, $S_{1234} = \{1\ 0\ 1\ 1\}$; (e) active state, $S_{1234} = \{1\ 0\ 0\ 1\}$ (*Source*: Sun 2014 [1]. Reproduced with permission of IEEE).

Table 13.1 Device currents and time intervals of single-phase qZSI in one switching cycle for positive half fundamental cycle. (*Source*: Sun 2014 [1]. Reproduced with permission of IEEE.)

Devices	Conduction Time	Conducting Current
S_1, S_4	$T_s \cdot (1+m)/2$	i_a
	$T_s \cdot D/2$	$2i_{L1}$
S_2, S_3	$T_s \cdot D/2$	$2i_{L1}$
D_2, D_3	$T_s \cdot [1-(1+m)/2]$	i_a
D	$T_s \cdot (1-D-m)$	$2i_{L1}$
	$T_s \cdot m$	$2i_{L1} - i_a$
L_1, L_2	T_s	i_{L1}
C_1, C_2	$T_s \cdot D$	$-i_{L1}$
	$T_s \cdot (1-D-m)$	i_{L1}
	$T_s \cdot m$	$i_{L1} - i_a$

$$P_{sw\_S\_th} = 2 \cdot V_{PN} \cdot 2i_{L1} \cdot \frac{tri+tfu}{2} \cdot f_s$$
$$+ 2 \cdot V_{PN} \cdot 2i_{L1} \cdot \frac{tru+tfi}{2} \cdot f_s \tag{13.25}$$

13.2.2 qZS Diode Power Loss

The qZS diode power loss consists of the conduction loss and the reverse recovery loss.
    The conduction loss of the qZS diode is

$$P_{con\_D} = \frac{2}{\pi} \int_0^\pi \left[u_{D0\_z} \cdot 2i_{L1} + R_{D\_z} \cdot (2i_{L1})^2 \right] \cdot (1-D-m) \, d\omega t$$
$$+ \frac{2}{\pi} \int_0^\pi \left[u_{D0\_z} \cdot (2i_{L1} - i_a) + R_{D\_z} \cdot (2i_{L1} - i_a)^2 \right] \cdot m \, d\omega t \tag{13.26}$$

where u_{D0_z} and R_{D_z} are the on-state zero-current voltage and the on-state resistance of the qZS network diode, respectively.
    The reverse recovery loss is given by

$$P_{rr\_D} = 2 \cdot Q_{rr\_z} \cdot V_{PN} \cdot f_s \tag{13.27}$$

where Q_{rr_z} is the reverse recovery charge of the qZS diode.

13.2.3 qZS Inductor Power Loss

The inductor losses consist of copper loss and core loss.

Copper loss is expressed by

$$P_{Cu} = i_{L1}^2 R_L \tag{13.28}$$

where R_L is the inductor's resistance.
  Core loss is expressed by

$$P_{Fe} = pV_e \tag{13.29}$$

where V_e is the core volume, and p is the loss coefficient. For example, high flux core 58090 gives

$$p = 492 B_{pk}^{2.22} f^{1.32},$$

B_{pk} is half of the AC flux swing, and f is the frequency.
  So the inductor loss is

$$P_L = 2P_{Cu} + P_{Fe} \tag{13.30}$$

13.2.4 qZS Capacitor Power Loss

Different operating states cause different capacitor currents, as shown in Table 13.1. The rms value of the capacitor current is calculated as

$$I_{C1} = \sqrt{ \begin{aligned} & \left(-i_{L1}\right)^2 \cdot D + \frac{1}{\pi}\int_0^\pi i_{L1}^2 \cdot (1-m-D)\,d\omega t \\ & + \frac{1}{\pi}\int_0^\pi \left(i_{L1} - i_a\right)^2 \cdot m\,d\omega t \end{aligned} } \tag{13.31}$$

The power loss of two capacitors is expressed by

$$P_{C1C2} = 2\left(I_{C1}\right)^2 R_{esr} \tag{13.32}$$

where R_{esr} is the capacitor's equivalent series resistance available in the capacitor datasheet.

13.3 Voltage and Current Stress

The voltage and current stresses depend on the modulation methods. Here, the voltage and current stress of the quasi-Z source inverter are analyzed as an example. The maximum constant boost control is considered in the analysis.

  The current flowing through the qZSI switches consists of two components: the load current in the non-shoot-through states, which is the same as the traditional voltage source inverter; the current in the shoot-through time interval, which is twice the quasi-Z-source inductor current, as the operating principle introduced in Chapter 2.

  Because of the symmetrical structure of the inverter, the current during shoot-through in terms of average is evenly distributed in three parallel paths in the maximum constant boost.

The current through the inverter during shoot-through is twice that of the inductor current. Therefore, the average current value in the shoot-through period through each switch is [2, 3]

$$I_{AV\_sh} = \frac{2}{3}I_L \tag{13.33}$$

where I_L is the average inductor current.
For the output power P_{out}, $I_L = P_{out}/V_{in}$, therefore

$$I_{AV\_sh} = \frac{2P_{out}}{3V_{in}} \tag{13.34}$$

where

$$P_{out} = 3V_{ac}I_{ac}\cos\varphi, \quad V_{ac} = \frac{v_{ac}}{\sqrt{2}} = \frac{MB}{2\sqrt{2}}V_{in} \tag{13.35}$$

V_{ac} and I_{ac} denote the AC phase-to-neutral RMS voltage and current; M and B denote modulation index and boost factor, respectively.
In the non-shoot-through states, the average current is

$$I_{AV\_non} = 2\frac{\sqrt{2}I_{ac}}{2\pi} = \frac{\sqrt{2}P_{out}}{3\pi V_{ac}\cos\varphi} = \frac{4P_{out}}{3\pi V_{in}MB\cos\varphi} \tag{13.36}$$

Then the average current flowing through each switch is

$$I_{AV\_ss} = DI_{AV\_sh} + (1-D)I_{AV\_non} \tag{13.37}$$

Using (13.34) and (13.36), (13.37) gives

$$I_{AV\_ss} = \frac{P_{out}}{V_{in}}\left[\frac{2D}{3} + \frac{4(1-D)}{3\pi MB\cos\varphi}\right] \tag{13.38}$$

The voltage stress of the inverter switch, equal to the dc-link peak voltage, is

$$V_s = BV_{in} = \frac{V_{in}}{\sqrt{3}M - 1} \tag{13.39}$$

From (13.38) and (13.39), the total average switch device power (SDP) can be expressed as

$$SDP_{AV} = 6V_s I_{AV\_ss} = 2P_{out}\left(\frac{2 - \sqrt{3}M}{\sqrt{3}M - 1} + \frac{2\sqrt{3}}{\pi\cos\varphi}\right) \tag{13.40}$$

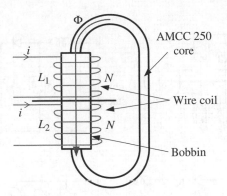

Figure 13.7 Coupled inductors layout.

13.4 Coupled Inductor Design

For the coupled inductor shown in Figure 13.7, the flux in the core is double, because both inductances are coupled and they carry the same current. This is an important factor to be considered when the coupled inductors are built [4].

For the coupled inductors, the flux is twice, and it is given by: $\varphi = 2 \times P \times N \times i$; P is a constant related to the core material and given by $\mu H/N^2$, N is the number of turns, and i is the current into the coil.

Now the inductance will be given by: $L = \dfrac{N\varphi}{i} = 2PN^2$. Hence, the number of turns needed is given by:

$$N = \sqrt{\frac{L}{2P}}, I_L = \frac{P_e}{V_{dc}}, \text{ampere-turns} = NIL$$

This ampere-turns value is useful to check if the designed inductor will go into saturation, to confirm the working point of the inductor, according to the inductance curves of the core. It is important to operate the inductor out of saturation, because after this point the inductance decreases rapidly. When the inductor goes into saturation the current will increase rapidly and it will no longer behave as an inductor.

13.5 Efficiency, Cost, and Volume Comparison with Conventional Inverter

13.5.1 Efficiency Comparison

An efficiency evaluation is an important task during inverter design. Efficiency comparison of the conventional voltage source inverter (VSI), two-stage dc-dc boosted VSI, and ZSI, which are shown in Figure 13.8, using the maximum constant boost control is based on the following assumptions of operation conditions:

- The rated output power is 50 kW and power factor is unity.
- The switching frequency for ZSI and both the boost converter and the VSI is 10 kHz.
- The output voltage and current of the inverter can be derived as in Table 13.2.

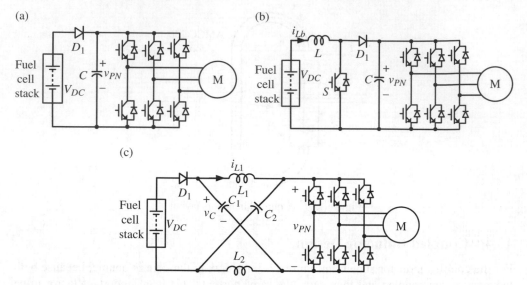

Figure 13.8 Three inverter systems configured for fuel cell vehicles with a PMSM as a load: (a) conventional PWM inverter, (b) dc-dc boosted PWM inverter, (c) ZSI.

Table 13.2 Operating conditions at different power (*Source*: Shen 2007 [2]. Reproduced with permission of IEEE)

Power rating		50 kW	40 kW	30 kW	20 kW	10 kW
Fuel cell voltage (V)		250	280	305	325	340
Motor phase voltage (V)	Conventional PWM inverter	101.7	81.4	61	40.7	20.3
	dc/dc boosted PWM inverter	170.8	136.6	102.4	68.3	34.2
	Z-source inverter	136.8	109.4	82.1	54.7	27.4
Motor current (A)	Conventional PWM inverter	182	182	182	182	182
	dc/dc boosted PWM inverter	108.4	108.4	108.4	108.4	108.4
	Z-sources inverter	135	135	135	135	135

Table 13.2 shows the derived operating conditions from the assumption. From the resultant voltage and current, the following devices are chosen for efficiency calculation: the switches for the main inverters of dc-dc boosted inverter and ZSI are Powerex IPM PM300CLA060, the switch for the dc-dc boost converter is Powerex PM300DSA060, the switch for the traditional VSI is Powerex IPM PM450CLA060, the input end diode of the traditional PWM inverter and the ZSI is IXYSMEO 500-06DA.

The efficiency comparison is shown in Figure 13.9 [2]. It can be seen that the ZSI provides the highest efficiency in most regions of the power range of the inverter itself [2, 5].

13.5.2 Cost and Volume Comparison

A cost and volume comparison can be done by studying the stored energy [6]. The results of the study in [5] have pointed out that both the volume and cost are increasing roughly with the stored energy of commercialized storage capacitors.

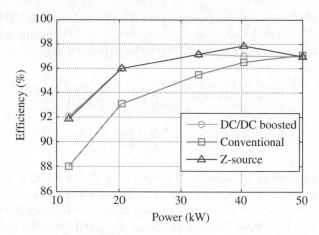

Figure 13.9 Comparison of inverter efficiencies (*Source*: Shen 2007 [2]. Reproduced with permission of IEEE).

Table 13.3 Symbol description

Symbols	Description
$L_{boost}, L_{ZSI}, L_{QZSI}$	Boost converter, ZSI, or qZSI inductors
$C_{boost}, C_{ZSI}, C_{QZSI}$	Boost converter, ZSI, or qZSI capacitors
V_{DC}, V_{PN}	DC power source and DC-link peak voltage
$D = \dfrac{1}{2}\left(1 - \dfrac{V_{DC}}{V_{PN}}\right)$	ZSI/qZSI duty cycle (2D for dc-dc boost converter)
v_C, v_{C1}, v_{C2}	ZSI/qZSI's capacitor voltage
$\Delta i_{Lb}, \Delta i_{LZ}, \Delta i_{Lq}$	Inductor current ripples
$\Delta v_{DC}, \Delta v_C, \Delta v_{C1}, \Delta v_{C2}$	Capacitor voltage ripples
V_{max}, I_{max}	Peak value of AC phase voltage and current
P_0	Inverter output power
$T = 1/f_s$	Switching period
I_{sc}	Short-circuit current during extra shoot-through zero states in ZSI/qZSI

The high-frequency ripples are therefore considered to design the capacitors. Analytical expressions are given in the following for three compared systems: (13.41)–(13.43). The parameters used are detailed in Table 13.3 and are the same as used in the dc-dc boosted inverter, ZSI, and qZSI.

1. dc-dc boosted VSI – For a given current ripple Δi_{Lb}, dc-link voltage ripple Δv_{PN}, and DC source voltage V_{DC}, both the electrostatic energy W_{Cboost} and the magnetic energy W_{Lboost} of the dc-dc boosted inverter in Figure 13.8(b) can be expressed as

$$W_{Lboost} = \frac{1}{2}L_{boost}i_{Lb}^2 = \frac{1}{2}\frac{V_{DC}}{\Delta i_{Lb}}\left(1-\frac{V_{DC}}{V_{PN}}\right)T\left(\frac{P_0}{V_{DC}}\right)^2$$

$$W_{Cboost} = \frac{1}{2}C_{boost}V_{PN}^2 = \frac{1}{2}\frac{I_{max}}{\Delta v_{PN}}\left(1-\frac{V_{DC}}{V_{PN}}\right)TV_{PN}^2$$

(13.41)

2. ZSI – Coupled inductors are considered for both the ZSI and qZSI. So both the stored electrostatic and magnetic energies in the case of coupled systems is accounted for with $L_{eq}=L_{ZSI}+M_{ZSI}\approx 2L_{ZSI}$, where M_{ZSI} represents the mutual inductance supposed to have a value close to L_{ZSI}. Following electrostatic energy W_{CZSI} and magnetic energy W_{LZSI} can be obtained for the ZSI shown in Figure 13.8(c):

$$W_{LZSI} = 2\cdot\frac{1}{2}L_{eq}i_{L1}^2 = 2\left[\frac{1}{8}\cdot\left(\frac{V_{PN}}{V_{DC}}+1\right)\cdot\frac{V_{DC}}{\Delta i_{LZ}}\cdot\left(1-\frac{V_{DC}}{V_{PN}}\right)\cdot\frac{T}{4}\right]\cdot\left(\frac{P_0}{V_{DC}}\right)^2,$$

$$W_{CZSI} = 2\cdot\frac{1}{2}C_{ZSI}v_C^2 = \frac{P_0}{V_{DC}}\cdot\frac{T}{4\Delta v_C}\cdot\left(1-\frac{V_{DC}}{V_{PN}}\right)\cdot\frac{1}{4}\cdot\left(\frac{V_{PN}}{V_{DC}}+1\right)^2 V_{DC}^2$$

(13.42)

3. qZSI – With regard to the qZSI system in Figure 13.1, the same considerations as previously mentioned for the ZSI are made. The only difference concerns the capacitive voltages which are changed by the source voltage. Thus, the mean value of the capacitive voltages are different for the two qZSI capacitors, and the stored electrostatic energy W_{CQZSI} and magnetic energy W_{LQZSI} are

$$W_{LQZSI} = W_{LZSI},$$

$$W_{CQZSI} = \frac{1}{2}C_{QZSI}v_{C1}^2 + \frac{1}{2}C_{QZSI}v_{C2}^2 = \frac{1}{2}\cdot\frac{P_0}{V_{DC}\Delta v_{C2}}\cdot\frac{T}{4}\cdot\left(1-\frac{V_{DC}}{V_{PN}}\right)\cdot\frac{1}{2}\cdot\left[\left(\frac{V_{PN}}{V_s}\right)^2+1\right]V_{DC}^2$$

(13.43)

Figure 13.10 shows the stored energy versus the ratio V_{PN}/V_{DC} at different powers [6]. The results show that in terms of stored energy and thus in volume, weight, and cost terms, both the ZSI and qZSI are competitive with the two-stage dc-dc boosted VSI, which is supposed to store more energy when the source voltage has to be increased (about two times in comparison with the qZSI system).

13.6 Conclusion

In this chapter, the design method for the inductance and capacitance of impedance source networks was presented. An example was shown based on the modeling and analysis of a single-phase qZSI. The power loss computation was also presented for the discussed qZSI.

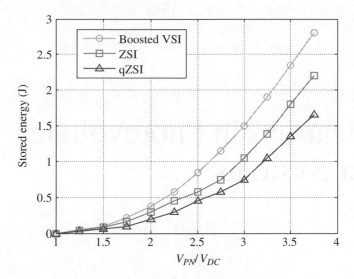

Figure 13.10   Stored energies in the three inverter systems at different powers (*Source*: Battiston 2014 [6]. Reproduced with permission of IEEE).

Comparison results of efficiency, cost, and volume among the traditional VSI, two-stage dc-dc boosted VSI, ZSI, and qZSI validated that the ZSI/qZSI were competitive solutions to the VSI and the two-stage dc-dc boosted VSI. The presented modeling, circuit analysis, and power loss evaluation provided an illustration for the future building and development of impedance source inverters/converters.

References

[1] D. Sun, B. Ge, X. Yan, D. Bi, H. Zhang, Y. Liu, H. Abu-Rub, L. Ben-Brahim, F. Z. Peng, "Modeling, impedance design, and efficiency analysis of quasi-Z source module in cascade multilevel photovoltaic power system," *IEEE Trans. Ind. Electron.*, vol.61, no.11, pp.6108–6117, 2014.

[2] M. Shen, A. Joseph, J. Wang, F. Z. Peng, D. J. Adams, "Comparison of traditional inverters and Z-source inverter for fuel cell vehicles," *IEEE Trans. Power Electron.*, vol.22, no.4, pp.1453–1463, July 2007.

[3] Y. Liu, B. Ge, H. Abu-Rub, F. Z. Peng, "Overview of space vector modulations for three-phase Z-source/quasi-Z-source inverters," *IEEE Trans. Power Electron.*, vol.29, no.4, pp.2098–2108, April 2014.

[4] M. Roldan, A. Barrado, J. Pleite, J. Vazquez, E. Olias, "Size and cost reduction of the energy-storage capacitors," in *Proc. 19th Annu. APEC*, 2004, pp.723–729.

[5] J. Li, J. Liu, J. Liu, "Comparison of Z-source inverter and traditional two-stage boost-buck inverter in grid-tied renewable energy generation," in *Proc. IEEE 6th International Power Electronics and Motion Control Conference (IPEMC)*, pp.1493–1497, 2009.

[6] Battiston, J.-P. Martin, E.-H. Miliani, B. Nahid-Mobarakeh, S. Pierfederici, F. Meibody-Tabar, "Comparison criteria for electric traction system using Z-source/quasi Z-source inverter and conventional architectures," *IEEE Journal of Emerging and Selected Topics in Power Electronics*, vol.2, no.3, pp.467–476, Sept. 2014.

14

Applications in Photovoltaic Power Systems

Z-source/quasi-Z-source inverters (ZSI/qZSI) have attracted considerable attention in photovoltaic (PV) power generation applications because of the single-stage power conversion, no dead time between switches of one bridge leg, and the ability to handle wide dc voltage variation from the PV. This chapter discusses this application, including the typical configurations, parameter design, maximum power point tracking (MPPT), and system control methods, as well as some examples.

14.1 Photovoltaic Power Characteristics

Separate PV power source is an array composed of identical PV panels in parallel and series. A typical PV application model, considering both solar radiation and PV panel temperature, is built as [1–3]

$$i_{PV} = I_{sc} \left\{ 1 - C_1 \left[\exp\left(\frac{v_{PV} - \Delta v_{PV}}{C_2 V_{oc}} \right) - 1 \right] \right\} + \Delta i_{PV} \tag{14.1}$$

where $C_1 = (1 - I_m/I_{sc}) \cdot \exp(-V_m/C_2 V_{oc})$, $C_2 = (V_m/V_{oc} - 1)/\ln(1 - I_m/I_{sc})$, $\Delta v_{PV} = -\beta \cdot \Delta T - R_s \cdot \Delta i_{PV}$, $\Delta i_{PV} = \alpha \cdot S/S_{ref} \cdot \Delta T + (S/S_{ref} - 1) \cdot I_{sc}$, $\Delta T = T - T_{ref}$; V_m, V_{oc}, I_m, and I_{sc} are the MPP voltage, open circuit voltage, MPP current, and short-circuit current of a PV array; v_{PV} and i_{PV} are the output voltage and current of a PV array; S_{ref} and T_{ref} are the solar irradiance and PV panel temperature references, at the standard test conditions (STC) of $1000 \, \text{W/m}^2$ and $25 \, °\text{C}$; S and T are the present irradiance and temperature of the PV array; α is the temperature coefficient of current change at reference irradiance (A/°C), β is temperature coefficient of

Impedance Source Power Electronic Converters, First Edition. Yushan Liu, Haitham Abu-Rub, Baoming Ge, Frede Blaabjerg, Omar Ellabban, and Poh Chiang Loh.

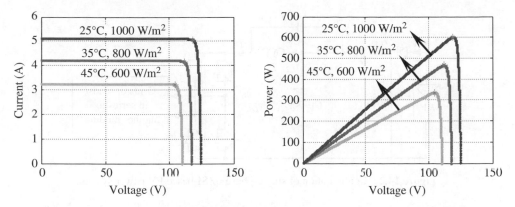

Figure 14.1   PV power characteristics of current and power versus voltage for a 600 W system.

voltage change at reference irradiance (V/°C); R_s is the series resistance of the PV panel; V_{oc} is the open-circuit voltage of the PV panel; and I_{sc} is the short-circuit current of the PV panel.

Figure 14.1 shows the PV power characteristics of current and power versus voltage of a 600 W system. It can be seen that PV panels are susceptible to temperature and irradiance variations and there are maximum power points at a specific condition.

14.2   Typical Configurations of Single-Phase and Three-Phase Systems

Figure 14.2 shows the single-phase qZSI-based PV power system, which couples the qZS impedance network to the H-bridge inverter's dc-link. The qZS network combines two inductors L_1 and L_2, two capacitors C_1 and C_2, and one diode D_1. The C_p is the PV panel's terminal capacitance. In addition, the single-phase qZSI power module could also cascade to form the qZS-CMI, as shown in Chapter 12.

Similarly, the three-phase qZSI-based PV power system is shown in Figure 14.3, the system can be connected to local loads or grids [4–6].

14.3   Parameter Design Method

For the three-phase system, the switching-frequency ripple of inductor current and dc-link voltage are the main concern in designing the qZS inductance and capacitance, which was introduced in Section 13.1.1. For the single-phase system, double-line-frequency ripple of the inductor current and capacitor voltage should also be taken into consideration [7–9]. Therefore, the qZS inductance and capacitance as well as the PV panel terminal capacitance design for single-phase qZSI-based PV system is shown as follows.

The PV panel voltage has the dynamic relationship with its current and internal resistance of [10]

$$\tilde{v}_{PV} = -R_s \tilde{i}_{PV} \tag{14.2}$$

where v_{PV} and i_{PV} represent the voltage and current of the PV panel; R_s is the PV panel's internal resistance; symbols with "~" being the ac dynamic variables.

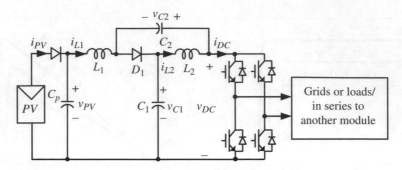

Figure 14.2 Configuration of single-phase qZSI-based PV power system.

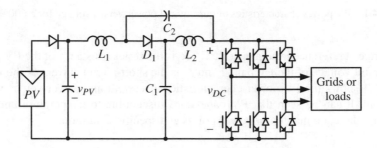

Figure 14.3 Configuration of three-phase qZSI-based PV power system.

The general steady-state relationship of the 2ω ripple components and qZS network parameters are

$$
\begin{cases}
\tilde{v}_{PV} = \dfrac{a_4 R_s}{2\omega C R_s a_5 - \left(1 + 2\omega C_p R_s\right) a_3} \tilde{i}_{DC} \\[3mm]
\tilde{i}_{L1} = -\dfrac{2\omega C a_5}{a_3} \tilde{v}_{PV} + \dfrac{(1-D)(1-2D)}{a_6} \tilde{i}_{DC} \\[3mm]
\tilde{i}_{L2} = \dfrac{a_2 D}{a_1(1-D)} \tilde{i}_{L1} - \dfrac{2\omega C D}{a_1(1-D)} \tilde{v}_{PV} + \dfrac{2D-1}{a_1} \tilde{i}_{DC} \\[3mm]
\tilde{v}_{C1} = \dfrac{2\omega L(1-D)}{2D-1} \tilde{i}_{L1} + \dfrac{2\omega L D}{2D-1} \tilde{i}_{L2} - \dfrac{1-D}{2D-1} \tilde{v}_{PV} \\[3mm]
\tilde{v}_{C2} = \dfrac{2\omega L D}{2D-1} \tilde{i}_{L1} + \dfrac{2\omega L(1-D)}{2D-1} \tilde{i}_{L2} - \dfrac{D}{2D-1} \tilde{v}_{PV}
\end{cases}
\tag{14.3}
$$

where, $a_1 = 2D - 1 - 4\omega^2 LC$, $a_2 = 2D - 1 + 4\omega^2 LC$, $a_3 = a_1^2(1-D)^2 - a_2^2 D^2$, $a_4 = (1-D)(2D-1)$ $[a_2 D + a_1(1-D)]$, $a_5 = a_2 D^2 + a_1(1-D)^2$, and $a_6 = (1-2D)^2 + 4\omega^2 LC$; the other circuit variables are as defined in Figure 14.2.

To limit the switching-frequency current ripple, the qZS inductance should be [4]

$$L = \frac{v_{C1}dt}{di_L} = \frac{V_{C1}\left(T_s D_{max}/k_{sh}\right)}{\left(P/V_{PV,min}\right)\Delta i_L^*} = \frac{D_{max}\left(1-D_{max}\right)V_{PV,min}^2}{k_{sh}f_s P\left(1-2D_{max}\right)\Delta i_L^*} \tag{14.4}$$

where k_{sh} is the shoot-through times per control period; $V_{PV,min}$ is the minimum PV panel voltage; D_{max} is the maximum available shoot-through duty cycle. For the simple boost control, D_{max} is given by

$$G_{max} = \frac{\sqrt{2}U_{AC}}{V_{PV,min}}, \quad D_{max} = \frac{G_{max}-1}{2G_{max}-1} \tag{14.5}$$

where G_{max} is the maximum voltage gain.

From (14.3), the qZS inductance and capacitance affect 2ω ripple of dc-link voltage and PV panel voltage in [10]

$$\left[4\omega^2 LC + \left(1-2D\right)^2\right]\tilde{v}_{DC} + 4\omega L\left(1-D\right)\tilde{i}_{DC} = \left(1-2D\right)\tilde{v}_{PV} \tag{14.6}$$

With the qZS inductance in (14.4) to filter the high frequency current ripple, the qZS capacitance is calculated from (14.6) by

$$C \geq \frac{\left(1-2D\right)\Delta v_{PV}^* V_{PV} - \left(1-2D\right)^2 \Delta v_{DC}^* V_{dc}}{4\omega^2 L \Delta v_{DC}^* V_{DC}} - \frac{\left(1-D\right)\hat{\tilde{i}}_{DC}}{\omega \Delta v_{DC}^* V_{DC}} \tag{14.7}$$

which will limit the PV panel 2ω voltage ripple within Δv_{PV}^*, and the dc-link voltage's 2ω ripple within Δv_{DC}^*.

After calculating the qZS inductance and capacitance in (14.4) and (14.7), respectively, for a PV panel terminal capacitance C_p that will be designed in the next subsection, from (14.3) the 2ω current ripple of the inductor L_1 is examined by

$$\tilde{i}_{L1} = \left\{\frac{\left(1-D\right)\left(1-2D\right)}{a_6} - \frac{2\omega C R_s a_5 a_4}{a_3\left[2\omega C R_s a_5 - \left(1+2\omega C_p R_s\right)a_3\right]}\right\}\tilde{i}_{DC} \tag{14.8}$$

if $\tilde{i}_{L1} \leq \Delta i_L^*$, where Δi_L^* is the 2ω current ripple limitation of inductor L_1, the obtained inductance L and capacitance C with C_p are the designed values used for the single-phase qZSI based PV power system; otherwise, we can increase inductance L and re-examine (14.8) until the 2ω current ripple is less than Δi_L^*.

It should be noted that the PV panel is not a constant dc voltage source, due to internal resistance and dynamic characteristic between PV voltage and current; moreover, a capacitor is usually paralleled to the PV panel to filter and stabilize its output, which will also affect the system performance. The method introduced in Chapter 13 regards the input voltage source of qZSI as a constant dc voltage. The above presented model in this section for the single-phase qZS PV inverter is a comprehensive one, which investigates the qZS network parameters considering the 2ω voltage and current ripple, PV panel's dynamic, and PV panel terminal capacitance.

14.4   MPPT Control and System Control Methods

Figure 14.4 illustrates a system control method for a three-phase qZSI-based PV power system. It mainly includes: (1) a dc-link peak voltage based shoot-through duty ratio to maintain the dc-link peak voltage at the command value $V_{dc}{}^{*}$, (2) the space vector modulation (SVM) of the qZSI, which substantially improves the voltage utilization and reduces the harmonic distortions, and (3) the synchronization of the inverter output ac current with the required grid voltage, where the reference voltages $u_{\alpha}{}^{*}$ and $u_{\beta}{}^{*}$ of SVM are from the d-q decoupling control that delivers the maximum power from the PV panels through using MPPT.

The double-loop shoot-through duty cycle control is introduced in Chapter 2, and the modulation technique of the ZSI is addressed in Chapter 4. The inverter output power is controlled by the MPPT algorithm. The perturb and observe (P&O) based MPPT can be used to maximize the active power from PV panels due to its easy implementation and reasonable tracking ability [11]. The PV voltage closed-loop will output the desired grid-injected d-axis current component to control the inverter output power tracking the PV panel's maximum power point.

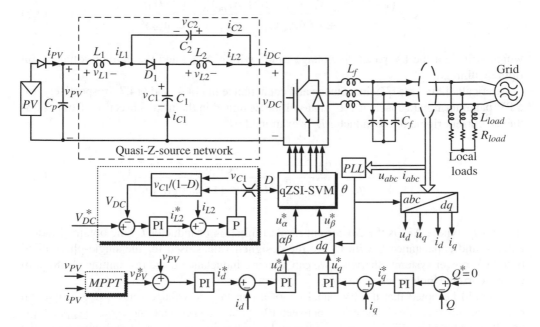

Figure 14.4   A control method for a three-phase qZSI-based PV power system.

Table 14.1 System specifications for single-phase qZSI PV system (*Source*: Liu 2014 [9]. Reproduced with permission of IEEE).

Parameters	Values
PV array voltage, v_{PV}	$350 \sim 600\,\text{V}$
Power rating, P_m	$21\,\text{kW}$
RMS phase-to-phase voltage of three-phase ac output, U_{AC}	$690\,\text{V}$
AC load frequency, f_0	$50\,\text{Hz}$
Switching frequency, f_s	$5\,\text{kHz}$
Desired high frequency and 2ω current ripple through the qZS inductors, Δi^*_L	25%
Desired 2ω ripple of PV panel voltage, Δv^*_{PV}	2%
Desired 2ω ripple of dc-link peak voltage, Δv^*_{DC}	6%

14.5 Examples Demonstration

14.5.1 Single-Phase qZS PV System and Simulation Results

A simulation example of a single-phase qZS PV inverter is shown in the following. The system specifications are listed in Table 14.1. Through the parameter design using (14.4)–(14.8), $3\,mH$ qZS inductance, $4.7\,mF$ qZS capacitance, and $1.1\,mF$ PV terminal capacitance are selected for the configuration of Figure 14.2. Figure 14.5 and Figure 14.6 show the calculated results and simulation results [9].

At 0.7 modulation index, $702\,\text{V}$ dc-link peak voltage is required to get $690\,\text{V}$ phase-to-phase voltage; hence, the shoot-through duty ratio is 0.286. From the principle of voltage-fed qZSI in Chapter 2, the average inductor current is $70\,\text{A}$, capacitor C_1 voltage is $501\,\text{V}$, and capacitor C_2 voltage is $201\,\text{V}$, theoretically.

As seen from Figure 14.5(a) and (b), (c) and (d), as well as Figure 14.6(a) and (b), (c) and (d), respectively, the 2ω ripple are introduced to the single-phase qZSI's dc-link voltage, qZS inductor currents, qZS capacitor voltages, and the PV panel's voltage and current. It is noticeable that the theoretical calculation from the built 2ω ripple model matches the simulations very well.

Using the parameter design method in Section 14.3, the peak-to-peak 2ω ripple components are limited within the desired ranges. For example, the Δv_{PV} is 1.33% seen from Figure 14.5(a) and (b), and Δv_{DC} equals 5.7% from Figure 14.5(c) and (d); Δi_{L1} and Δi_{L2} are within 25%, seen from Figure 14.6(a) and (b). Moreover, the results of the 2ω ripple model match the simulation results well.

14.5.2 Three-Phase qZS PV Power System and Simulation Results

The three-phase qZS PV power system in configuration of Figure 14.4 is simulated. The results are shown in Figure 14.7. Figure 14.7(a) shows the PV characteristics of working conditions. As shown in Figure 14.7(b), the PV panel output voltage varies from $105\,\text{V}$ to $117\,\text{V}$, and then to $111\,\text{V}$. Accordingly, the active power injected to the grid varies from $480\,\text{W}$, to $430\,\text{W}$, then to $460\,\text{W}$. At the same time, the shoot-through duty ratio is regulated to keep a constant dc-link peak voltage in Figure 14.7(c).

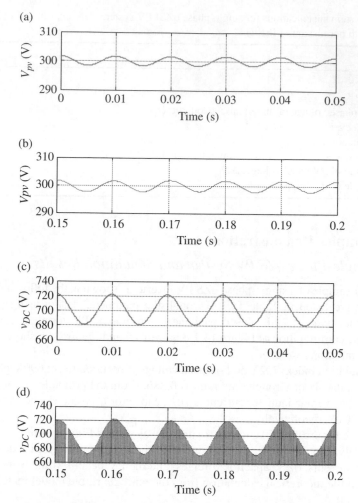

Figure 14.5 Calculated and simulated results of PV panel voltage and dc-link voltage in single qZS H-bridge module: (a, b) for PV panel voltage v_{pv} from 2ω ripple model and simulation, (c, d) for dc-link peak voltage from 2ω ripple model and dc-link voltage from simulation (*Source*: Liu 2014 [9]. Reproduced with permission of IEEE).

14.5.3   1 MW/11 kV qZS CMI Based PV Power System and Simulation Results

A 1 MW/11 kV three-phase qZS-CMI PV power system is simulated on the basis of the qZSI module shown in Figure 14.2. As Figure 14.8 shows, the 1 MW/11 kV qZS-CMI PV power system consists of 16 single-phase qZSI modules in cascade per phase. Simulation results using simple boost modulation are shown in Figure 14.9. From Figure 14.9(b), it can be seen that the qZS-CMI output voltage has low harmonics when the 16 modules are in series. Pure sinusoidal phase voltage is obtained only with a 1 mH filter inductance. The results also demonstrate a transformerless connection to the medium voltage of the qZS-CMI.

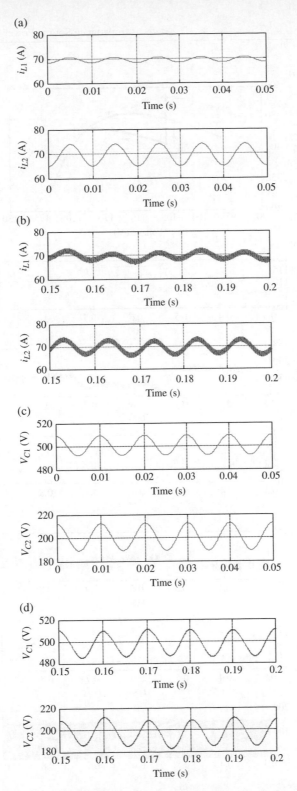

Figure 14.6 Calculated and simulated results of inductor currents and capacitor voltages in single qZS H-bridge module: (a) inductor currents i_{L1} and i_{L2} from 2ω ripple model, (b) inductor currents i_{L1} and i_{L2} from simulation, (c) capacitor voltages v_{C1} and v_{C2} from 2ω ripple model, and (d) capacitor voltages v_{C1} and v_{C2} from simulation (*Source*: Liu 2014 [9]. Reproduced with permission of IEEE).

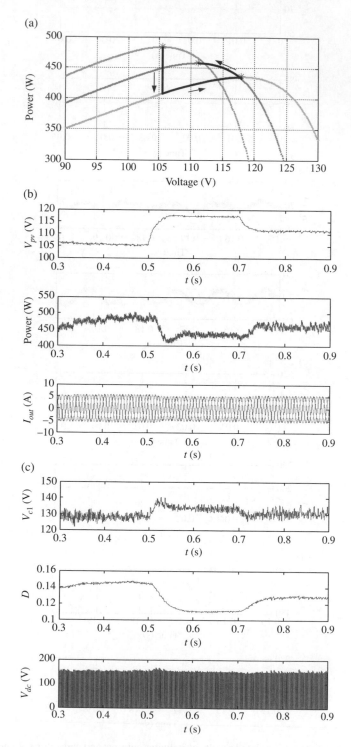

Figure 14.7   Simulation results of the three-phase qZS PV power system: (a) sketch map of working conditions, (b) PV-panel voltage, grid-connected power and current, (c) capacitor C_1 voltage, shoot-through duty ratio, and dc-link voltage.

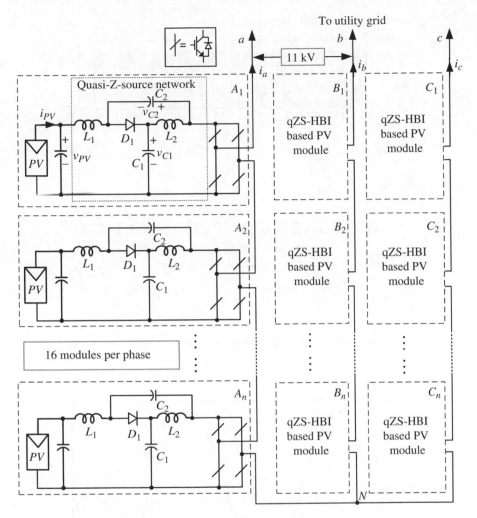

Figure 14.8   System configuration of 1 MW/11 kV qZS CMI system.

14.6   Conclusion

The ZSI/qZSI applied to PV power systems have been presented in this chapter, and the configurations of single-phase/three-phase/cascade-multilevel qZSI based PV systems were demonstrated. The design method of impedance parameters was shown by the single-phase qZSI, based on the dc-side low-frequency ripple analysis. A closed-loop control method was addressed for the three-phase qZSI PV system, which could also be extended to other ZSI/qZSI PV systems. Example simulation results were illustrated for the three configurations. The wide DC voltage handling ability of the ZSI/qZSI in single-stage power conversion provided a wide-ranging investigation into current and future PV applications.

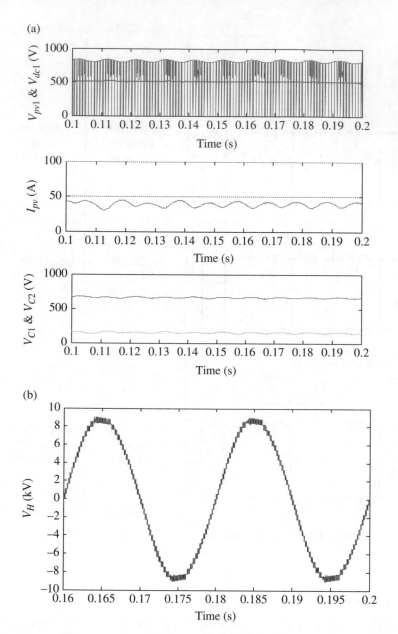

Figure 14.9 Simulation results of 1 MW/11 kV qZS CMI PV power system: (a) one module's PV-panel voltage, dc-link voltage, PV-panel current, and qZS-capacitor voltages, (b) stepwise phase voltage of qZS-CMI, (c) one-phase load voltage, (d) one-phase load current, (e) system output power.

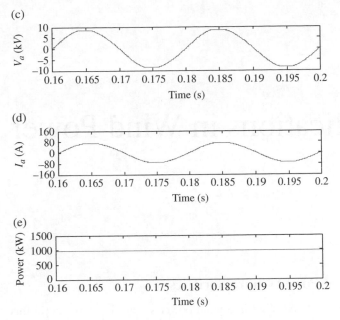

Figure 14.9   (*Continued*)

References

[1] L. Castaner, S. Silvestre, *Modelling Photovoltaic Systems Using PSpice*, Wiley & Sons Ltd, 2002.

[2] S. B. Kjaer, J. K. Pedersen, F. Blaabjerg, "A review of single-phase grid-connected inverters for photovoltaic modules," *IEEE Trans. Ind. Appl.*, vol.41, no.5, pp.1292–1306, Sept.–Oct. 2005.

[3] D. Meneses, F. Blaabjerg, O. García, J. A. Cobos, "Review and comparison of step-up transformerless topologies for photovoltaic ac-module application," *IEEE Trans. Power Electron.*, vol.28, no.6, pp.2649–2663, June 2013.

[4] Y. Li, J. Anderson, F. Z. Peng, D. Liu, "Quasi-Z-Source inverter for photovoltaic power generation systems," in *Proc. 2009 IEEE Applied Power Electronics Conference and Exposition (APEC)*, pp.918–924, 2009.

[5] B. Ge, H. Abu-Rub, F. Peng, Q. Lei, A. de Almeida, F. Ferreira, D. Sun, Y. Liu, "An energy stored quasi-Z-source inverter for application to photovoltaic power system," *IEEE Trans. Ind. Electron.*, vol.60, no.10, pp.4468–4481, Oct. 2013.

[6] H. Abu-Rub, A. Iqbal, S. Moin Ahmed, F. Z. Peng, Y. Li, G. Baoming, "Quasi-z-source inverter-based photovoltaic generation system with maximum power tracking control using ANFIS," *IEEE Trans. Power Electron.*, vol.4, no.1, pp.11–20, Jan. 2013.

[7] Y. Yu, Q. Zhang, X. Liu, S. Cui, "DC-link voltage ripple analysis and impedance network design of single-phase Z-source inverter," in *Proc. 2011–14th European Conference on Power Electronics and Applications (EPE 2011)*, pp.1–10, 2011.

[8] D. Sun, B. Ge, X. Yan, D. Bi, H. Abu-Rub, F. Z. Peng, "Impedance design of quasi-Z source network to limit double fundamental frequency voltage and current ripples in single-phase quasi-Z source inverter," in *Proc. 2013 IEEE Energy Conversion Congress & Exposition (ECCE)*, pp.2745–2750, 2013.

[9] Y. Liu, H. Abu-Rub, B. Ge, F. Z. Peng "Impedance design of 21 kW quasi-Z-source H-bridge module for MW-scale medium-voltage cascaded multilevel photovoltaic inverter," in *Proc. 2014 International Symposium on Industrial Electronics (ISIE)*, pp.2486–2491, Istanbul, Turkey.

[10] Y. Liu, B. Ge, H. Abu-Rub, D. Sun, "Comprehensive modeling of single-phase quasi-Z-source photovoltaic inverter to investigate low-frequency voltage and current ripple," *IEEE Trans. Ind. Electron.*, vol.62, no.7, pp.4194–4202, July 2015.

[11] M. A. Elgendy, B.Zahawi, D. J. Atkinson, "Assessment of perturb and observe MPPT algorithm implementation techniques for PV pumping applications," *IEEE Trans. Sustain. Energy*, vol.3, no.1, pp.21–33, Jan. 2012.

15

Applications in Wind Power

15.1   Wind Power Characteristics

The application of ZSI/qZSI in a wind power system is addressed in this chapter, including their typical configurations, parameters design, maximum power point tracking (MPPT), and also system control methods. The contents are based on the authors' publication [1].

The power captured from the wind energy through wind turbine can be generally expressed as

$$P_W = 0.5C_p\left(\lambda\right)\pi\rho R^2 v^3, \tag{15.1}$$

where ρ is the air density in kg/m³, R is the blade radius in m, v is the wind speed in m/s, λ is the tip speed ratio, defined as $\lambda = \omega R/v$, ω is the angular speed of wind turbine in rad/s, C_p is the power coefficient of the wind turbine, which is related to the pitch angle β and the tip speed ratio λ, with an expression of [2]

$$C_p = 0.22\left(116/\gamma - 0.4\beta - 5\right)e^{-12.5/\gamma}, \tag{15.2}$$

$$\gamma = 1/\left(\frac{1}{\lambda + 0.08\beta} - \frac{0.035}{\beta^3 + 1}\right) \tag{15.3}$$

where γ is an intermediate variable.

When β remains unchanged, C_p depends only on λ. Therefore, the maximum utilization factor C_{pmax} can be reached if the wind turbine is operating at optimal tip speed ratio λ_{opt}, which will produce the highest conversion efficiency of the wind turbine. Thus, the curves of power versus rotor speed with the wind speed as a parameter can be obtained, as shown in Figure 15.1.

Impedance Source Power Electronic Converters, First Edition. Yushan Liu, Haitham Abu-Rub, Baoming Ge, Frede Blaabjerg, Omar Ellabban, and Poh Chiang Loh.
© 2016 John Wiley & Sons, Ltd. Published 2016 by John Wiley & Sons, Ltd.

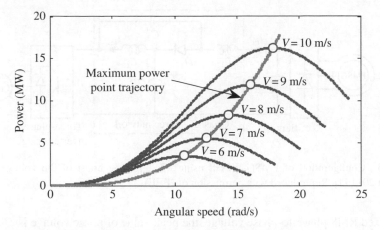

Figure 15.1 Example of wind power characteristics of power versus rotor speed.

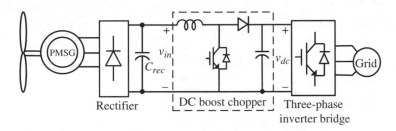

Figure 15.2 Traditional PMSG-WPGS with dc-dc boost converter-based two-stage inverter.

15.2   Typical Configurations

Figure 15.2 shows the traditional permanent magnet synchronous generator (PMSG) based wind power generation system (WPGS) using a dc-dc boost converter-based two-stage inverter [3, 4]. While in the qZSI based PMSG-WPGS, the dc boost chopper is replaced by the Z-source/quasi-Z-source network, as using the qZS network illustrated in Figure 15.3 [1, 5–11]. Through the MPPT control of wind turbine, the maximum power P_{Wmax} generated by the wind turbine is delivered to the direct-drive PMSG. The three-phase ac is converted to dc as the dc source voltage v_{in} of qZSI by a diode rectifier.

15.3   Parameter Design

A qZSI-based wind power generation system is taken into consideration. When the qZS-network input dc voltage range is confirmed, the maximum qZS inductor current is

$$I_L = \frac{P}{v_{in,min}} \tag{15.4}$$

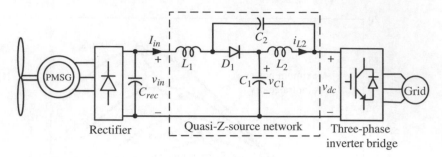

Figure 15.3 Configuration of PMSG-WPGS using qZSI for the boost of the voltage to the grid (*Source*: Liu 2011 [1]. Reproduced with permission of IEEE).

At the fixed RMS phase-to-phase voltage, the peak value of phase voltage is

$$v_{ac} = V_{RMS}\sqrt{\frac{2}{3}} \tag{15.5}$$

Maximum gain of the inverter is

$$G_{max} = \frac{v_{ac}}{v_{in,min}/2} \tag{15.6}$$

Minimum modulation index is

$$M_{min} = \frac{G_{max}}{\sqrt{3}G_{max}-1} \tag{15.7}$$

Maximum duty ratio is

$$D_{max} = 1 - \frac{\sqrt{3}M_{min}}{2} \tag{15.8}$$

The switching frequency is supposed to be relatively high, so the time at which it is conducting is not very short. Therefore, twice the switching frequency f_s for active states, the maximum interval for shoot-through will be

$$T_{0max} = \frac{2-\sqrt{3}M_{min}}{2f_s} \tag{15.9}$$

The current ripple of the inductors is limited to $r_i=20\%$, according to the ZSI/qZSI parameter design method in Chapter 12, we have

$$L = \frac{1}{2} \cdot T_{0max} \cdot \frac{M_{min}v_{in}}{r_i I_{in}} \tag{15.10}$$

The voltage ripple of the capacitors is chosen to be $r_v = 1\%$, the capacitance is

$$C = 2T_{0max} \cdot \frac{I_{in}}{r_v \hat{v}_{dc}} = \frac{2T_{0max} I_{in} (1 - 2D_{max})}{r_v v_{in}} \tag{15.11}$$

According to the system power and voltage ratings, the qZS inductance and capacitance could be selected.

15.4 MPPT Control and System Control Methods

Figure 15.4 shows a system control method of qZSI-based PMSG-WPGS [1]. The control system mainly contains shoot-through duty cycle control, the PWM of qZSI, and the dc-link and AC-output voltage control.

To capture the maximum wind energy so as to obtain optimum dc input voltage of qZSI, the widely used hill-climbing algorithm is adopted to realize MPPT control. The speed ω is adjusted in time to maintain maximum power of wind turbine when the wind speed changes. Then the optimum power is delivered from the wind turbine to the PMSG. Therefore, the obtained rectifier voltage v_{in}, which is the dc source voltage of qZSI, is highest in this way.

The shoot-through states are inserted in the traditional zero vector states without having any effects on the inverter output voltage. The dc-link peak voltage should be controlled to keep it constant. However, the dc-link voltage of qZSI contains the shoot-through zero state, and it is difficult to directly detect its peak value. The relationship between capacitor voltage V_{C1} and the dc-link peak voltage is

$$\hat{v}_{dc} = \frac{1}{1 - D} V_{C1}, \tag{15.12}$$

with the relationship between V_{C1} and V_{in}, the command signal of V_{C1} is

$$V_{C1}{}^* = \frac{1}{2} \left(V_{in}{}^* + \hat{v}_{dc}{}^* \right), \tag{15.13}$$

where $\hat{v}_{dc}^*$ is set by the grid requirements, while V_{in}^* depends on the MPPT control as mentioned before. Consequently, V_{C1} is used to control $\hat{v}_{dc}$ indirectly, as shown in Figure 15.4.

If the control reference frame is oriented along the grid voltage, the power equations in the synchronous reference frame are expressed as

$$\begin{cases} P = \dfrac{3}{2} u_d i_d \\ Q = -\dfrac{3}{2} u_d i_q \end{cases}, \tag{15.14}$$

where P and Q are active and reactive power, respectively, u is the grid voltage, and i is the grid current. The subscripts "d" and "q" stand for direct and quadrature components, respectively.

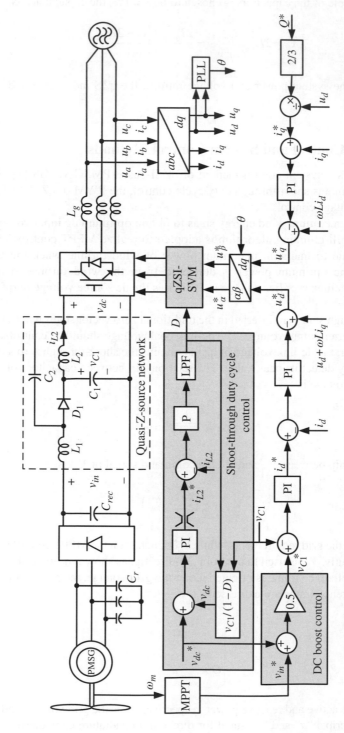

Figure 15.4 System control of PMSG-WPGS using qZSI (*Source:* Liu 2011 [1]. Reproduced with permission of IEEE).

From this, it is known that the independent control of P and Q can be achieved by controlling the direct and quadrature components of the grid currents i_d and i_q. The decoupling algorithm, seen in Figure 15.4, is

$$
\begin{cases}
u_d^* = -\left(i_d^* - i_d\right)\left(K_P + \dfrac{K_I}{s}\right) + u_d + \omega L i_q \\[2mm]
u_q^* = -\left(i_q^* - i_q\right)\left(K_P + \dfrac{K_I}{s}\right) + u_q - \omega L i_d
\end{cases}
\tag{15.15}
$$

Through the decoupling control, the direct and quadrature components of the voltage u_d^* and u_q^* can be obtained. Then they are converted to the stationary reference frame by Clark inverse transform. The PLL ensures the tracking of the grid voltage frequency [12]. In this way, the AC-side grid-connected control is achieved, as demonstrated in Figure 15.4.

15.5 Simulation Results of a qZS Wind Power System

Based on the aforementioned PMSG-WPGS, the relevant simulations are performed in MATLAB/Simulink [1]. The parameters of the wind turbine are as follows: air density $1.225\,kg/m^3$, wind wheel radius $38.8\,m$. The other parameters are shown in Table 15.1. The command signal of dc-link peak voltage is set to 1500 V. The operation conditions are: $0 < t \leq 0.6\,s$, the wind speed is 8 m/s; $0.6\,s < t \leq 1.4\,s$, the wind speed is 7 m/s; $1.4\,s < t \leq 2\,s$, the wind speed is 7.5 m/s. The simulated results are shown in Figure 15.5 and Figure 15.6. It works for a wider speed as well, but the system is just simulated in these conditions due to settling time and memory limits.

When the wind speed changes from 8 m/s to 7 m/s, then to 7.5 m/s, obviously the output power of generator in Figure 15.5(b) is tracking the maximum power according to Figure 15.1. And the rotor speed of the generator in Figure 15.5(c), which is also the wind turbine speed due to the direct-drive PMSG, is the optimum speed. The corresponding results verify the correctness of the designed MPPT control.

When the wind speed varies, the rectifier voltage V_{in}, which is also the dc source voltage of qZSI, is 1110 V, 980 V, 1050 V, respectively. According to the theoretical calculations of qZSI's

Table 15.1 System parameters of a qZSI-based wind power system (*Source*: Liu 2011 [1]. Reproduced with permission of IEEE)

PMSG and Rectifier		qZSI and Grids	
Rated capacity	2.5 MW	QZSI inductors	$4\,mH$
Rated frequency	15.88 Hz	Parasitic resistance of qZSI Inductors	$0.005\,\Omega$
Stator resistance	$0.001\,\Omega$	Q-ZSI capacitors	$1\,mF$
Inductance	$1.5\,mH$	Series resistance of Q-ZSI capacitors	$0.05\,\Omega$
Pole pairs	40	Switching frequency	10 kHz
Input capacitors (C_r)	$10\,\mu F$	Circuit inductance	$10\,mH$
Series resistance of C_r	$0.1\,\Omega$	Grid line RMS voltage	690 V
Rectifier capacitor (C_{rec})	$10\,mF$	Grid frequency	50 Hz

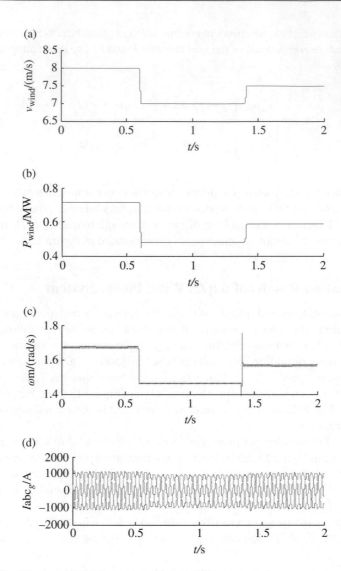

Figure 15.5 Simulated results of generator side in qZSI based wind power system: (a) wind speed, (b) power of wind turbine, (c) rotor speed of generator, (d) generator currents (*Source*: Liu 2011 [1]. Reproduced with permission of IEEE).

operating principle, the capacitor voltage V_{C1} should be 1305 V, 1240 V, 1275 V; the shoot-through duty cycle D should be 0.13, 0.1733, and 0.15. It can be seen from Figure 15.6(b) and (c) that the simulated results are identical to the theoretic results. On the other hand, when the wind speed varies, the rectifier voltage V_{in} also changes. However, the dc-link peak voltage can maintain the command value after a transient regulating process, referring to the results in Figure 15.6(d) and (e). Consequently, the correctness and validity of the designed closed-loop shoot-through duty cycle control are confirmed.

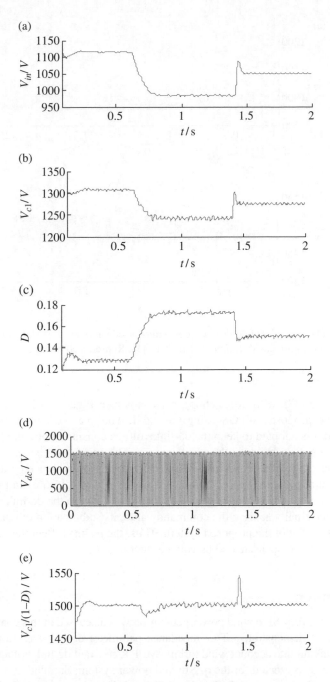

Figure 15.6 Simulated results of qZSI based wind power system: (a) rectifier voltage, (b) capacitor voltage of quasi-Z-source network, (c) shoot-though duty cycle, (d) dc-link voltage of qZSI; (e) dc-link peak voltage of qZSI (*Source*: Liu 2011 [1]. Reproduced with permission of IEEE).

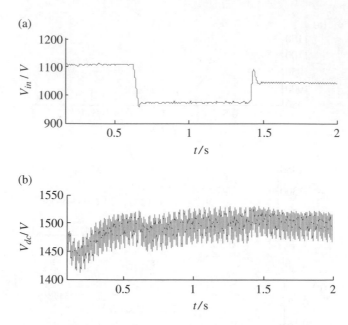

Figure 15.7 Simulated results of dc-dc boost converter-based wind power system: (a) rectifier voltage, (b) dc-link voltage of traditional VSI (*Source*: Liu 2011 [1]. Reproduced with permission of IEEE).

The traditional WPGS with boost chopper, as shown in Figure 15.2, is also simulated to compare with the proposed WPGS using the qZSI. The average current-mode control of the dc-dc converter is adopted to keep the dc-link voltage constant. For comparison, the operation conditions are the same as in the previous case. The simulated results of dc-link voltage are shown in Figure 15.7. It can be seen that the rectifier voltage of the generator in Figure 15.7-(a) is the same as that of qZSI in Figure 15.6-(a). Comparing Figure 15.6(e) and Figure 15.7-(b), it can be seen that when the rectifier voltage v_{in} changes, although the dc-link voltage of traditional WPGS could finally achieve the command value, its stabilization time is longer and the ripple is larger than that of the proposed system. Thus, the results reflect the superiority of the WPGS using qZSI to compensate dc-bus voltage fluctuations.

15.6 Conclusion

The application of ZSI/qZSI to wind power generation was discussed in this chapter, and typical configurations were demonstrated. The impedance parameter design was shown. The system control method, taking into account wind turbine MPPT, constant dc-link voltage, and grid-side synchronization, was presented for the qZSI wind power system. Simulation results were illustrated and compared when using qZSI and a traditional dc-dc boost converter based two-stage inverter, demonstrating the wide dc-link voltage handling capability of the ZSI/qZSI in single-stage power conversion topology.

References

[1] Y. Liu, B. Ge, F. Z. Peng, H. Abu-Rub, A. T. De Almeida, F. J. T. E. Ferreira, "Quasi-Z-source inverter based PMSG wind power generation system," in *Proc. IEEE Energy Conversion Congress and Exposition (ECCE)*, pp.291–297, 2011.

[2] T. Ackermann, *Wind Power in Power Systems*, Wiley & Sons Ltd, 2012.

[3] F. Blaabjerg, M. Ke, "Future on power electronics for wind turbine systems," *IEEE Journal of Emerging and Selected Topics in Power Electronics*, vol.1, no.3, pp.139–152, Sept. 2013.

[4] C. Zhe, J. M. Guerrero, F. Blaabjerg, "A review of the state of the art of power electronics for wind turbines," *IEEE Trans. Power Electron.*, vol.24, no.8, pp.1859–1875, Aug. 2009.

[5] Q. Sun, Y. Wang, "On control strategy of Z-source inverter for grid integration of direct-driven wind power generator," in *Proc. 2012 31st Chinese Control Conference (CCC)*, pp.6720–6723, 2012.

[6] X. Wang, D. M. Vilathgamuwa, K. J. Tseng, C. J. Gajanayake, "Controller design for variable-speed permanent magnet wind turbine generators interfaced with Z-source inverter," in *Proc. 2009 International Conference on Power Electronics and Drive Systems (PEDS)*, pp.752–757, 2009.

[7] S. M. Dehghan, M. Mohamadian, A. Y. Varjani, "A new variable-speed wind energy conversion system using permanent-magnet synchronous generator and Z-source inverter," *IEEE Trans. Energy Convers.*, vol.24, no.3, pp.714–724, Sept. 2009.

[8] U. Supatti, F. Z. Peng, "Z-source inverter with grid connected for wind power system," in *Proc. IEEE Energy Conversion Congress and Exposition (ECCE)*, pp.398–403, 2009.

[9] T. Maity, H. Prasad, V. R. Babu, "Study of the suitability of recently proposed quasi Z-source inverter for wind power conversion," in *Proc. International Conference on Renewable Energy Research and Application (ICRERA)*, pp.837–841, 2014.

[10] W.-T. Franke, M. Mohr, F. W. Fuchs, "Comparison of a Z-source inverter and a voltage-source inverter linked with a DC/DC-boost-converter for wind turbines concerning their efficiency and installed semiconductor power," in *Proc. IEEE Power Electronics Specialists Conference (PESC)*, pp.1814–1820, 2008.

[11] B. K. Ramasamy, A. Palaniappan, S. M.Yakoh, "Direct-drive low-speed wind energy conversion system incorporating axial-type permanent magnet generator and Z-source inverter with sensorless maximum power point tracking controller," *IET Renew. Power Gen.*, vol.7, no.3, pp.284–295, May 2013.

[12] L. Hadjidemetriou, E. Kyriakides, F. Blaabjerg, "A new hybrid PLL for interconnecting renewable energy systems to the grid," *IEEE Trans. Ind. Appl.*, vol.49, no.6, pp.2709–2719, 2013.

16

Z-Source Inverter for Motor Drives Application: A Review

This chapter presents an overview for the Z-source inverter (ZSI) as an emerging topology for power electronics dc/ac converters in general purpose motor drive applications. Different ZSI topologies, different motor types with different phases, and different control algorithms will be reviewed as a part of the ZSI motor drive system. In addition, various applications of the ZSI in the automotive field will be highlighted. The ZSI is a very promising power electronics converter topology for motor drive applications as it provides many advantages compared with the traditional variable speed drive system configuration, such as improving the drive system reliability and extending the output voltage range from the converter.

16.1 Introduction

The traditional motor drive system that is based on the voltage source inverter (VSI), which consists of a diode rectifier front end, a dc-link capacitor (C), a dc inductor (L), and inverter bridge, as shown in Figure 16.1, suffers from some limitations and problems, such as the following [1–8]:

- The obtainable output is limited some way below the input line voltage.
- Voltage sags can interrupt an adjustable speed drive (ASD) system and shut down critical loads and processes.
- Inrush and harmonic current from the diode rectifier can pollute the line.
- Performance and reliability are compromised by the VSI structure (mis-gating, dead time, and common-mode voltage).

Impedance Source Power Electronic Converters, First Edition. Yushan Liu, Haitham Abu-Rub, Baoming Ge, Frede Blaabjerg, Omar Ellabban, and Poh Chiang Loh.
© 2016 John Wiley & Sons, Ltd. Published 2016 by John Wiley & Sons, Ltd.

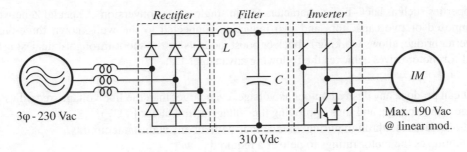

Figure 16.1 Traditional variable speed drive system configuration with induction motor (IM) (*Source*: Peng 2003 [1]. Reproduced with permission of IEEE).

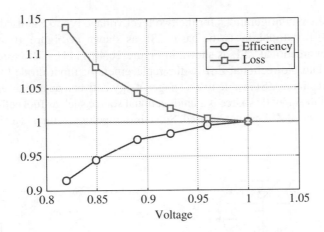

Figure 16.2 Effect of under-voltage on losses and efficiency of an 11 kW induction motor (measurements, all values are referred to their rated values) [4,5]. (*Source*: Sack 2008 [4]. Reproduced with permission of IEEE.)

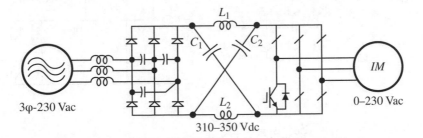

Figure 16.3 Z-source inverter ASD system configuration (*Source*: Peng 2005 [2]. Reproduced with permission of IEEE).

In addition, the efficiency of the motor drive system is strongly affected if its terminal voltage is lowered compared to its rated value, as shown in Figure 16.2 for an induction motor [4].

The Z-source inverter (ZSI) is one of the most promising power electronics converter topologies suitable for motor drive applications, as shown in Figure 16.3. The ZSI has interesting properties such as buck–boost characteristics in single-stage conversion. A special Z-network composed of two capacitors and two inductors connected to the well-known three-phase inverter bridge, allows working in buck or boost mode using the shoot-through state. Using the ZSI for motor drives achieves the following advantages [2]:

- It can produce any desired output ac voltage, even larger than the line voltage, regardless of the input voltage, and thus improving the voltage utility factor (VUF).
- It provides ride-through during voltage sags without any additional circuits.
- Minimizes the motor ratings to deliver a required power.
- It improves the power factor and reduces harmonic current and common-mode voltage of the line.

There are different configurations for the two level voltage type ZSIs used for motor drive applications. The first topology is the basic ZSI, as shown in Figure 16.4(a). The second topology is the bidirectional ZSI, as shown in Figure 16.4(b) [11]. The basic version of ZSI can be changed into bidirectional ZSI by the replacement of input diode by a bidirectional switch. The bidirectional ZSI is able to exchange energy between ac and dc energy storage in both directions. If the input dc source is a unidirectional source such as fuel cells or photovoltaic arrays, the bidirectional ZSI cannot be used. So, a third topology, which is a high performance

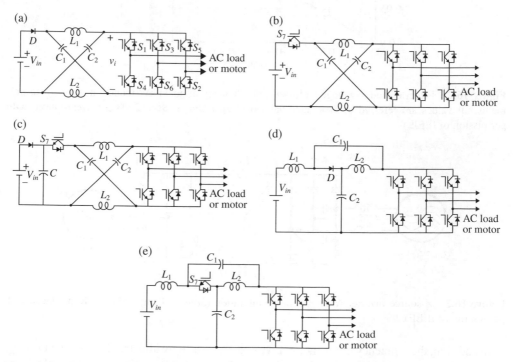

Figure 16.4 Different ZSI topologies used in motor drive applications: (a) ZSI, (b) BZSI [11], (c) HP-ZSI [12], (d) QZSI [13], (e) BQZSI (*Source*: Guo 2013 [14]. Reproduced with permission of IEEE).

ZSI, as shown in Figure 16.4(c), can be used [12]. The high performance ZSI can operate at wide load range with small Z-network inductor and eliminate the possibility of the dc-link voltage drops, and simplify the Z-network inductor design and system control. The fourth topology is the quasi-Z-source inverter (QZSI) as shown in Figure 16.4(d), which has several advantages over the basic ZSI topology, including; lower component rating (Z-network capacitor voltages are lower than in case of the basic ZSI topology), the joint earthing of the input power source and the dc-link bus, which reduce the common-mode noise and continuous input current [13]. The fifth topology is the bidirectional QZSI (BQZSI) topology, as shown in Figure 16.4(e) [14]. The contents of this chapter are based on the authors' publication of [15].

16.2   Z-Source Inverter Feeding a Permanent Magnet Brushless DC Motor

Due to its compact and simple structure, reliable operation, higher power density, higher efficiency, low maintenance costs, simple control algorithms, the permanent magnet brushless dc motor (PMBDCM) has been widely used in industry. Accordingly, a PMBDCM drive system fed by a ZSI has the advantages of all of them, and offers a potential candidate for ASD system application. Figure 16.5 shows the main circuit of the ZSI based PMBDCM drive system [16–20].

In [21–24], the ZSI was used to supply a brushless dc motor from a photovoltaic (PV) array for a water pumping system. Figure 16.6 describes the basic building blocks of this brushless dc motor drive system [21, 22]. The drive system consists of the speed controller, reference current generator, hysteresis current controller, three-phase ZSI, and the BLDC. The control algorithm is composed of two main parts: the shoot-through (ST) duty ratio control for the ZSI, and the current control of the brushless dc motor. In the first part, a fuzzy-logic incremental-conductance maximum power point tracking (FL-IC MPPT) algorithm is used to intelligently determine the step change in the ST duty ratio (ΔD) based on a fuzzy controller, and to absorb the maximum power from the PV array. In order to reduce torque ripples, a current shaping technique is used in the second part of the control algorithm. As indicated in Figure 16.6, the current shaping control consists of a PID speed controller, a reference current generator, and hysteresis current controllers. The speed of the motor is compared with its reference value, and the speed error is processed in the PID speed controller, tuned by the particle swarm optimization (PSO) technique. The output of this controller is considered as the reference torque. The reference current shaping block generates square reference current waveforms, using the electrical rotor position. In [23, 24], the direct torque control (DTC) scheme was used to provide faster torque response and reduced torque ripple for driving the brushless dc motor, as indicated in Figure 16.7. In [25], a sensorless BLDCM was proposed based on the back emf method, as indicated in Figure 16.8, which is one of the most popular control methods for the sensorless BLDCM control algorithm, the core of which is to detect the zero-crossing point (ZCP) of the back emf. This control algorithm can operate within a wide speed range without switching the detection points and the reference levels, which improves the overall performance of the ZSI based PMBDCM drive system. To sum up, the ZSI improves the stability and safety of a BLDCM drive system under complex conditions.

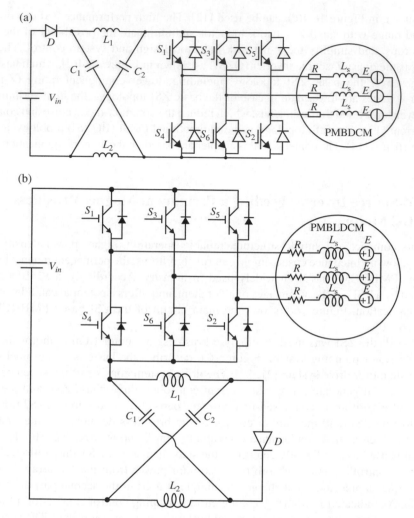

Figure 16.5 Main circuit of (a) Z-source inverter based PMBDCM (*Source*: Wang 2007 [16]. Reproduced with permission of IEEE) and (b) improved ZSI based PMBDCM (*Source*: Peter 2014 [19]. Reproduced with permission of Taylor & Francis).

16.3 Z-Source Inverter Feeding a Switched Reluctance Motor

Switched reluctance motor (SRM) drives are under consideration in various applications requiring high performance. This is certainly due to its numerous advantages such as simple and robust construction, low inertia, high speed, high-temperature performance, low costs, and fault tolerance control capabilities [26]. In [27], a closed loop control of switched reluctance motor drives using Z-source inverter with the simplified rule base of fuzzy logic controller has been proposed, as shown in Figure 16.9. The proposed system combines the advantages of both the ZSI and the SRM. The proposed control algorithm shows very good stability and robustness against speed and load variations over a wide range of operating conditions.

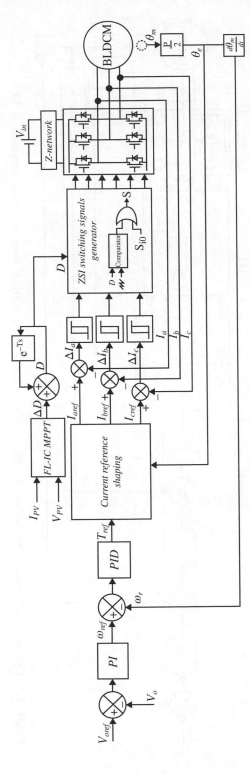

Figure 16.6 Block diagram of the BLDC motor drive and MPPT of PV array (*Source*: Hosseini 2010 [21]. Reproduced with permission of IEEE).

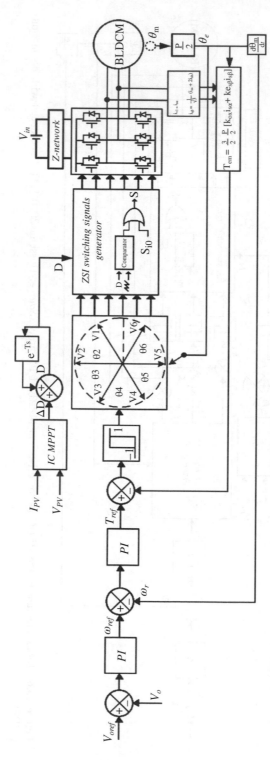

Figure 16.7 Overall block diagram of the DTC of a BLDC motor drive (*Source*: Feyzi 2010 [23]. Reproduced with permission of IEEE).

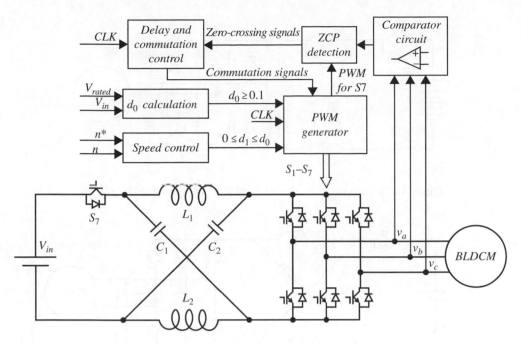

Figure 16.8 The complete block diagram of a sensorless BLDCM control (*Source*: Xia 2015 [25]. Reproduced with permission of IEEE).

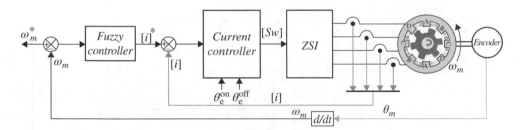

Figure 16.9 Z-source inverter based switched reluctance motor drive system.

16.4 Z-Source Inverter Feeding a Permanent Magnet Synchronous Motor

Permanent-magnet synchronous motors (PMSM) are capable of providing high torque-to-current ratios, high power-to-weight ratios, high efficiency, and robustness. Owing to these advantages, PMSMs are widely used in modern variable speed ac drives. In [28, 29], a bidirectional ZSI (BZSI) is applied to a PMSM drive system, as shown in Figure 16.10, as a single-stage converter instead of a two-stage converter (bidirectional dc/dc converter and VSI). Figure 16.11 shows the control scheme of the PMSM drive system with a BZSI for an electric vehicle application. The ZSI provides an adjustable boost dc-link voltage when the PMSM needs to operate in the high-speed region. The application of a BZSI to PMSM drive system can improve drive system reliability. Furthermore, the adjustable dc-link

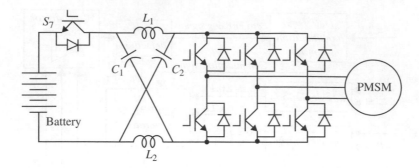

Figure 16.10   Schematic diagram of a PMSM drive system fed by a BZSI.

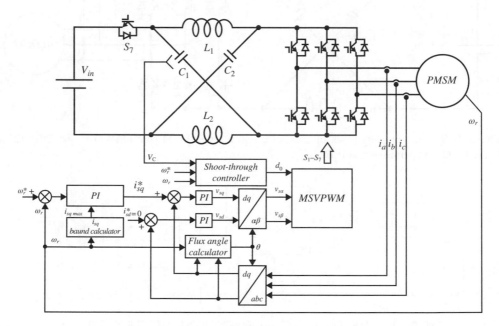

Figure 16.11   Control scheme of a PMSM drive system fed by a BZSI for electric vehicle applications (*Source*: Liu 2011 [29]. Reproduced with permission of Elsevier).

voltage leads to an additional degree of freedom in the loss minimization control structure [30]. In [31, 32] the sliding-mode control (SMC) method is used to control the ZSI feeding a PMSM, as shown in Figure 16.12, whereas a flatness-based control has been proposed to drive an actuator and generate the peak dc bus voltage reference, which resulted in an improvement of global efficiency of the drive system.

In [33], a flux-weakening algorithm has been developed to achieve constant output power with an interior permanent magnet synchronous motor (IPMSM). The principle of the boosted voltage flux-weakening (BVFW) control is that the output power of IPMSMs remains constant during this operation and the current limit is still a constraint. As the speed increases, the current space vector rotates counterclockwise in the *d-q* plane, which

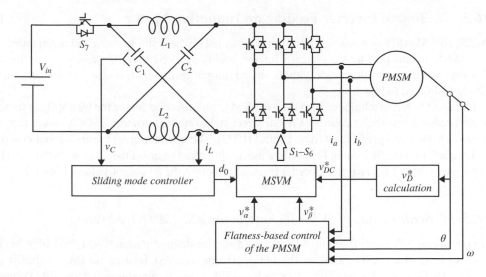

Figure 16.12   Sliding-mode control of a PMSM fed by a BZSI (*Source*: Battiston 2014 [32]. Reproduced with permission of IEEE).

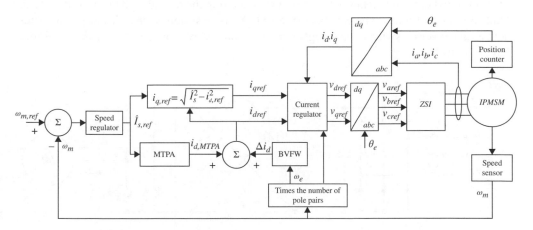

Figure 16.13   Block diagram of an IPMSM-ZSI drive system with flux weakening control (*Source*: Lin 2014 [33]. Reproduced with permission of IEEE).

produces a stronger demagnetizing current component to reduce the output torque of such IPMSMs. The voltage level, which could be much higher than the dc bus voltage of such drives, depends on the speed and the d, q axis currents. Figure 16.13 illustrates a block diagram of an IPMSM-drive system with a closed-loop speed control and the BVFW control. The IPMSM shown operates in the constant torque region with maximum torque per ampere (MTPA) control and in the constant power region with BVFW control. Compared with the conventional flux-weakening strategies, the BVFW control allows an IPMSM to largely extend its constant power speed range (CPSR) and operate at higher speed with higher torque.

16.5 Z-Source Inverter Feeding an Induction Motor

A ZSI for IM ASD systems was first proposed in [34], where the features and advantages of the ZSI fed induction motor system over the traditional VSI based system were outlined. The regenerative operation mode without any additional components is also possible using the ZSI as indicated in [4, 6].

Different control techniques have been applied to an induction motor fed by a VSI for motor drive applications: scalar control (V/F), indirect field oriented control (IFOC), direct torque control (DTC), and predictive torque control (PTC). The application of the above four control techniques for the ZSI feeding IM needs some modification to insert the ST state in the switching signal within the zero state in order to avoid affecting the ac output voltage [34–52].

16.5.1 Scalar Control (V/F) Technique for ZSI-IM Drive System

An open-loop V/F control combined with pulse amplitude modulation PWM (PAM/PWM) is used to control an ASD based on the high-performance ZSI feeding an IM, as shown in Figure 16.14 [34]. The traditional open-loop V/F control maintains the dc-link voltage constant, only by controlling the modulation index (PWM) to follow the V/F law. Combining the pulse amplitude modulation (PAM) and PWM, named PAM/PWM control strategy with the traditional control V/F law, which results in changing the dc-link voltage according to the load conditions, brings some benefits to the ZSI-IM drive system such as lower voltage stress across the inverter switches and higher modulation index, compared with the traditional PWM control.

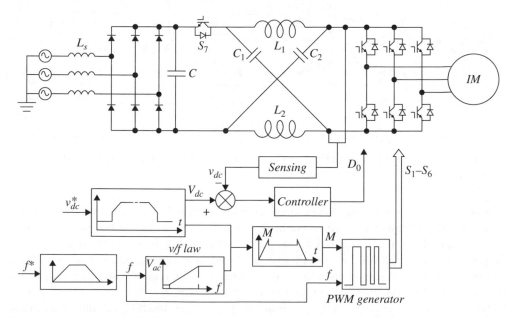

Figure 16.14 PAM/PWM based open-loop V/F control of a ZSI-IM based ASD system (*Source*: Ding 2007 [34]. Reproduced with permission of IEEE).

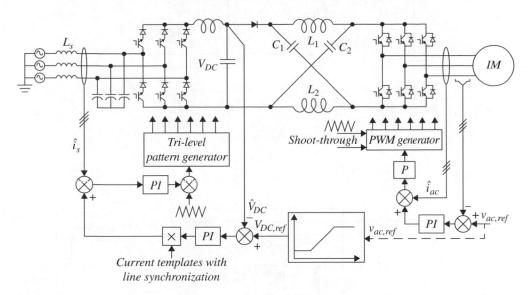

Figure 16.15 Control block diagram of the PWM CSR-ZSI-IM ASD system (*Source*: Ding 2007 [35]. Reproduced with permission of IEEE).

Reference [35] presented a second example of a V/F controlled ZSI-IM drive system by investigating the complete system including a current source rectifier (CSR). The control of the current source rectifier ZSI-IM adjustable speed drive (CSR-ZSI-IM ASD) system, as indicated in Figure 16.15, is similar to the traditional converters except for the control of the variable dc-bus voltage and the ZSI. This control algorithm has many advantages: it provides a bidirectional power flow, maintains the unity power factor, reduces the input line current harmonics, provides ride-through during the voltage sags and voltage swells without any additional circuits, and keeps the inverter modulation index high during low-speed operation.

In [36], a four-switch three-phase ZSI was used to feed an induction motor with V/F control. The four-switch ZSI is shown in Figure 16.16(a), where one of the capacitors in the Z-network is split into two and the middle point is connected to one phase of the IM. The algorithm to control the dc boost, split capacitor voltage balance, and the ac output voltage of the four-switch three-phase ZSI feeding an induction motor drive is shown in Figure 16.16(b). In [37], a four-switch single-phase to three-phase quasi Z-source inverter, which is a more efficient, more reliable, and cost efficient converter, has been used for driving a three-phase induction motor, as shown in Figure 16.17. The proposed quasi Z-source single-phase to three-phase converter reduces the cost of the system, the switching losses, and the complexity of the control method.

A voltage mode integrated control technique (VM-ICT) was utilized in [38] for a ZSI fed IM drive using two control loops, as illustrated in Figure 16.18. An outer voltage loop controls the motor line voltage through a PI controller, and an inner current loop regulates the motor phase current through a PI controller and a limiter to provide the reference voltage for the modified space vector modulation (MSVM) block, where there is no speed or torque control for the induction motor.

(a)

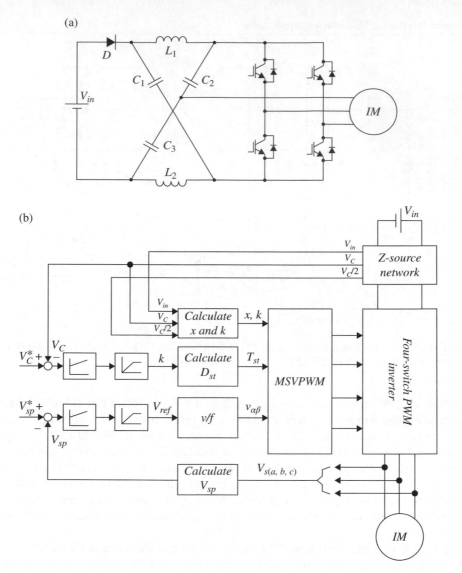

(b)

Figure 16.16 Four-switch three-phase ZSI-IM drive system: (a) topology configuration, (b) V/F control algorithm (*Source*: Baba 2014 [36]. Reproduced with permission of IEEE).

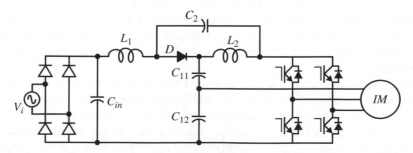

Figure 16.17 Single-phase to three-phase QZSI feeding IM (*Source*: Khosravi 2011 [37]. Reproduced with permission of IEEE).

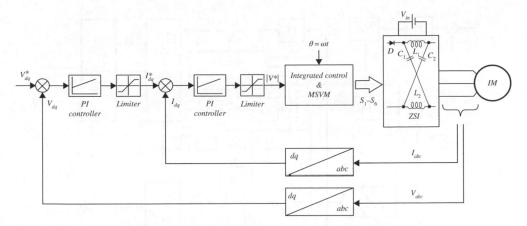

Figure 16.18   Voltage mode integrated control technique (VM-ICT) for ZSI fed IM drive.

16.5.2   Field Oriented Control Technique for ZSI-IM Drive System

In order to achieve a high dynamic performance in an induction motor drive application, vector control is often applied. The principal aim of vector control is to independently control the flux and torque in the induction motor. In the indirect field oriented control (IFOC) method, the reference frame is rotating at synchronous angular velocity, ω_e. This reference frame allows the phase currents to be viewed as two dc quantities under steady state conditions. The q axis component is responsible for the torque-producing current, i_{qs}, and the d axis is responsible for the field-producing current, i_{ds}. These two vectors are orthogonal so that the field current and the torque current can be controlled independently. Figure 16.19 shows a block diagram of the IFOC technique for an induction motor. The entire closed loop system, as shown in Figure 16.20, contains: input battery, ZSI, capacitor voltage control, and IFOC speed control, where the capacitor voltage control generates the ST duty ratio and the IFOC generates the modulation index according to the operating conditions [39–44].

16.5.3   Direct Torque Control (DTC) Technique for ZSI-IM Drive System

Direct torque control (DTC) is one of the advanced control schemes for ac drives. It is characterized by a simple control algorithm, an easy digital implementation, and robust operation. In this section, the principle of the classical DTC scheme and space vector modulation direct torque control (SVM-DTC) scheme are introduced.

The basic idea of the conventional DTC algorithm, as shown in Figure 16.21(a), is to select, directly without a PWM modulator, the most suitable voltage vector for the inverter, which produces the desired torque and flux in the motor. In the DTC, the instantaneous values of the flux and torque are calculated from the stator variables (current and voltage). Selecting the optimum inverter switching state can control them directly and independently. The selection is made to restrict the error of the flux and torque within the hysteresis bands and to obtain the fastest torque response at every instant [45].

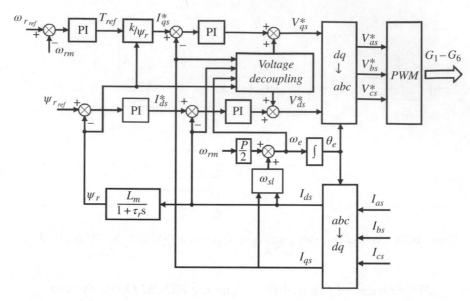

Figure 16.19   Block diagram of the IFOC of a three-phase IM (*Source*: Ellabban 2011 [39]. Reproduced with permission of IEEE).

The conventional DTC scheme has many drawbacks such as variable switching frequency, high current and torque ripples, starting and low-speed operation problems, and high sampling frequency needed for digital implementation of the hysteresis controllers. To overcome these problems, the space vector modulation is combined with the conventional DTC scheme for induction motor drives to provide a constant inverter switching frequency. In the SVM-DTC scheme, as shown in Figure 16.21(b), the torque and flux hysteresis comparators are replaced by PI controllers to regulate the flux and torque magnitudes respectively. The output of these PI controllers generates the d and q components of the reference voltage command for inverter control, which ensures that the inverter switching frequency is fixed. So the inverter switching frequency is significantly increased, and the associated torque ripple and current harmonics can be dramatically reduced, in comparison with the conventional switching table based DTC scheme [46].

Figure 16.22 shows the complete block diagram of the closed loop DTC-SVM speed controlled IM fed by the high performance ZSI. The complete system has five controllers: PI speed controller, PI torque controller, PI flux controller, voltage, and current controllers for peak dc-link voltage control [47, 48].

In [49], there is a comparative study of the most significant control methods: V/F, IFOC, and DTC for an induction motor fed by a ZSI for automotive applications. The three control techniques are implemented using PWM voltage modulation. The comparison was based on various criteria including motor dynamic performance, speed and torque ripples, ac current harmonic content, control algorithm implementation complexity, ZSI performance, and overall system efficiency. The three control techniques were compared based on a simulated benchmark. The main results of that work are summarized in Table 16.1, which indicates that the IFOC seems to be the most suitable control technique for controlling an induction motor fed by a high-performance ZSI for automotive applications.

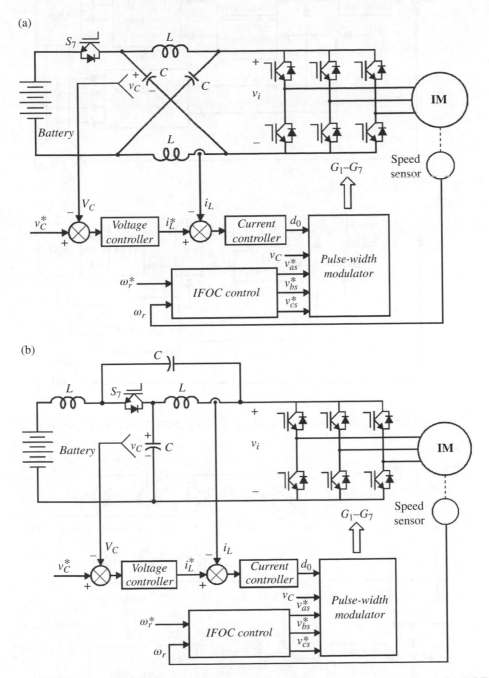

Figure 16.20 Closed-loop speed control of an induction motor fed by: (a) BZSI and (b) BQZSI (*Source*: Ellabban 2012 [40]. Reproduced with permission of IEEE).

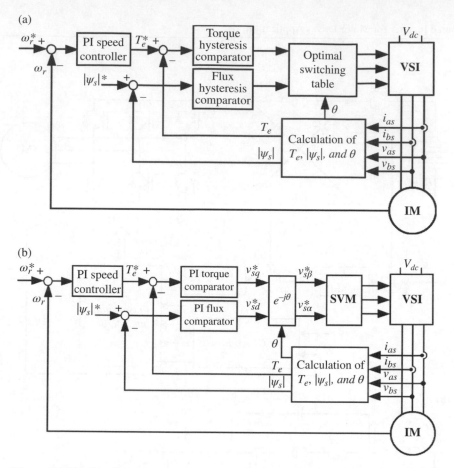

Figure 16.21 Block diagram of: (a) classical DTC and (b) DTC-SVM based IM drive.

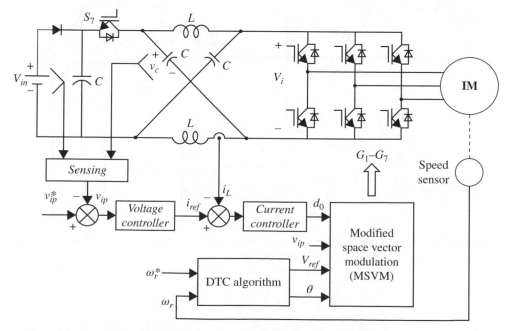

Figure 16.22 DTC-SVM closed-loop speed control of three-phase induction motor fed by a high-performance ZSI (*Source*: Ellabban 2011 [47]. Reproduced with permission of IEEE).

Table 16.1 Summary of the performance comparison of different control techniques for IM fed by a ZSI (*Source*: Ellabban 2011 [49]. Reproduced with permission of IEEE)

Comparison Criterion	V/F	IFOC	SVM-DTC
Dynamic response	Poor	Very good	Good
Low speed response	Very poor	Good	Very good
Torque ripples	Large	Small	Small
Speed error	Large	Small	Medium
Current THD	Low	High	Medium
ZSI performance	Good	Good	Poor
Complexity	Low	High	High
Efficiency	Medium	High	Low

16.5.4 Predictive Torque Control for ZSI-IM Drive System

Recently, predictive control was successively applied to conventional motor drives (IM fed by a VSI) with superior performance compared to the vector control or the DTC [50]. These results lead us to use it in preference to other power electronics converter topologies, such as the BQZSI. Since the BQZSI based drive has a higher number of state variables compared to a typical VSI, its control is a rather difficult task and requires a cascaded controller structure on both sides of the converter. Therefore, applying the predictive torque control algorithm, which is characterized by its single structure, superior performance, and its ability to deal simultaneously with any number of variables, to control an IM fed by a BQZSI, a high degree of flexibility will be obtained. Furthermore, the interesting advantages of the PTC algorithm are combined with the advantages of the BQZSI.

In predictive torque control (PTC), the predictions for the future values of the controlled variables (motor stator flux, torque, and BQZSI capacitor voltage) are calculated. Hence, the reference condition, which is implemented by a cost function, considers the future behavior of these variables. Predictions are calculated for every actuating possibility, and the cost function selects the voltage vector that optimizes the reference tracking. The block diagram and the flowchart of the PTC algorithm are shown in Figure 16.23 [51, 52].

16.6 Multiphase Z-Source Inverter Motor Drive System

Multiphase machines offer higher efficiency, reduced torque pulsations, higher torque density, and greater fault tolerance. Furthermore, in a multiphase drive the energy delivered through each phase of the inverter is less than in a three-phase system, giving advantages for the design of inverters for high power machines. Multiphase induction machines have been receiving much attention in the literature, but much of this has been related to a specific number of phases [53].

The combination of the multiphase machine concept and the ZSI with the vector control concept to produce a very high reliable adjustable speed drive (ASD) system was proposed in [54–57]. Figure 16.24 shows the complete block diagram of a five-phase IM fed by a ZSI system containing: input dc source, HP-ZSI, dual loop capacitor voltage control, IFOC speed control, and five-phase induction motor with speed and current measurements. Figure 16.25 shows a six-phase Z-source inverter feeding an IM, where a neural network

(a)

(b)

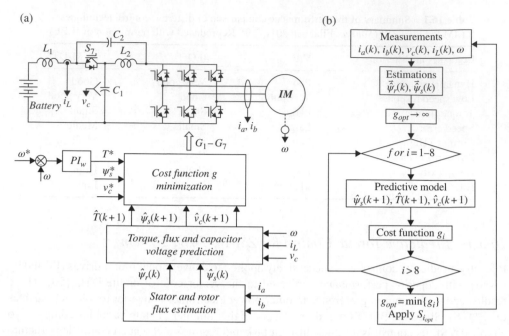

Figure 16.23 Predictive torque control for a BQZSI feeding an IM: (a) block diagram, (b) implementation flowchart (*Source*: Ellabban 2013 [51]. Reproduced with permission of IEEE).

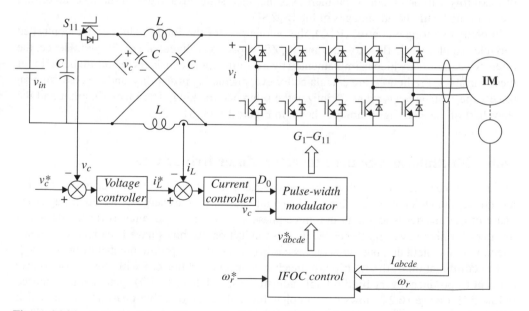

Figure 16.24 Complete block diagram of a five-phase ZSI-IM ASD system (*Source*: Ellabban 2013 [54]. Reproduced with permission of IEEE).

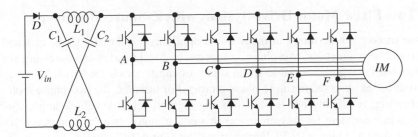

Figure 16.25   Six-phase Z-source inverter feeding an IM.

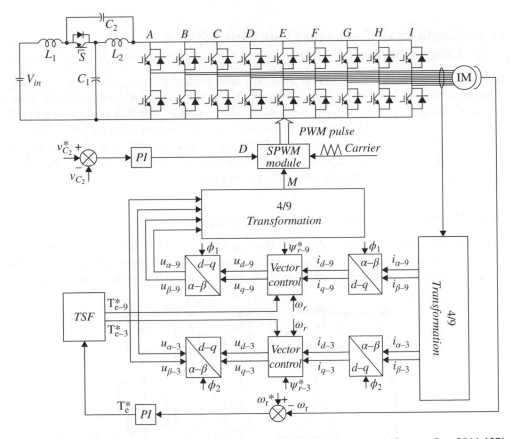

Figure 16.26   The control scheme for a nine-phase QZSI IM drive system (*Source*: Sun 2011 [57]. Reproduced with permission of IEEE).

(NN) classification is used for SVPWM calculations; the NN classification reduces the required computational efforts and makes it possible to increase the switching frequency [55]. Figure 16.26 shows the complete control scheme of a nine-phase QZSI IM drive system. It consists of the qZS network capacitor voltage controller and double vector controller of the nine-phase IM.

16.7   Two-Phase Motor Drive System with Z-Source Inverter

Two-phase motors can be found in many home appliances, industrial tools, or small power applications. The symmetrical two-phase motors can be considered as an interesting alternative for a fractional horsepower variable-speed drive, owing to its characteristic of no pulsating torque produced as observed in single-phase motors. For instance, the two-phase motor can be used for heating, ventilating, and air conditioning in hybrid electric vehicle applications [58].

In [59], a drive system for a symmetrical two-phase motor using a Z-source inverter was proposed, as shown in Figure 16.27. Furthermore, the hybrid PWM strategy was applied to the two-phase ZSI motor drive system to guarantee the boost operation of the ZSI and the control of the two-phase IM.

16.8   Single-Phase Induction Motor Drive System Using Z-Source Inverter

Single-phase induction motors (SPIM) are widely used for low power applications, especially in appliances where three-phase supply is not available. Some examples of these applications are water pump, air conditioner, and washing machine. In [60], a ZSI is applied to a SPIM drive system, so that the disadvantages of low VUF in conventional three-phase inverter are eliminated. The schematic diagram of a SPIM drive system with a ZSI is shown in Figure 16.28(a) and Figure 16.28(b) shows the control block diagram of a ZSI using a V/F method to generate the switching pulses and operate the SPIM.

16.9   Z-Source Inverter for Vehicular Applications

In the literature, there are a lot of ZSI vehicular applications [61]. In [62] is a comparison of three different inverters: traditional voltage source inverter (VSI), boosted VSI, and ZSI for fuel cell vehicle (FCV) applications, with a conclusion that the ZSI offers higher efficiency

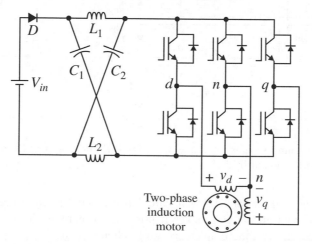

Figure 16.27 Two-phase motor drive system with Z-source converter (*Source*: Santos 2010 [59]. Reproduced with permission of IEEE).

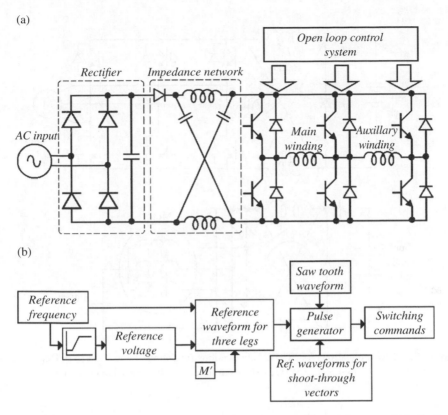

Figure 16.28   SPIM drive system using ZSI: (a) schematic diagram and (b) control block diagram. (*Source*: Rajaei 2010 [59]. Reproduced by permission of the Institution of Engineering & Technology. Full acknowledgment to the Author, Title, and date of the original work.)

with fewer switching devices and higher passive components requirement. The ZSI can also be applied to fuel cell hybrid electric vehicle (FCHEV) with slight modification to include a battery; two configurations were proposed for FCHEVs as shown in Figure 16.29(a), (b) [63, 64]. The first configuration uses a battery connected in parallel with one of the Z-network capacitors. This configuration has some disadvantages such as a high voltage battery must be used (the battery voltage equal to the Z-network capacitor voltage) and the dc-link voltage is twice the battery voltage during regenerative braking when disconnecting the fuel cell stack from the converter input terminals, which may damage the inverter switches. The second configuration uses a battery connected to the motor neutral point, but this also has some disadvantages such as some dc current flows through the traction motor windings, which increases the motor copper loss, therefore the motor needs to be oversized. A bidirectional Z-source nine-switch inverter (BZS-NSI), as shown in Figure 16.30, was proposed in [65] to replace the conventional series HEV configuration with a bidirectional dc/dc converter and two VSIs for bidirectional power transfer between a battery, an electric motor, and an electric generator. This BZS-NSI has some disadvantages such as lower reliability, since a failure in one switch will disable the complete system; high current switch stress; and complicated

(a)

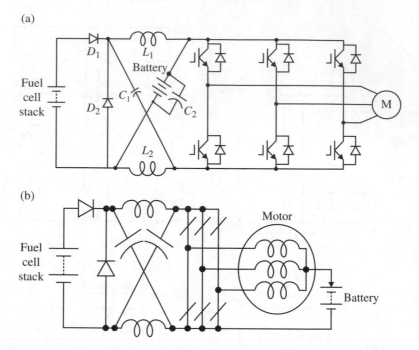

(b)

Figure 16.29 ZSI applications for FCHEV applications (a: *Source*: Peng 2007 [63]. Reproduced with permission of IEEE, b: *Source*: Shen 2007 [64]. Reproduced with permission of IEEE).

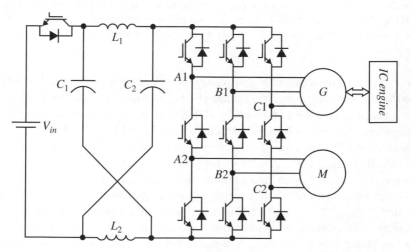

Figure 16.30 Bidirectional Z-source nine-switch inverter (BZS-NSI) for HEV applications (*Source*: Dehghan 2010 [65]. Reproduced with permission of IEEE).

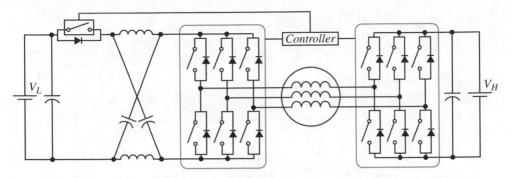

Figure 16.31 Bidirectional Z-source inverter in double-ended inverter drive system for HEV [66].

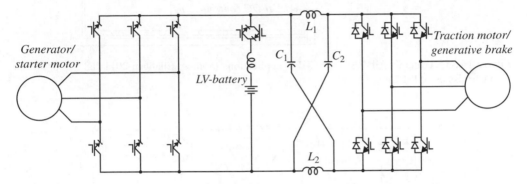

Figure 16.32 Back-to-back Z-source topology in HEV applications (*Source*: Rogers 2011 [67]. Reproduced with permission of IEEE).

control algorithm. In [66], a BZSI is used in double-ended inverter drive system, as shown in Figure 16.31, where the BZSI can be operated in boost mode, buck mode as a normal VSI, or a charger mode. The back-to-back ZSI topology was presented in [66], as shown in Figure 16.32, where the Z-network is inserted into the dc-link, so that the voltage level at the traction inverter is controlled independently of that at the generator inverter. A fuel cell hybrid electric vehicle (FCHEV) system supplied by a fuel cell system and a super-capacitor module using the high-performance Z-source inverter was presented in [68], as shown in Figure 16.33. The super-capacitor module delivers the transient and instantaneous peak power demands and absorbs the deceleration and regenerative braking energies while the fuel cell system delivers the load mean power. In [69], the Z-source inverter has been used in place of voltage source inverter and investigates the performance of locomotive drives, as shown in Figure 16.34.

16.10   Conclusion

This chapter gave a review of the use of the Z-source inverter in motor drive applications. Different ZSI topologies: ZSI, BZSI, HP-ZSI, QZSI, and BQZSI; different motor types: PMBDCM, SRM, PMSM, and IM for different phase numbers: single, two, three, five, and nine phases and different control algorithms: V/F, FOC, DTC, and PTC have been reviewed.

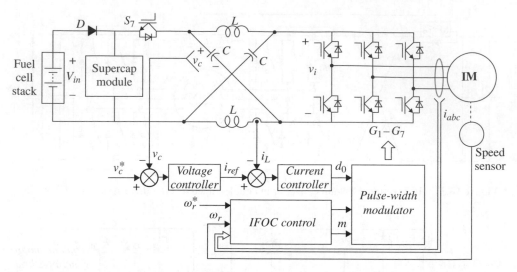

Figure 16.33   Using the HP-ZSI for FCHEV applications (*Source*: Ellabban 2011 [68]. Reproduced with permission of IEEE).

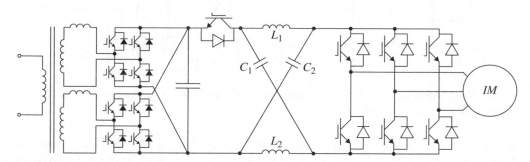

Figure 16.34   Z Source inverter fed locomotive drives (*Source*: Vasanthi 2012 [69]. Reproduced with permission of IEEE).

This chapter also presented different examples of utilizing the Z-source inverter for various vehicular applications. In summary, the Z-source inverter ASD system has several unique advantages that are very desirable for many ASD applications, such as ride-through capability during voltage sags, reduced line harmonics, improved power factor and reliability, and extending the output voltage range.

References

[1] F. Z. Peng, X. Yuan, X. Fang, Z. Qian, "Z-source inverter for adjustable speed drives," *IEEE Power Electron Lett.*, vol.1, no.2, pp.33–35, June 2003.

[2] F. Z. Peng, A. Joseph, J. Wang, M. Shen, L. Chen, Z. Pan, E. Ortiz-Rivera, Y. Huang, "Z-source inverter for motor drives," *IEEE Trans. Power Electron.*, vol.20, no.4, pp.857–863, July 2005.

[3] F. Z. Peng, M. Shen, A. Joseph, "Z-source inverters, control, motor drive applications," *KIEE International Transactions on Electrical Machinery and Energy Conversion System*, vol.5-B, no.1, pp.6–12, 2005.

[4] L. Sack, B. Piepenbreier, M. V. Zimmermann, "Z-source inverter for general purpose drives in motoring and regenerating operation," in *Proc. International Symposium on Power Electronics, Electrical Drives, Automation and Motion (SPEEDAM)*, pp.766–771, 2008.

[5] L. Sack, B. Piepenbreier, M. von Zimmermann, "Dimensioning of the Z-source inverter for general purpose drives with three-phase standard motors," in *Proc. IEEE Power Electronics Specialists Conference (PESC)*, 15–19 June 2008, pp.1808–1813.

[6] M. V. Zimmermann, S. Labusch, B. Piepenbreier, "Bidirectional AC-AC Z-source inverter with active rectifier and feedforward control," in *Proc. IEEE Energy Conversion Congress and Exposition (ECCE)*, pp.3180–3186, 12–16 Sept. 2010.

[7] Y. P. Siwakoti, G. E. Town, "Common-mode voltage reduction techniques of three-phase quasi Z-source inverter for AC drives," in *Proc. Twenty-Eighth Annual IEEE Applied Power Electronics Conference and Exposition (APEC)*, pp.2247–2252, 17–21 March 2013.

[8] D. Kumar, Z. Husain, "A comparative study of Z-source inverter fed three-phase IM drive with CSI and VSI fed IM," *International Journal of Power Electronics and Drive System (IJPEDS)*, vol.3, no.3, September 2013, pp.259–270.

[9] O. Ellabban, J. V. Mierlo, P. Lataire, "Comparison between different PWM control methods for different Z-source inverter topologies," in *Proc. 13th European Conference on Power Electronics and Applications*, pp.1–11 Sept. 2009.

[10] O. Ellabban, J. V. Mierlo, P. Lataire, "Experimental study of the shoot-through boost control methods for the Z-source inverter," *European Power Electronics Journal (EPE J)*, vol.21, no.2, pp.18–29, June 2011.

[11] J. Rabkowski, "The bidirectional Z-source inverter as an energy storage/grid interface," in *Proc. International Conference on "Computer as a Tool" (EUROCON)*, pp.1629–1635, 9–12 Sept. 2007.

[12] X. Ding, Z. Qian, S. Yang, B. Cui, F. Z. Peng, "A high-performance Z-source inverter operating with small inductor at wide-range load," *Proc. Twenty Second Annual IEEE Applied Power Electronics Conference (APEC)*, pp.615–620.

[13] J. Anderson, F. Z. Peng, "Four quasi-Z-source inverters," in *Proc. IEEE Power Electronics Specialists Conference (PESC)*, pp.2743–2749, 15–19 June 2008.

[14] F. Guo, L. Fu, C. Lin, C. Li, W. Choi, J. Wang, "Development of an 85 kW bidirectional quasi-Z-source inverter with DC-link feed-forward compensation for electric vehicle applications," *IEEE Trans. Power Electron.*, vol.28, no.12, pp.5477–5488, Dec. 2013.

[15] Omar Ellabban, Haitham Abu-Rub, An overview for the Z-Source Converter in motor drive applications, Renewable and Sustainable Energy Reviews, vol. 61, August 2016, pp. 537–555.

[16] J. Wang, L. Zhou, G. Tao, J. Shi, "Modeling and simulation of a permanent magnet brushless DC motor fed by PWM Z-source inverter," in *Proc. International Conference on Electrical Machines and Systems (ICEMS)*, pp.834–838, 8–11 Oct. 2007.

[17] G. Buja, R. Keshri, R. Menis, "Characteristics of Z-source inverter supply for permanent magnet brushless motors," in *Proc. 35th Annual Conference of IEEE Industrial Electronics (IECON)*, pp.1234–1239, 3–5 Nov. 2009.

[18] G. Buja, R. Keshri, R. Menis, "Comparison of DBI and ZSI supply for PM brushless DC drives powered by fuel cell," in *Proc. IEEE International Symposium on Industrial Electronics (ISIE)*, pp.165–170, 27–30 June 2011.

[19] P. G. Peter, M. Rajaram "An enhanced Z-source inverter topology based permanent magnet brushless DC motor drive speed control," *International Journal of Electronics*, pp.1–17, 2014.

[20] J. Shang, Z. Liu, Y. Zhang, "Analysis and Experimental Validation of Performance of Z-source Inverter-fed Permanent Magnet Brushless DC Motor Drive," *Advanced Materials Research*, vol.383–390, pp.7063–7068, 2012.

[21] S. H. Hosseini, F. Nejabatkhah, S. A. K. Mozafari Niapoor, S. Danyali, "Supplying a brushless dc motor by Z-source PV power inverter with FL-IC MPPT," in *Proc. International Conference on Green Circuits and Systems (ICGCS)*, pp.485–490, 21–23 June 2010.

[22] S. A. K. Mozafari Niapoor, S. Danyali, M. B. B. Sharifian, "PV power system based MPPT Z-source inverter to supply a sensorless BLDC motor," in *Proc. 1st Power Electronic & Drive Systems & Technologies Conference (PEDSTC)*, pp.111–116, 17–18 Feb. 2010.

[23] M. R. Feyzi, S. A. K. Mozafari Niapoor, S. Danyali, M. Shafiei, "Supplying a brushless DC motor by Z-source PV power inverter with FLC-IC MPPT by DTC drive," in *Proc. International Conference on Electrical Machines and Systems (ICEMS)*, pp.694–699, 10–13 Oct. 2010.

[24] S. A.K. Mozafari Niapoor, S. Danyali, M. B. B. Sharifian, M. R. Feyzi, "Brushless DC motor drives supplied by PV power system based on Z-source inverter and FL-IC MPPT controller," *Energy Conversion and Management*, vol.52, pp.3043–3059, 2011.

[25] C. Xia, X. Li, "Z-source inverter-based approach to the zero-crossing point detection of back EMF for sensorless brushless DC motor," *IEEE Trans. Power Electron.*, vol.30, no.3, pp.1488–1498, March 2015.

[26] O. Ellabban, H. Abu-Rub, "Torque control strategies for a high performance switched reluctance motor drive system," in *Proc. 7th IEEE GCC Conference and Exhibition (GCC)*, pp.257–262, 17–20 Nov. 2013.

[27] M. H. Prabhu, "SRM drives using Z-source inverter with the simplified fuzzy logic rule base," *International Electrical Engineering Journal (IEEJ)*, vol.5, no.2, pp.1280–1286, 2014.

[28] P. Liu, H. Liu, "Permanent-magnet synchronous motor drive system for electric vehicles using bidirectional Z-source inverter," *IET Electrical Systems in Transportation*, vol.2, no.4, pp.178–185, December 2012.

[29] P. Liu, H. Lu, "Application of Z-source inverter for permanent-magnet synchronous motor drive system for electric vehicles," *Procedia Engineering*, vol.15, pp.309–314, 2011.

[30] S. Tenner, S. Gunther, W. Hofmann, "Loss minimization of electric drive systems using a Z-source inverter in automotive applications," in *Proc. 15th European Conference on Power Electronics and Applications (EPE)*, pp.1–8, 2–6 Sept. 2013.

[31] A. Battiston, J.-P. Martin, E.-H. Miliani, B. Nahid-Mobarakeh, S. Pierfederici, F. Meibody-Tabar, "Control of a PMSM fed by a quasi Z-source inverter based on flatness properties and saturation schemes," in *Proc. 15th European Conference on Power Electronics and Applications (EPE)*, pp.1–10, 2–6 Sept. 2013.

[32] A. Battiston, E.-H. Miliani, J.-P. Martin, B. Nahid-Mobarakeh, S. Pierfederici, F. Meibody-Tabar, "A control strategy for electric traction systems using a PM-motor fed by a bidirectional Z-source inverter," *IEEE Trans. Veh. Technol.*, vol.63, no.9, pp.4178–4191, Nov. 2014.

[33] M. Li, J. He, N. A. O. Demerdash, "A flux-weakening control approach for interior permanent magnet synchronous motors based on Z-source inverters," in *Proc. IEEE Transportation Electrification Conference and Expo (ITEC)*, pp.1–6, 15–18 June 2014.

[34] X. Ding, Z. Qian, S. Yang, B, Cui, F. Z. Peng, "A new adjustable-speed drives (ASD) system based on high-performance Z-source inverter," in *Proc. 42nd IEEE Industry Applications Annual Meeting*, pp.2327–2332, 2007.

[35] X. Ding, J. Liu, Y. Lu, B. Yu, "A novel adjustable-speed system based on Z-source inverter," in *Proc. 30th Chinese Control Conference (CCC)*, pp.3523–3527, 22–24 July 2011.

[36] M. Baba, C. Lascu, I. Boldea, F. Blaabjerg, "Parallel and series 4 switch Z-source converters in induction motor drives," in *Proc. International Conference on Optimization of Electrical and Electronic Equipment (OPTIM)*, pp.639–646, 22–24 May 2014.

[37] F. Khosravi, N. A. Azli, A. Kaykhosravi, "A new single-phase to three-phase converter using quasi Z-source network," *IEEE Applied Power Electronics Colloquium (IAPEC)*, pp.46–50, 18–19 April 2011.

[38] S. Thangaprakash, A. Krishnan, "Current mode integrated control technique for Z-source inverter fed induction motor drives," *Journal of Power Electronics*, vol.10, no.3, pp.285–292, May 2010.

[39] O. Ellabban, J. V. M., P. Lataire, "A new closed loop speed control of induction motor fed by a high performance Z-source inverter," in *Proc. the IEEE Electrical Power and Energy Conference (EPEC)*, August 25–27, Halifax, Canada.

[40] O. Ellabban, H. Abu-Rub, "Indirect field-oriented control of an induction motor fed by a bidirectional quasi-Z-source inverter," in *Proc. 38th Annual Conference of IEEE Industrial Electronics (IECON)*, Montreal, Canada, pp.5297–5302, 25–28 Oct 2012.

[41] U. Flisar, D. Vončina, P. Zajec, "Voltage sag independent operation of induction motor based on Z-source inverter," in *Proc. International Journal for Computation and Mathematics in Electrical and Electronic Engineering (COMPEL)*, vol.31 no.6, 2012, pp.1931–1944.

[42] A. Usman, T. Izhar, "Z-source inverter based indirect field oriented control with synchronous current injection and feed-forward CEMF compensation for an induction motor drive," in *Proc. IEEE Student Conference on Research and Development (SCOReD)*, pp.40–45, 19–20 Dec.

[43] A. A. Kumar, D. Singh, "DSP based IFO control of HEV fed through impedance source inverter," in *Proc. Annual IEEE India Conference (INDICON)*, pp.1–6, 13–15 Dec. 2013.

[44] A. K. Akkarapaka, D. Singh, "The IFOC based speed control of induction motor fed by a high performance Z-source inverter," in *Proc. International Conference on Renewable Energy Research and Application (ICRERA)*, pp.539–543, 19–22 Oct. 2014.

[45] O. S. Ellabban, H. A. A. Fattah, H. M. Emara, A. F. Sakr, "Particle swarm optimized direct torque control of induction motors," in *Proc. 32nd Annual Conference on Industrial Electronics (IECON)*, 6–10 Nov. 2006, pp.1586–1591.

[46] S. A. Zaid, O. A. Mahgoub, K. A. El-Metwally, "Implementation of a new fast direct torque control algorithm for induction motor drives," *IET Electr. Power Appl.*, 2010, vol.4, no.5, pp.305–313.

[47] O. Ellabban, J. V. Mierlo, P. Lataire, "Direct torque controlled space vector modulated induction motor fed by a Z-source inverter for electric vehicles," in *Proc. International Conference on Power Engineering, Energy and Electrical Drives (POWERENG)*, pp.1–7, 11–13 May 2011.

[48] C.-T. Pham, A. W. Shen, P. Q. Dzung, N. B. Anh, L. H. Viet, "Self-tuning fuzzy PI-type controller in Z-source inverter for hybrid electric vehicles," *International Journal of Power Electronics and Drive System (IJPEDS)*, vol.2, no.4, pp.353–363, December 2012.

[49] O. Ellabban, J. V. Mierlo, P. Lataire, "A comparative study of different control techniques for induction motor fed by a Z-source inverter for electric vehicles," in *Proc. International Conference on Power Engineering, Energy and Electrical Drives (POWERENG)*, pp.1–7, 11–13 May 2011.

[50] R. Kennel, J. Rodriguez, J. Espinoza, M. Trincado, "High performance speed control methods for electrical machines: An assessment," in *Proc. IEEE International Conference on Industrial Technology*, pp.1793–1799.

[51] O. Ellabban, H. Abu-Rub, J. Rodríguez, "Predictive torque control of an induction motor fed by a bidirectional quasi Z-source inverter," in *Proc. 39th Annual Conference of the IEEE Industrial Electronics Society (IECON)*, pp.5854–5859, 10–13 November 2013.

[52] S. A. Davari, D. Arab Khaburi, "Using predictive control and q-ZSI to drive an induction motor supplied by a PV generator," in *Proc. 5th Power Electronics, Drive Systems and Technologies Conference (PEDSTC)*, pp.61–65, 5–6 Feb. 2014.

[53] E. Levi, "Multi-phase electric machines for variable-speed applications," *IEEE Trans. Ind. Electron.*, vol.55, no.5, pp.1893–1909, May 2008.

[54] O. Ellabban, H. Abu-Rub, "Field oriented control of a five phase induction motor fed by a Z-source inverter," in *Proc. IEEE International Conference on Industrial Technology (ICIT)*, pp.1624–1629, 2013.

[55] S. M. J. Rastegar Fatemi, J. Soltani, N. R. Abjadi, G. R. A. Markadeh, "Space-vector pulse-width modulation of a Z-source six-phase inverter with neural network classification," *IET Power Electron.*, vol.5, no.9, pp.1956–1967, November 2012.

[56] A. Kouzou, H. Abu-Rub, "Multi-phase Z-source inverter using maximum constant boost control," *Archives of Control Sciences*, vol.23, no.1, pp.107–26, 2013.

[57] D. Sun, B. Ge, W. Wu, D. Bi, F. Z. Peng, H. Abu-Rub, "Quasi-Z source inverter based pole-phase modulation machine drive system," in *Proc. International Conference on Electrical Machines and Systems (ICEMS)*, pp.1–6, 20–23 Aug. 2011.

[58] F. Blaabjerg, F. Lungeanu, K. Skaug, M. Tonnes, "Two-phase induction motor drives," *IEEE Ind. Appl. Mag.*, vol.10, pp.24–32, July-Aug 2004.

[59] E. C. dos Santos, M. Pacas, M. G. Molina, "Two-phase motor drive systems with Z-source inverter and hybrid PWM," in *Proc. IEEE Energy Conversion Congress and Exposition (ECCE)*, pp.3877–3882, 12–16 Sept. 2010.

[60] A. H. Rajaei, M. Mohamadian, S. M. Dehghan, A. Yazdian, "Single-phase induction motor drive system using Z-source inverter," *IET Electric Power Appl.*, vol.4, no.1, pp.17–25, January 2010.

[61] O. Ellabban, J. V. Mierlo, P. Lataire, P. V. den Bossche, "Z-source inverter for vehicular applications," in *Proc. IEEE Vehicle Power and Propulsion Conference (VPPC)*, pp.1–6, 6–9 September 2011.

[62] M. Shen, A. Joseph, J. Wang, F. Z. Peng, D. J. Adams "Comparison of traditional inverters and Z-source inverter for fuel cell vehicles," *IEEE Trans. Power Electron.*, vol.22, no.4, pp.1453–1463, July 2007.

[63] F. Z. Peng, M. Shen, K. Holland, "Z-source inverter control for traction drive of fuel cell – battery hybrid vehicles," *IEEE Trans. Power Electron.*, vol.22, no.3, May/June, 2007, pp.1054–1061.

[64] M. Shen, S. Hodek, F. Z. Peng, "Control of the Z-source inverter for FCHEV with the battery connected to the motor neutral point," in *Proc. IEEE Power Electronics Specialist Conference (PESC)*, 2007, pp.1485–1490.

[65] S. M. Dehghan, M. Mohamadian, A. Yazdian, "Hybrid electric vehicle based on bidirectional Z-source nine-switch inverter," *IEEE Trans. Veh. Technol.*, vol.59, no.6, pp.2641–2653, July 2010.

[66] B. A. Welchko, J. M. Nagashima, G. S. Smith, S. Chakrabarti, M. Perisic, G. John, "Double ended inverter system with an impedance source inverter subsystem," United States Patent US7956569, 06/07/2011.

[67] C. B. Rogers, F. Z. Peng, "Back to back Z-source inverter topology for the series hybrid electric bus," in *Proc. IEEE Energy Conversion Congress and Exposition (ECCE)*, pp.2353–2357, 17–22 Sept. 2011.

[68] O. Ellabban, V. Joeri, P. Lataire, P. V. Bossche, "Z-source inverter for vehicular applications," in *2011 IEEE Vehicle Power and Propulsion Conference (VPPC)*, pp.1–6, 6–9 Sept. 2011.

[69] V. Vasanthi, S. Ashok, "Performance evaluation of Z source inverter fed locomotive drives with different control topologies," in *Proc. IEEE International Conference on Power Electronics, Drives and Energy Systems (PEDES)*, pp.1–4. 2012.

17

Impedance Source Multi-Leg Inverters

Sertac Bayhan[1,2] and Haitham Abu-Rub[2]

[1] Department of Electronics and Automation, Gazi University, Ankara, Turkey
[2] Electrical and Computer Engineering Program, Texas A&M University at Qatar, Qatar Foundation, Doha, Qatar

17.1 Impedance Source Four-Leg Inverter

17.1.1 Introduction

Standalone distribution generation (DG) systems have recently received much attention for their advantages in supplying power to remote customers. They constitute the most promising alternative to the expensive and complicated grid connection. A variety of standalone systems are possible which includes single home, small and large communities, islands, satellite stations, aircraft, ship propulsion systems, and large-scale computer systems. However, these DG systems require power converters (dc-dc and/or dc-ac) to generate suitable power with specific voltage and frequency. Therefore, power electronic converters play an important role in providing clean power to the loads in standalone power systems. The loads are of single-/three-phase, balanced/unbalanced, and linear/non-linear nature [1].

In most cases, traditional three-leg voltage source inverters (VSIs) are proven to be the best candidates for supplying three-phase balanced loads. This approach is easy to implement. However, it suffers from two major drawbacks [2, 3].

The first drawback is insufficient input voltage due to the intermittent and stochastic nature of renewable energy sources (RESs). The VSIs must have an input voltage that is greater than the maximum value of the line-to-line output voltage in order to guarantee reliable and uninterruptible power for the loads, which is a major challenge in such systems. To overcome this drawback, frequently controllable dc-dc converters are used as an input stage [4]. By using dc-dc converters in the input stage, the output voltage of the

Impedance Source Power Electronic Converters, First Edition. Yushan Liu, Haitham Abu-Rub, Baoming Ge, Frede Blaabjerg, Omar Ellabban, and Poh Chiang Loh.
© 2016 John Wiley & Sons, Ltd. Published 2016 by John Wiley & Sons, Ltd.

VSI can be regulated as desired. However, this solution brings some problems such as a more complex power circuit and control structure, lower reliability, and higher system cost. Recently, a number of new power converter topologies have been proposed to cope with these problems [5, 8]. One of these topologies is the voltage fed Z-source inverter (ZSI) that has appeared as an emerging power converter topology for RES-based DG systems [9, 11]. The main advantage of this topology is that the number of switches is lower than the conventional two-stage power conversion systems (dc-dc and dc-ac) and it does not require protecting leg switches to prevent short-circuit. These ZSIs replace the dc-dc input stage with a simple LC network, which allows the input voltage of the VSI to be varied as desired [12].

Another disadvantage of three-phase VSI topology is that most of these inverters are designed to supply balanced three-phase voltage to the loads. However, unbalanced load conditions are common for DG systems, where the power is delivered to the local loads. The single- or three-phase, (un)balanced and/or (non)linear loads are supplied by the three-phase four-wire systems in many commercial and industrial applications. The neutral connection to the loads should be provided by the VSI or by an additional star-delta or zig-zag transformer connection [13]. The transformerless neutral connection can be accomplished by connecting the neutral point of the load to the mid-point of split dc-link capacitors of a three-leg VSI or to the mid-point of the additional fourth (neutral) leg. With the practice of split dc-link capacitors, the three-leg VSI acts as three single-phase half-bridge inverters and causes poor utilization of the dc-link [13]. Since the unbalanced neutral current flows through the capacitors, a larger capacitor is also required to maintain an acceptable voltage ripple. Moreover, the step change in the load currents causes a surge, which will destroy the dc-link capacitors. On the other hand, the three-phase four-leg inverters are promising candidates for three-phase four-wire applications. In this configuration, the load neutral is connected to the fourth-leg instead of dc-link capacitor mid-point. The four-leg inverter provides enhanced dc-link utilization (15% higher compared to the three-leg inverter), a lower ripple on the dc-link voltage, and smaller size for dc-link capacitors [14, 15].

To eliminate the drawbacks of the two-stage dc-dc converters and three-phase VSI, the Z-source four-leg inverter is examined in this chapter, starting with an analysis of the unbalanced load, and the effects of an unbalanced load. Then, the different inverter topologies for mitigation of unbalanced conditions are introduced. The Z-source four-leg inverter topology is then presented, along with a description of the topology, model, and switching schemes for control. Also, the control methods of the Z-source inverters are presented with the three most popular boost control methods.

17.1.2   Unbalanced Load Analysis Based on Fortescue Components

To quantify an unbalance in the voltage or current of a three-phase system, the so-called Fortescue components or symmetrical components are used [16]. The three-phase system is decomposed into a so-called positive-sequence, negative-sequence, and zero-sequence system, indicated by subscripts p, n, 0.

According to the theory, an asymmetrical three-phase signal (either voltage or current) can be represented as a sum of positive, negative, and zero-sequence components [13–17]. The subscripts a, b, c indicate the different phases. The transformation is expressed as

$$\begin{bmatrix} X_p \\ X_n \\ X_0 \end{bmatrix} = \frac{1}{3} \begin{bmatrix} 1 & a & a^2 \\ 1 & a^2 & a \\ 1 & 1 & 1 \end{bmatrix} \begin{bmatrix} X_a \\ X_b \\ X_c \end{bmatrix} \tag{17.1}$$

where X could be V (voltage) or I (current) and the rotation operator a is given by

$$a = e^{j2\pi/3} \tag{17.2}$$

These transformations are energy-invariant, so any power quantity calculated with the original or transformed values will result in the same value. The inverse transformation is given by

$$\begin{bmatrix} X_a \\ X_b \\ X_c \end{bmatrix} = \begin{bmatrix} 1 & 1 & 1 \\ a^2 & a & 1 \\ a & a^2 & 1 \end{bmatrix} \begin{bmatrix} X_p \\ X_n \\ X_0 \end{bmatrix} \tag{17.3}$$

The positive system is associated with a positively rotating field whereas the negative system yields a negatively rotating field. Zero components have identical phase angles and simply oscillate at a particular frequency. In systems without neutral conductors zero currents can obviously not flow, but significant voltage differences between the "zero voltages" at the neutral points of the Y-connections in the supply system and the loads may arise.

Since the neutral current in a three-phase four-wire system is defined as in (17.4), it is clear that the neutral current equals three times the zero-sequence current.

$$I_n = -\left(I_a + I_b + I_c \right) \tag{17.4}$$

According to [13] and [18], the IEC gives a definition of the degrees of the unbalance, the unbalance factor is described in terms of the negative-sequence unbalance factor, and the zero-sequence unbalance factor expressed in the following equations

$$Unbal\_N\% = \frac{X_n}{X_p} \times 100$$
$$\tag{17.5}$$
$$Unbal\_0\% = \frac{X_0}{X_p} \times 100$$

where X_p, X_n, X_0 are positive, negative, and zero-sequence components, respectively. Figure 17.1 shows the decomposition of an unbalanced three-phase signal.

17.1.3 Effects of Unbalanced Load Condition

A three-phase power system is called balanced if the three-phase voltages and currents have the same amplitude and they are phase shifted by 120° with respect to each other. If either or

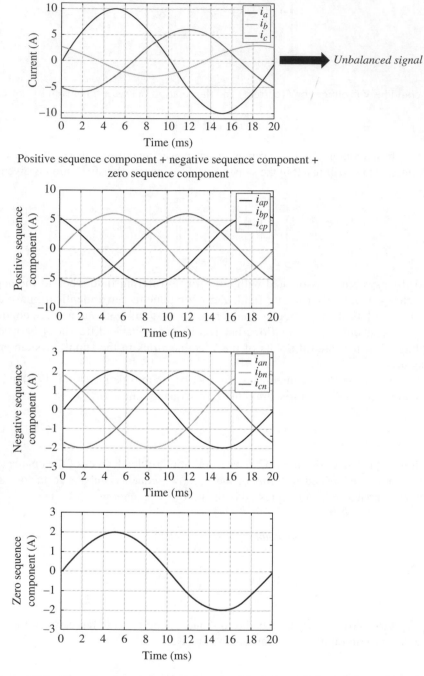

Figure 17.1 Example of a symmetrical decomposition of an unbalanced three-phase signal.

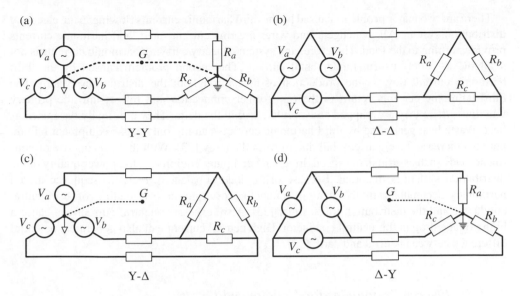

Figure 17.2 Possible connections between the power source and the load: (a) Y-Y connection, (b) Δ-Δ connection, (c) Y-Δ connection, (d) Δ-Y connection.

both of these conditions are not met, the system is called unbalanced. There are four types of connection between the load side and the source side of a power system.

A Y-Y connection is shown in Figure 17.2(a), a negative-sequence current will circulate at twice the fundamental frequency between the load side and source side through the three phase conductors. The zero-sequence current will travel at a fundamental frequency between the neutral points of the source and load through the neutral line. Since the neutral line is generally designed for balanced conditions, the neutral current will cause excess heat in the line, and it is even worse when the load is non-linear [13, 19].

For the Δ-Δ connection shown in Figure 17.2(b), there is no neutral point on either the source side or the load side. The effect of the zero-sequence current is eliminated because the zero-sequence current normally circulates in the Δ configuration of the load side. The negative-sequence current will circulate at twice the fundamental frequency of the two sides, causing the voltage potential to shift constantly at both sides. The potential difference of both sides, through parasitic capacitance, could lead to a common-mode current, thereby causing EMI issues [20].

For Y-Δ and Δ-Y connections as shown in Figure 17.2(c), (d), apart from the fact that negative-sequence current will be the same as in Δ-Δ connections, there is a zero-sequence current circulating in the Δ configuration side.

In the three-phase four-wire system, loads are connected from line to the neutral of the three phases. In an ideal balanced sinusoidal three-phase power system, the neutral current is the vector sum of the three phase currents and should be equal to zero. Under normal operating conditions, some phase unbalance occurs resulting in a small neutral current. Non-linear loads draw highly distorted phase current. In the non-linear circuits, the tripled harmonic current adds arithmetically in the neutral line resulting in high harmonic distortion in the neutral [21]. The third harmonic is the main contributor of harmonic current [22].

There are two major problems caused by the third harmonic currents flowing in the electrical distribution system. All switchgear and wires are rms current rated, and harmonic currents provide nothing to the load. Therefore, any system capacity carrying harmonic currents is not available to supply useful fundamental current. The second problem caused by the third harmonic current flow is the production of heat throughout the distribution system. The third harmonic currents return to the transformer connection and reflect into the primary winding, where they are trapped and circulate within the delta. They are finally dissipated as heat. Waste heat generated by third harmonic currents can not only cause equipment failure, but also increase the electricity bill due to wasted energy [23]. With the growing use of non-linear loads such as adjustable speed drives and computer equipment, the power quality in the distribution system is distorted. Because of the load unbalances, the zero sequence tripled harmonics accumulate in the neutral conductor resulting in overloading of the neutral conductor and the distribution transformer [23]. Also, excessive neutral current can cause a high voltage drop in the neutral conductor. High neutral current can also generate a potential difference between neutral and earth [13].

17.1.4   Inverter Topologies for Unbalanced Loads

There are many converter topologies available for three-phase four-wire systems. In the following, the main types and variations of these topologies are described in details.

17.1.4.1   A Three-Phase Four-Wire VSI with Split DC-Link Capacitors

A very simple approach to connecting the neutral conductor is to connect the star point of the loads to the mid-point of the dc-link capacitors as shown in Figure 17.3. This connection arrangement is easy to install. However, it has two major drawbacks.

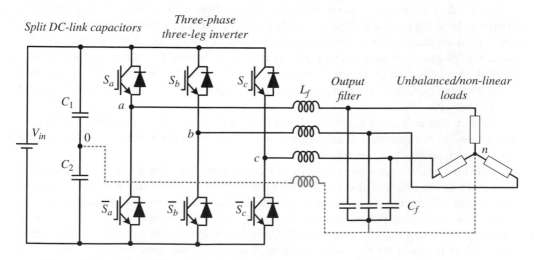

Figure 17.3   A three-phase four-wire VSI with split dc-link capacitors and an output filter.

The first drawback is the low utilization of the dc-link voltage. By connecting the start point and mid-point of the dc-link capacitors together, the potential difference between the two points is then forced to be zero, because the variation of the voltage between the mid-point of dc-link and the ground is slow and small [23]. This means the star point of the load is not free to float regardless of the switching schemes that are used. In this case, the maximum phase voltage is 0.5 V_{DC}, which shows that dc-link utilization is limited to $\sqrt{3}/2$. This is 15% less than would be achieved with a three-phase three-leg VSI using the SVM scheme [13].

Another disadvantage of this topology is that two large capacitors (C_1 and C_2) are needed to deal with the current ripples caused by both negative-sequence current and zero-sequence current [21]. The choice of the dc-link capacitors is a complex task, and the size of the capacitor depends on the ripple level [21]. As a result, this topology will increase the cost of the system dramatically, especially for high power applications.

17.1.4.2 A Three-Phase Four-Wire VSI with a Δ-Y transformer

An alternative approach for connecting the neutral point is to connect the star point of the loads to the secondary side of a Δ-Y transformer. As can be seen in Figure 17.4, owing to the strong transformer coupling effect and by keeping the inductors on the primary side of the transformer, the zero-sequence current will only flow in the primary side windings [24]. This prevents the zero-sequence current from returning to the dc-link, and it is widely used in UPS applications [24]. However, this topology will not solve the unbalanced/non-linear load conditions caused by the negative-sequence current. Another disadvantage is the size of the transformer, which makes the whole system more bulky and expensive.

A similar topology uses a zig-zag transformer instead of a Δ-Y transformer [25, 26]. This is based on the idea that the zero-sequence current can be canceled out by using the phase shift of each phase of the transformer. The size and cost are still a disadvantage for this topology.

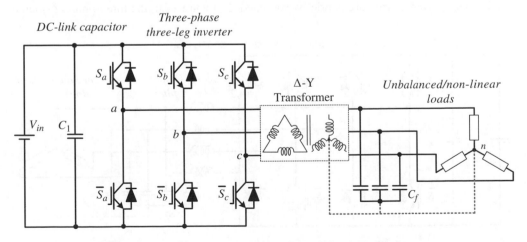

Figure 17.4 A three-phase four-wire VSI with a Δ-Y transformer.

17.1.4.3   A Three-Phase Four-Leg VSI

A three-phase four-leg VSI has been introduced to handle the neutral current caused by unbalanced and/or non-linear load conditions [27, 28]. As shown in Figure 17.5, the load star point is connected to the mid-point of a fourth leg of the converter, and by doing so, the neutral point is now controllable through the fourth leg. The advantages of this topology are as follows [13]:

- Full dc-link voltage utilization. Unlike with the three-phase four-wire VSI with split dc-link capacitors, the star point of the loads is now free to vary by controlling the fourth leg. Therefore maximum unity dc-link utilization can be realized by adopting switching schemes such as 3D SVM [29], and carrier-based PWM [30]. These will be presented in more detail in the following sections.
- The dc-link capacitors can be smaller compared to the split dc-link option. Since the star point of the loads is now tied to the mid-point of the fourth leg, there is no need to use two large capacitors for the dc-link.
- A transformerless converter. Compared to the converters that are either use a Δ-Y transformer or a zig-zag transformer, this topology does not need such a large or heavy transformer. An extra pair of IGBT/diode is a comparatively small size/weight increase compared with a transformer. Of course, a more complicated switching scheme and control method are needed for this topology, but with the availability of the modern controller system, this should not present a problem.
- EMI and common-mode voltage reduction. In any fast switching power electronic system, EMI is a potential problem that increases with the voltage levels. For a three-phase system, common-mode voltage switching noise is a significant EMI source. By controlling the fourth leg, the common-mode voltage can be reduced to a certain level [13].

17.1.5   Z-Source Four-Leg Inverter

17.1.5.1   Overview of the Topology

To describe the operating principle and control of the Z-source four-leg inverter in Figure 17.6, the Z-source inverter structure is briefly examined. In Figure 17.6, the three-phase Z-source

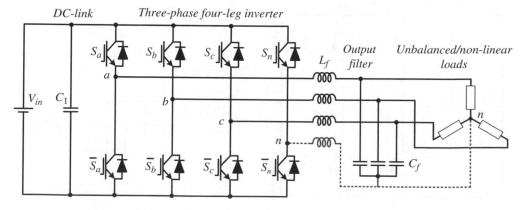

Figure 17.5   A three-phase four-leg VSI supplying an unbalanced load.

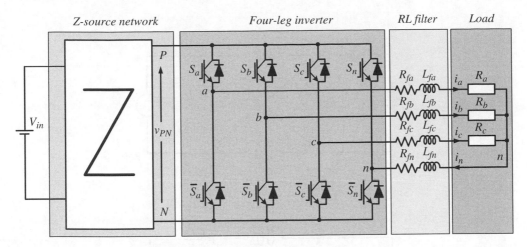

Z-source network   Four-leg inverter   RL filter   Load

Figure 17.6 Impedance source three-phase four-leg inverter topology (*Source*: Sertac 2016 [47]. Reproduced with permission of IEEE).

four-leg inverter bridge has 17 ($2^4 + 1$) permissible switching states (vectors), unlike the traditional three-phase voltage-source four-leg inverter that has 16 (2^4). The traditional three-phase four-leg inverter has 14 active vectors when the dc voltage is impressed across the load and two zero vectors when the load terminals are shorted through either the lower or upper three devices, respectively. However, the three-phase Z-source four-leg inverter bridge has one extra zero state (or vector) when the load terminals are shorted through both the upper and lower devices of any one phase leg (i.e. both devices are gated on), any two phase legs, or all three phase legs. This shoot-through zero state (or vector) is forbidden in the traditional inverter because it would cause a short-circuit. This is called the third zero state (vector), the shoot-through zero state (or vector), which can be generated in 15 different ways: shoot-through via any phase leg, combinations of any two phase legs, and all three phase legs. The Z-source network makes the shoot-through zero state possible. This shoot-through zero state provides the unique buck–boost feature of the inverter.

Furthermore, the load neutral point is connected to the mid-point of the inverter fourth leg to allow for zero sequence current/voltage control, as can be seen in Figure 17.6. However, the addition of an extra leg makes the switching schemes more complex compared with a traditional three-leg voltage source inverter. Nevertheless, using an extra leg improves the inverter capability and reliability. The four-leg inverter can be used under balanced/unbalanced and/or linear/non-linear load conditions to compensate for that.

17.1.5.2 Impedance Source Network Models

The Z-source and quasi-Z-source networks will be addressed in this section.

Z-source network model

The equivalent circuits of the Z-source network with its non-shoot-through state and shoot-through states are illustrated in Figure 17.7(a)–(c) [31], assuming that the inductors (L_1 and L_2) and capacitors (C_1 and C_2) have the same inductance and capacitance, respectively.

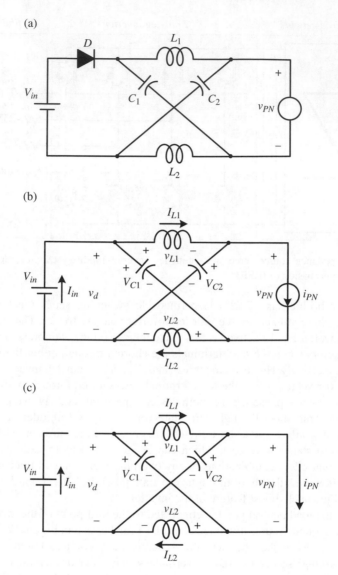

Figure 17.7 Model of Z-source inverter: (a) equivalent circuit of the Z-source network, (b) in the non-shoot-through state, (c) in the shoot-through state (*Source*: Sertac 2016 [47]. Reproduced with permission of IEEE).

Thus, the Z-source network becomes symmetrical. From the equivalent circuits, we have

$$V_{C1} = V_{C2} = V_C, \quad v_{L1} = v_{L2} = v_L \tag{17.6}$$

During the non-shoot-through state (T_0) the four-leg inverter model is represented by a constant current source, as can be seen from Figure 17.7(b). From the equivalent circuit, Figure 17.7(b), the inductor voltage (v_L), dc-link voltage (v_{PN}), and diode voltage (v_d) are written as

$$v_L = V_0 - V_C, \quad v_d = V_{in}, \quad v_{PN} = V_C - v_L = 2V_C - V_{in} \tag{17.7}$$

where V_{in} is the dc source voltage.

During the shoot-through state (T_0) the four-leg inverter model is represented by a short-circuit, as can be seen from Figure 17.7(c). By applying Kirchhoff's voltage law to Figure 17.7(c), inductor voltage (v_L), dc-link voltage (v_{PN}), and diode voltage (v_d) are written as

$$v_L = V_C, \quad v_{diode} = 2V_C, \quad v_{PN} = 0 \tag{17.8}$$

The average voltage of the inductors over one switching period $(T = T_0 + T_1)$ should be zero in steady state, so from (17.7) and (17.8), we obtain

$$V_C = \frac{T_1}{T_1 - T_0} V_{in} \tag{17.9}$$

Similarly, the average dc-link voltage across the inverter bridge can be found as

$$v_{PN} = \frac{T_1}{T_1 - T_0} V_{in} = V_C \tag{17.10}$$

The peak dc-link voltage across the inverter bridge is expressed in (17.7) and can be rewritten as

$$\hat{v}_{PN} = V_C - v_L = 2V_C - V_{in} = \frac{T}{T_1 - T_0} V_{in} = BV_{in} \tag{17.11}$$

where B is the boost factor resulting from the shoot-through zero state.

The output peak phase voltage from the inverter can be expressed as

$$\hat{v}_{ac} = M \frac{\hat{v}_{PN}}{2} \tag{17.12}$$

where M is the modulation index. Using (17.6) and (17.7) it can be further expressed as

$$\hat{v}_{ac} = MB \frac{V_{in}}{2} \tag{17.13}$$

This equation shows that the output voltage can be stepped up and down by choosing an appropriate voltage gain

$$G = MB = (0 \rightarrow \infty) \tag{17.14}$$

where G is the voltage gain that is determined by the modulation index M and the boost factor B. The boost factor as expressed in (17.6) can be controlled by the duty cycle (i.e. interval ratio) of the shoot-through zero state over the non-shoot-through states of the inverter PWM. Note that the shoot-through zero state does not affect the PWM control of the inverter, because it equivalently produces the same zero voltage to the load terminal. The available shoot-through period is limited by the zero-state period, which is determined by the modulation index.

Quasi-Z-source network model

The equivalent circuits of the qZS network with its non-shoot-through state and shoot-through states are illustrated in Figure 17.8(a)–(c) [32, 33]. All voltages and currents are defined in these figures, and the polarities are shown with arrows.

During the non-shoot-through state the four-leg inverter model is represented by a constant current source, as can be seen from Figure 17.8(b). By applying Kirchhoff's voltage law to

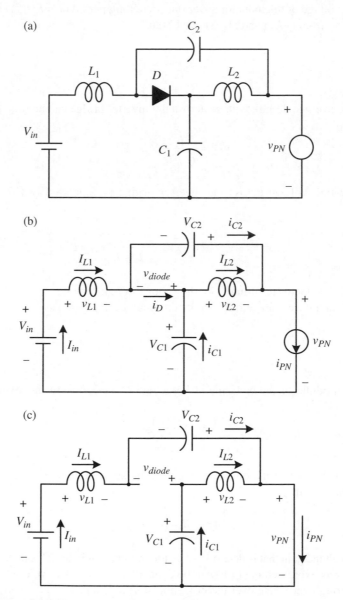

Figure 17.8 Model of qZ-source inverter: (a) equivalent circuit of the quasi-Z-source network, (b) in the non-shoot-through state, (c) in the shoot-through state (*Source*: Sertac 2016 [48]. Reproduced with permission of IEEE).

Figure 17.8(b), inductor voltages (v_{L1} and v_{L2}), dc-link voltage (v_{PN}), and diode voltage (v_d) are written as

$$v_{L1} = V_{in} - V_{C1}, \quad v_{L2} = -V_{C2} \tag{17.15}$$

$$v_{PN} = V_{C1} - v_{L2} = V_{C1} + V_{C2}, \quad v_d = 0 \tag{17.16}$$

During the shoot-through state the four-leg inverter model is represented by a short-circuit, as can be seen from Figure 17.8(c). By applying Kirchhoff's voltage law to Figure 17.8(c), the inductor voltages (v_{L1} and v_{L2}), dc-link voltage (v_{PN}), and diode voltage (v_d) are written as

$$v_{L1} = V_{in} + V_{C2}, \quad v_{L2} = V_{C1} \tag{17.17}$$

$$v_{PN} = 0, \quad v_d = V_{C1} + V_{C2} \tag{17.18}$$

At steady state, the average voltage of the capacitors over one switching cycle is

$$V_{C1} = \frac{T_1}{T_1 - T_0} V_{in} = m V_{in},$$

$$V_{C2} = \frac{T_0}{T_1 - T_0} V_{in} = n V_{in}. \tag{17.19}$$

where T_0 is the duration of the shoot-through state and T_1 is the duration of the non-shoot-through state, and V_{in} is the input dc voltage. Here, $m > 1$ and $m - n = 1$.

From (17.16), (17.18), and (17.19), the peak dc-link voltage across the inverter bridge is

$$v_{PN} = V_{C1} + V_{C2} = \frac{T}{T_1 - T_0} V_{in} = B V_{in} \tag{17.20}$$

where T is the switching cycle, and B is the boost factor of the qZSI. If $B \leq 1$ then the inverter works in buck conversion mode and if $B > 1$, the inverter works in boost conversion mode. From (17.19), the boost factor can be rewritten as $m + n = B$, and $1 < m < B$.

The average current of the inductors L_1, and L_2 can be calculated by the system power P

$$I_{L1} = I_{L2} = I_{in} = \frac{P}{V_{in}} \tag{17.21}$$

According to Kirchhoff's current law and (17.21), we can also get

$$i_{C1} = i_{C2} = i_{PN} - I_{L1} \tag{17.22}$$

The voltage gain (G) of the qZSI can be expressed as

$$G = \hat{v}_{in} / 0.5 v_{PN} \tag{17.23}$$

where M is the modulation index, $\hat{v}_{in}$ is the peak ac phase voltage.

In summary, there are some unique merits of the qZSI when comparing it with the ZSI [9]:

1. The two capacitors in ZSI sustain the same high voltage, while in qZSI the voltage on capacitor C_2 is lower, which means a lower capacitor rating.
2. The ZSI has discontinuous input current in the boost mode, but the input current of the qZSI is continuous owing to the input inductor L_1, which will significantly reduce the input stress.
3. For the qZSI, there is a common dc rail between the source and the inverter, which is easier to assemble and causes less EMI.

17.1.5.3 Four-Leg Inverter Model

The equivalent circuit of the four-leg inverter with the output RL filter is shown in Figure 17.9, where the L_{fj} is the filter inductance, R_{fj} is the filter resistance, and R_j is the load resistance ($j = a, b, c$) [14].

For a three-phase four-leg inverter, the addition of the fourth leg makes the switching states 16 (2^4). The valid switching states with the corresponding phase and line voltages for the traditional four-leg inverter are shown in Table 17.1.

The four control signals, named S_a, S_b, S_c, and S_n, form a total of 16 (2^4) switching states of the converter. The voltages in each leg of the inverter, measured from the negative point of dc-link N, can be expressed as

$$\begin{bmatrix} v_{aN} \\ v_{bN} \\ v_{cN} \\ v_{nN} \end{bmatrix} = \begin{bmatrix} S_a \\ S_b \\ S_c \\ S_n \end{bmatrix} v_{dc} \tag{17.24}$$

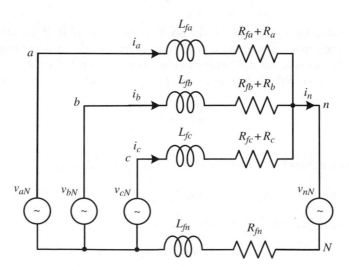

Figure 17.9 Equivalent circuit of the three-phase four-leg inverter (*Source*: Sertac 2016 [48]. Reproduced with permission of IEEE).

Table 17.1 Switching states of the three-phase four-leg inverter

Switching state	S_a	S_b	S_c	S_n	v_{aN}	v_{bN}	v_{cN}	v_{nN}	v_{an}	v_{bn}	v_{cn}
1	1	0	0	0	v_{dc}	0	0	0	v_{dc}	0	0
2	1	1	0	0	v_{dc}	v_{dc}	0	0	v_{dc}	v_{dc}	0
3	0	1	0	0	0	v_{dc}	0	0	0	v_{dc}	0
4	0	1	1	0	0	v_{dc}	v_{dc}	0	0	v_{dc}	v_{dc}
5	0	0	1	0	0	0	v_{dc}	0	0	0	v_{dc}
6	1	0	1	0	v_{dc}	0	v_{dc}	0	v_{dc}	0	v_{dc}
7	1	1	1	0	v_{dc}	v_{dc}	v_{dc}	0	v_{dc}	v_{dc}	v_{dc}
8	0	0	0	0	0	0	0	0	0	0	0
9	1	0	0	1	v_{dc}	0	0	v_{dc}	0	$-v_{dc}$	$-v_{dc}$
10	1	1	0	1	v_{dc}	v_{dc}	0	v_{dc}	0	0	$-v_{dc}$
11	0	1	0	1	0	v_{dc}	0	v_{dc}	$-v_{dc}$	0	$-v_{dc}$
12	0	1	1	1	0	v_{dc}	v_{dc}	v_{dc}	$-v_{dc}$	0	0
13	0	0	1	1	0	0	v_{dc}	v_{dc}	$-v_{dc}$	$-v_{dc}$	0
14	1	0	1	1	v_{dc}	0	v_{dc}	v_{dc}	0	$-v_{dc}$	0
15	1	1	1	1	v_{dc}	v_{dc}	v_{dc}	v_{dc}	0	0	0
16	0	0	0	1	0	0	0	v_{dc}	$-v_{dc}$	$-v_{dc}$	$-v_{dc}$

which can be written as

$$v_{jn} = S_j v_{dc}, \quad j = a,b,c,n. \tag{17.25}$$

The voltage applied to the output RL filter, in terms of these voltages, is

$$\begin{bmatrix} v_{an} \\ v_{bn} \\ v_{cn} \end{bmatrix} = \begin{bmatrix} S_a - S_n \\ S_b - S_n \\ S_c - S_n \end{bmatrix} v_{dc} \tag{17.26}$$

The above expression can be simplified to

$$v_{kn} = v_{kN} - v_{nN} = \left(S_k - S_n\right)v_{dc}, \quad k = a,b,c \tag{17.27}$$

The inverter voltages are then given as

$$\begin{bmatrix} v_{aN} \\ v_{bN} \\ v_{cN} \\ v_{nN} \end{bmatrix} = \begin{bmatrix} R_{fa} & 0 & 0 & 0 \\ 0 & R_{fb} & 0 & 0 \\ 0 & 0 & R_{fc} & 0 \\ 0 & 0 & 0 & R_{fn} \end{bmatrix} \begin{bmatrix} i_a \\ i_b \\ i_c \\ i_n \end{bmatrix} + \begin{bmatrix} L_{fa} & 0 & 0 & 0 \\ 0 & L_{fb} & 0 & 0 \\ 0 & 0 & L_{fc} & 0 \\ 0 & 0 & 0 & L_{fn} \end{bmatrix} \frac{d}{dt} \begin{bmatrix} i_a \\ i_b \\ i_c \\ i_n \end{bmatrix}$$

$$+ \begin{bmatrix} R_a & 0 & 0 & 0 \\ 0 & R_b & 0 & 0 \\ 0 & 0 & R_c & 0 \\ 0 & 0 & 0 & R_n \end{bmatrix} \begin{bmatrix} i_a \\ i_b \\ i_c \\ i_n \end{bmatrix} + \begin{bmatrix} 1 \\ 1 \\ 1 \\ 1 \end{bmatrix} v_{nN} \tag{17.28}$$

which can be written as

$$v_{jn} = \left(R_{fj} + R_j\right)i_j + L_{fj}\frac{d_{ij}}{dt} + v_{nN}, \quad j = a,b,c,n \tag{17.29}$$

The derivative of the output current vector can be obtained from (17.29)

$$\frac{d_{ij}}{dt} = \frac{1}{L_{fj}}\left[\left(v_{jn} - v_{nN}\right) - \left(R_{fj} + R_j\right)i_j\right], \quad j = a,b,c. \tag{17.30}$$

Based on (17.25) and (17.29), the load neutral voltage v_{nN} can be expressed as

$$v_{nN} = L_{eq}v_{dc}\sum_{j=a,b,c,n}\frac{S_j}{L_{fj}} - L_{eq}\sum_{j=a,b,c,n}\frac{R_{fj} + R_j}{L_{fj}}i_j \tag{17.31}$$

with

$$L_{eq} = \left(\frac{1}{L_{fa}} + \frac{1}{L_{fb}} + \frac{1}{L_{fc}} + \frac{1}{L_{fn}}\right)^{-1} \tag{17.32}$$

$$i_a + i_b + i_c + i_n = 0 \tag{17.33}$$

17.1.6　Switching Schemes for Three-Phase Four-Leg Inverter

For a three-phase four-leg VSI, the addition of a fourth leg brings the switching states to $2^4 = 16$. This means the switching schemes that can be applied on the four-leg inverter are more complex than the conventional switching schemes implemented on the three-leg inverter [34]. The following sections list the most popular switching schemes that are implemented on the three-phase four-leg VSI.

17.1.6.1　Three-Dimensional Space Vector Modulation

The concept of three-dimensional space vector modulation (3D SVM) was invented by Richard Zhang and was first published in [35]. Since the day of its invention, this scheme has been regarded as the best switching scheme for a three-phase four-leg VSI under unbalanced/non-linear load conditions [36]. There are 16 switching states of the converter, as shown in Figure 17.10. After transforming the switching vectors from *a-b-c* coordinates to the α-β-γ coordinates, they are located into a three-dimensional space, as illustrated in Figure 17.11, which is where the name "3D" comes from. The transformation between the *a-b-c* and α-β-γ is given by

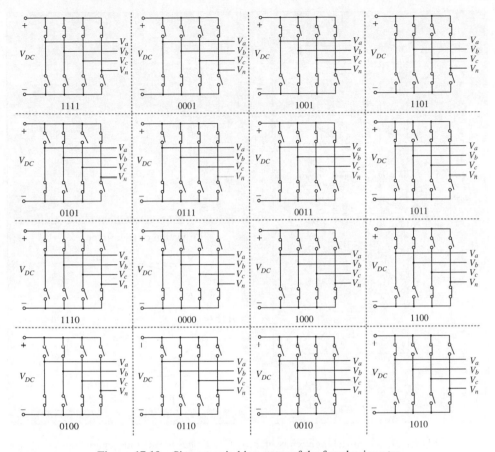

Figure 17.10   Sixteen switching states of the four-leg inverter.

$$
\begin{bmatrix} X_\alpha \\ X_\beta \\ X_\gamma \end{bmatrix} =
\begin{bmatrix} \dfrac{2}{3} & -\dfrac{1}{3} & -\dfrac{1}{3} \\ 0 & \dfrac{1}{\sqrt{3}} & -\dfrac{1}{\sqrt{3}} \\ \dfrac{1}{3} & \dfrac{1}{3} & \dfrac{1}{3} \end{bmatrix}
\begin{bmatrix} X_a \\ X_b \\ X_c \end{bmatrix}
\tag{17.34}
$$

$$
\begin{bmatrix} X_a \\ X_b \\ X_c \end{bmatrix} =
\begin{bmatrix} 1 & 0 & 1 \\ -\dfrac{1}{2} & \dfrac{\sqrt{3}}{2} & 1 \\ -\dfrac{1}{2} & -\dfrac{\sqrt{3}}{2} & 1 \end{bmatrix}
\begin{bmatrix} X_\alpha \\ X_\beta \\ X_\gamma \end{bmatrix}
\tag{17.35}
$$

where X could be a voltage (V) or a current (I) for the three-phase power system.

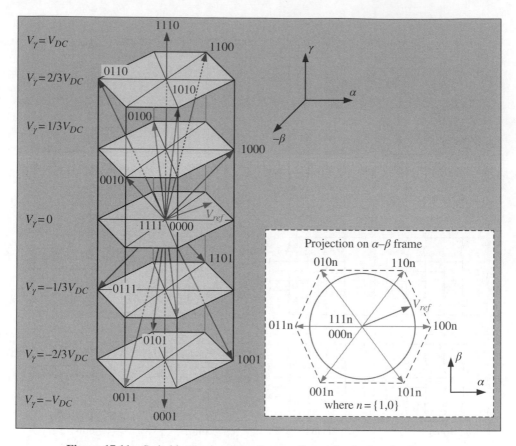

Figure 17.11 Switching vectors in α-β-γ coordinates for the four-leg inverter.

In Figure 17.11, it is shown that among all the 16 switching vectors, there are two zero switching vectors (1111, 0000) which are located in the middle layer of the space. The other 14 non-zero switching vectors are located at different layers according to the different voltage levels of $V\gamma$. It should be noted that $V\gamma$ is the zero-sequence component and related to the neutral current [29]. There are altogether seven different voltage levels of $V\gamma$ and, therefore, seven layers of the space. The projection of the switching vectors on the α-β plane is also shown in Figure 17.11. As can be seen, the projection of V_α and V_β form the same circle as that of the traditional SVM.

Synthesis of the rotating reference vector in the α-β-γ coordinates takes the following steps [13, 29]:

1. Prism and tetrahedron identification
2. Projection of the reference vector
3. Sequencing of the selected switching vectors
4. Generation of the modulation index

1) *Prism and tetrahedron identification*

For the conventional SVM (2D), the synthesis of the rotating reference vector is straightforward since the adjacent switching vectors are always chosen. For the same reason, adjacent switching vectors in the α-β-γ coordinates have to be selected. However, the adjacent switching vectors in the three-dimensional space are not always easy to identify. It takes two steps to choose the adjacent switching vectors: prism identification and tetrahedron identification [29]. As can be seen in Figure 17.12, six prisms can be identified in the space, like the six sectors in the conventional SVM. It should be noted that each prism is rotated by 60° from the previous prism.

The knowledge of V_α and V_β of the rotating reference vector will determine in which prism the reference vector is located. It is can also be seen that in each prism, there are six non-zero switching vectors and two zero switching vectors that can be used. After the prism

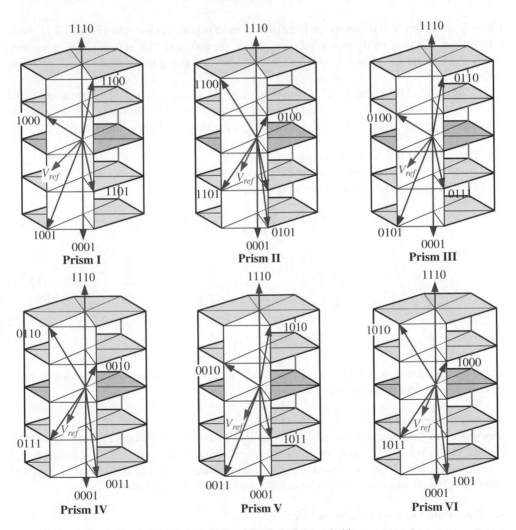

Figure 17.12 Prism identification of the switching vectors.

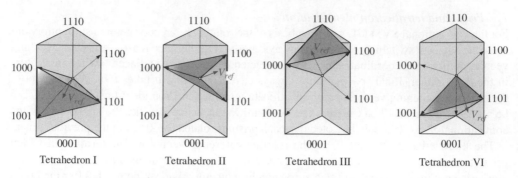

Figure 17.13 Tetrahedron identification in Prism 1.

identification, four tetrahedrons are identified in each prism, as shown in Figure 17.13. Each tetrahedron consists of three non-zero switching vectors and two zero switching vectors. These vectors are the adjacent switching vectors that are going to be used to synthesize the reference vector.

To decide in which tetrahedron the rotating reference vector locates, the information of the phase voltages in *A-B-C* coordinates is needed, as stated in [29], and the tetrahedron can be selected by directly comparing the relative sizes of the phase voltages and zero.

2)   *Projection of the reference vector*
The duration of each selected switching vector can be calculated by projecting the reference vector on the switching vectors. The calculations needed are

$$\vec{V}_{ref} = d_1\vec{V}_1 + d_2\vec{V}_2 + d_3\vec{V}_3 \tag{17.36}$$

$$\begin{bmatrix} d_1 \\ d_2 \\ d_3 \end{bmatrix} = \frac{1}{V_{DC}} G \begin{bmatrix} V_{\alpha\_ref} \\ V_{\beta\_ref} \\ V_{\gamma\_ref} \end{bmatrix} \tag{17.37}$$

$$d_z = 1 - d_1 - d_2 - d_3 \tag{17.38}$$

where *d* is the duration of each selected switching vector and *G* is the matrix needed to compute the duty ratios; it can be found in [35].

3)   *Sequencing of the selected switching vectors*
Sequencing of the selected switching vectors follows the same rule as the conventional SVM. In [36], two classes of the switching schemes are compared and analyzed. It is concluded in [35] that the Class II center-aligned is the best compromise between the switching loss and harmonic content and is normally adopted in a real-time implementation.

4)   *Generation of the modulation index*
The waveform of modulation index can be derived from the duty ratios of the switching vectors d_1, d_2, d_3, and d_z and the sequencing scheme used [35]. For demonstration purposes, a

balanced load condition is studied here. The modulation index waveform is obtained by supplying a balanced three-phase sinusoidal reference signal as

$$
\begin{bmatrix} V_{ab\_ref} \\ V_{bc\_ref} \\ V_{ca\_ref} \end{bmatrix} = MV_{DC} \begin{bmatrix} \sin(\omega t) \\ \sin(\omega t - 120^o) \\ \sin(\omega t + 120^o) \end{bmatrix}
\tag{17.39}
$$

where M is the modulation index. Based on (17.34), the trajectory of the rotating reference vector is a circle only based on the layer of $V\gamma = 0$. The rotating reference vector in α-β-γ coordinates is then given by

$$
\begin{bmatrix} V_{\alpha\_ref} \\ V_{\beta\_ref} \\ V_{\gamma\_ref} \end{bmatrix} = \frac{M}{\sqrt{3}} V_{DC} \begin{bmatrix} \cos(\omega t) \\ \sin(\omega t) \\ 0 \end{bmatrix}
\tag{17.40}
$$

17.1.6.2   Carrier-based PWM for four-leg VSI

The three-dimensional SVM is based on the concept of switching vectors corresponding to the switching states of the converter. Without using the concept of switching vectors, a carrier-based PWM scheme can be used in the scalar implementation for four-leg VSI [23, 36]. It can be proved to be equivalent to 3D SVM under open loop operation [36]. The modulation index waveforms for the first three phases follow the same rule as CBPWM for three-leg VSI. An offset voltage signal is then calculated for the fourth leg. A block diagram of the carrier-based PWM is shown in Figure 17.14. The variations of the modulation index waveforms depend on the offset modulation calculation. The fourth leg modulation index should be within the envelope of the minimum and maximum function. In [37], a similar idea was presented, but the offset modulation index was called the injected signal in this case. The calculation is different from that in [36], therefore the boundary of the injected signal waveform is different.

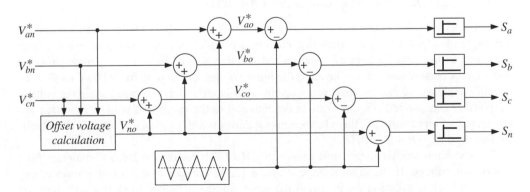

Figure 17.14   Block diagram of the carrier-based PWM for four-leg VSI.

As can be seen from Figure 17.14, the terminal phase reference voltages are given by

$$
\begin{bmatrix} V_{ao}^* \\ V_{bo}^* \\ V_{co}^* \end{bmatrix} = \begin{bmatrix} V_{an}^* \\ V_{bn}^* \\ V_{cn}^* \end{bmatrix} + V_{no}^*
\tag{17.41}
$$

The offset voltage calculation is then given by

$$
V_{fo}^* = \begin{cases} (0.5-\xi)V_{DC} - \xi V_{min}^* & , V_{max}^* < 0 \\ (0.5-\xi)V_{DC} + (\xi-1)V_{max}^* & , V_{min}^* > 0 \\ (0.5-\xi)V_{DC} + (\xi-1)V_{max}^* - \xi V_{min}^* & , elsewhere \end{cases}
\tag{17.42}
$$

where $V_{max}^* = \max(V_{an}^*, V_{bn}^*, V_{cn}^*)$, $V_{min}^* = \min(V_{an}^*, V_{bn}^*, V_{cn}^*)$, and ξ is called the zero-state partitioning function, and is defined as

$$
\xi = \frac{t_{0000}}{t_{0000} + t_{1111}}
\tag{17.43}
$$

where t_{0000} is the time of the switching state where the four bottom switches are on and t_{1111} is the time of the switching state where four top switches are on. Thus, the ζ-based offset voltage calculation establishes the unification of the carrier-based PWM switching scheme for the four-leg VSI.

The largest advantage of the CBPWM for the four-leg VSI is its simplicity and, therefore, the lower cost of implementation. However, the performance under an unbalanced load condition is not clear. Another drawback of this switching scheme is a controller design based on *d-q*-0 coordinates, or the synchronous reference frame is not possible since there is no rotating reference vector.

17.1.7 *Buck/Boost Conversion Modes Analysis*

Experiments have been performed to verify the theoretical analysis and confirm the control technique of qZS four-leg inverters that can operate both buck or boost conversion modes according to the input voltage and the desired output voltage. To test the voltage gain and boost factor performance of the inverter, the input voltage is changed from 180 V to 80 V. To simplify analysis, we assume that output voltage v_{ln} is equal to load voltage v_R, and the voltage on the filter is neglected (Figure 17.6). In order to ensure the same load voltage $v_{ln} = 50\sqrt{3}$ at a wide range of input voltages, the reference output currents are (i_a^*, i_b^*, i_c^*) set to 7 A and loads are balanced ($R_a = R_b = R_c = 10\,\Omega$).

The minimum input voltage must be $V_{in} = 2v_{ln}/M = 123$ V (with $M = 2/\sqrt{3}$) to maintain 50 V rms output voltage. If the input voltage is above 123 V, the qZ-source four-leg inverter can operate in buck conversion mode, and if the input voltage is below 123 V, the qZS four-leg inverter can operate in boost conversion mode.

Case 1: $V_{in} = 180\,V$, $M = 0.8$

Experimental results of this case are shown in Figure 17.15(a). Here, $V_{in} > 123\,V$, so, the qZSI works in buck conversion mode. Thus, the boost factor $B = 1$ and the voltage gain is $G = BM \approx 0.8$. The maximum output line-to-line is

$$v_{ab} = v_{bc} = v_{ca} = \sqrt{3}GV_{in}/2 \cong 123\,(V)$$

As can be seen from Figure 17.15(a), the voltage on C_1 is equal to the input voltage $180\,V$ and the voltage on C_2 is $0\,V$. It can be noted that a pure dc current flows through an inductor due to the voltage on L_1 being zero.

(a)

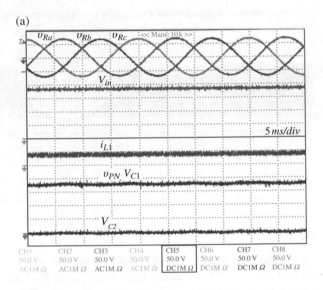

(b)

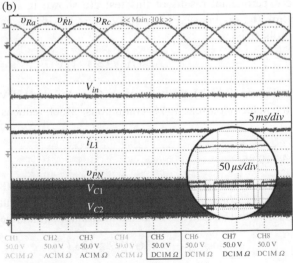

Figure 17.15 Experimental results with the same output voltage at (a) $V_{in} = 180\,V$, $M = 0.8$, (b) $V_{in} = 100\,V$, $M = 1$; (c) $V_{in} = 80\,V$, $M = 0.85$ (*Source*: Sertac 2016 [48]. Reproduced with permission of IEEE).

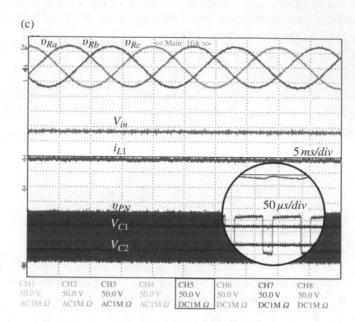

Figure 17.15   (*Continued*)

Case 2: $V_{in} = 100\,V$, $M = 1$

In order to maintain constant output voltage, qZSI works in boost conversion mode because $V_{in} < 123\,V$. From (17.48) and (17.49), we can get the boost factor $B \approx 1.42$ and the voltage gain $G \approx 1.42$. The experimental results of this test are shown in Figure 17.15(b). The voltage on the dc-link (v_{PN}) is boosted from 100 V to 142 V. In this case, the maximum output line-to-line voltage is

$$v_{ab} = v_{bc} = v_{ca} = \sqrt{3}GV_{in}/2 \cong 123\,(V)$$

As can be seen from Figure 17.15(b), the voltages on C_1 and C_2 are 120 V and 22 V, respectively. Notice that the inductor current (i_{L1}) is continuous, which reduces the input stress.

Case 3: $V_{in} = 80\,V$, $M = 0.85$

Figure 17.15(c) shows the experimental results in this case. From (17.48) and (17.49), we can get the boost factor $B = 2.08$ and the voltage gain $G = 1.768$. The voltage on the dc-link (v_{PN}) is boosted from 80 V to 166 V. In this case, the maximum output line-to-line voltage is

$$v_{ab} = v_{bc} = v_{ca} = \sqrt{3}GV_{in}/2 \cong 123\,(V)$$

As can be seen from Figure 17.15(c), the voltages on C_1 and C_2 are 124 V and 42 V, respectively. Experimental results show that the qZS four-leg inverter can provide constant output voltage under various input voltages without using a dc/dc converter or a transformer.

17.2 Impedance Source Five-Leg (Five-Phase) Inverter

Multiphase (more than three phases) motor drives have attracted huge attention in recent years and a large number of research papers have been published. The major attractions of multiphase systems come from the fact that they offer some inherent advantages when compared to standard three-phase drives, such as reduction in the amplitude and increase in the frequency of torque pulsation, reduction in the rotor current harmonics, reduction in the dc-link current harmonics, reduction in the current per phase without increasing the voltage per phase leading, and increasing the torque per ampere for the same volume machine. Although general purpose drive applications are still not envisaged for multiphase motor drives, they are considered more attractive in niche applications such as mining, "more electric aircraft," ship propulsion, and electric and hybrid vehicles [38].

Multiphase motors are supplied using power electronic converters since multiphase is not available from the grid. The most common choice is either two-level or multilevel multiphase voltage source inverters (VSIs). A new class of power converter known as impedance source and quasi-impedance source inverters has been introduced recently which has source voltage boosting capability. As mentioned in the first part of this section, the ZSI and qZSI are an emerging topology of power electronics converters with very interesting properties such as buck–boost characteristics and single-stage conversion. These can produce the desired ac output voltage, even if it is greater than the input voltage; provide ride-through during voltage sags without additional circuits; improve the power factor; and reduce the harmonic current and the common-mode voltage [39]. The above features make the ZSI and qZSI fed adjustable speed drive (ASD) systems highly desirable and reliable when compared to VSI fed ASDs [39].

17.2.1 Five-Phase VSI Model

The equivalent circuit of the five-phase inverter with an output RL load is shown in Figure 17.16. Each of the two-level inverter legs has two switching states. Thus for the five-phase inverter, there are 32 possible switching combinations. Among the 32 switching states, 30 states are

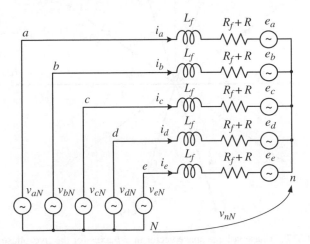

Figure 17.16 Equivalent circuit of the five-phase inverter.

active and two states are zero switching states (vectors). However, in addition to these switching states, for this application, one extra switching state is required in order to ensure the shoot-through state. Therefore, a total of 33 switching states are used for this application.

At any instant of time, the inverter can produce only one space vector. The space vector of phase voltages can be defined using power variant transformations such as

$$v_{\alpha\beta} = \frac{2}{5}\left(v_a + \mathbf{a}v_b + \mathbf{a}^2 v_c + \mathbf{a}^{*2} v_d + \mathbf{a}^* v_e\right) \tag{17.44}$$

where $\mathbf{a} = e^{\frac{j2\pi}{5}}$, $\mathbf{a}^2 = e^{\frac{j4\pi}{5}}$, $\mathbf{a}^* = e^{\frac{-j2\pi}{5}}$, $\mathbf{a}^{*2} = e^{\frac{-j4\pi}{5}}$, and * stands for a complex conjugate. The phase voltage space vectors thus obtained in the α-β plane are shown in Figure 17.17. Since it is a five-phase system, the transformation is further done in order to obtain space vectors in the x-y plane using equation (17.44) and the resulting space vectors are shown in Figure 17.18 [40].

$$v_{xy} = \frac{2}{5}\left(v_a + \mathbf{a}^2 v_b + \mathbf{a}^4 v_c + \mathbf{a}^6 v_d + \mathbf{a}^8 v_e\right) \tag{17.45}$$

The vectors are numbered by decimal values when converted to their equivalent binary and will give the switching state of each leg, where 1 corresponds to the upper switch being ON and 0 represents the lower switch being ON. Moreover the operation of each inverter leg is complementary in order to protect the dc source. As can be seen from Figure 17.17, the outer decagon space vectors of the α-β plane map into the inner decagon of the x-y plane

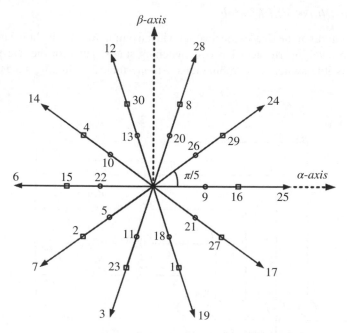

Figure 17.17 Phase voltage space vector in α-β axes for the five-phase inverter.

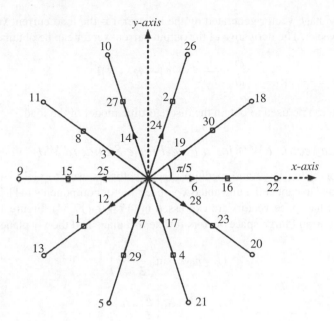

Figure 17.18 Phase voltage space vector in *x-y* axes for the five-phase inverter.

(Figure 17.18), the innermost decagon of the d-q plane forms the outer decagon of the *x-y* plane, and the middle decagon space vectors map into the same region. Further, it is observed from the above mapping that the phase sequence *a,b,c,d,e* of the *α-β* plane corresponds to *a,c,e,b,d* sequence of the *x-y* plane. It is also observed that there are three sets of vectors with three different lengths, referred to as large, medium, and small. The length of the large vectors is $\frac{1}{5}\cos\frac{\pi}{5}V_{dc}$, the medium vectors $\frac{2}{5}V_{dc}$, and the small vectors $\frac{1}{5}\cos\frac{2\pi}{5}V_{dc}$.

Taking into account the definitions of variables from the circuit shown in Figure 17.16, the equations for load current dynamics for each phase can be written as

$$
\begin{bmatrix} v_a \\ v_b \\ v_c \\ v_d \\ v_e \end{bmatrix} = \begin{bmatrix} R+R_f & 0 & 0 & 0 & 0 \\ 0 & R+R_f & 0 & 0 & 0 \\ 0 & 0 & R+R_f & 0 & 0 \\ 0 & 0 & 0 & R+R_f & 0 \\ 0 & 0 & 0 & 0 & R+R_f \end{bmatrix} \begin{bmatrix} i_a \\ i_b \\ i_c \\ i_d \\ i_e \end{bmatrix} + \begin{bmatrix} L_f & 0 & 0 & 0 & 0 \\ 0 & L_f & 0 & 0 & 0 \\ 0 & 0 & L_f & 0 & 0 \\ 0 & 0 & 0 & L_f & 0 \\ 0 & 0 & 0 & 0 & L_f \end{bmatrix} \frac{d}{dt} \begin{bmatrix} i_a \\ i_b \\ i_c \\ i_d \\ i_e \end{bmatrix} + \begin{bmatrix} e_a \\ e_b \\ e_c \\ e_d \\ e_e \end{bmatrix}
$$

(17.46)

where *R* is the load resistance, and R_f and L_f are the filter resistance and inductance, respectively. This equation can be written as

$$
\mathbf{v} = \left(R_f + R \right)\mathbf{i} + L_f \frac{d\mathbf{i}}{dt} + \mathbf{e}
$$

(17.47)

where **v** is the voltage vector generated by the inverter, **i** is the load current vector, and **e** the load back-emf vector. The derivative of the output current vector can be obtained from (17.47)

$$\frac{d\mathbf{i}}{dt} = \frac{1}{L_f}\left[(\mathbf{v}-\mathbf{e})-\left(R_f+R\right)\mathbf{i}\right] \tag{17.48}$$

This equation can be used to obtain the discrete-time model of the load.

17.2.2 Space Vector PWM for a Five-Phase Standard VSI

The combination of the large and medium space vectors is used to yield sinusoidal output phase voltages and to cancel the unwanted x-y space vector components [41]. The application time intervals of the space vectors are expressed in (17.49)–(17.54). Figure 17.19 illustrates the small, medium, and large space vectors for the α-β plane and the x-y plane.

$$t_{al} = m\cdot k\cdot\sin\left(\frac{\pi}{5}-\theta\right) \tag{17.49}$$

$$t_{am} = m\cdot\sin\left(\frac{\pi}{5}-\theta\right) \tag{17.50}$$

$$t_{bl} = m\cdot k\cdot\sin\left(\theta\right) \tag{17.51}$$

$$t_{bm} = m\cdot\sin\left(\theta\right) \tag{17.52}$$

$$t_0 = t_s - t_{al} - t_{bl} - t_{am} - t_{bm} \tag{17.53}$$

$$m = \frac{5\left|V^*_s\right|t_s}{2V_{dc}\cdot\sin\left(\frac{\pi}{5}\right)\cdot\left(1+k^2\right)} \tag{17.54}$$

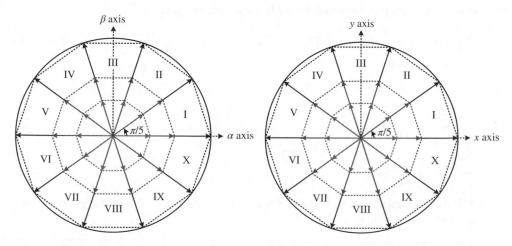

Figure 17.19 Space vectors for five-phase VSI (*Source*: Abduallah 2015 [42]. Reproduced with permission of IEEE).

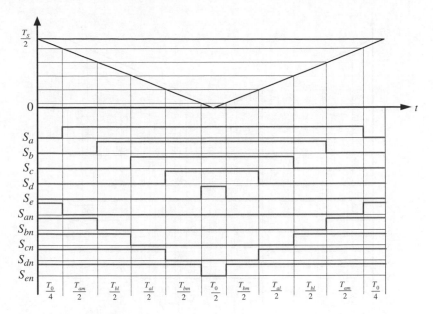

Figure 17.20 Traditional SVM switching pattern for a five-phase VSI while applying large and medium vectors (*Source*: Abduallah 2015 [42]. Reproduced with permission of IEEE).

Figure 17.20 illustrates the switching pattern while using the large and medium space vector combination, k is the ratio of the large and medium space vectors and is set equal to 1.618. Each switching device switches on and off once per cycle. As a result, a total of 20 switching states is produced per switching cycle [41]. The time interval of the zero states is the subtraction of the time of application of the large and medium vectors from the time interval of the switching period. The maximum achievable output voltage using this scheme is $0.5257\,V_{dc}$. This SVM method produces a sinusoidal output phase voltage, and the fundamental component of the phase voltage is equal to the reference voltage. This is due to the total cancellation of the unwanted x-y space vector components. The x-y components are cancelled by the suitable subdivision of the time intervals of the large and medium space vectors.

17.2.3 Space Vector PWM for Five-Phase qZSI

This section is based on the authors' publication [42]. The traditional SVM for qZSI introduces and inserts the shoot-through state in the zero vector interval to obtain a sinusoidal output using a combination of the large and medium voltage vectors as presented in Figure 17.21. In this scheme, four active voltage vectors are applied, and the shoot-through is introduced at several places within the switching period. In Figure 17.21 the shoot-through is applied in all five legs of the inverter and the shoot-through is divided into ten periods equally between all legs [42]. The shoot-through is divided equally between the phases. The method gives sinusoidal output phase voltages. It is important to note that the shoot-through time depends upon the number of legs in which shoot-through is introduced. So in SVM of qZSI,

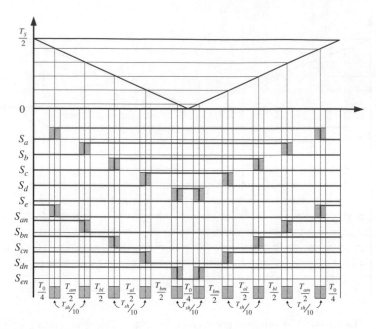

Figure 17.21 SVM for five-phase qZSI with large and medium vectors (*Source*: Abduallah 2015 [42]. Reproduced with permission of IEEE).

denoted as SVQ, the shoot-through scheme is divided into ten equal periods per switching cycle and each period divided equally between the upper and lower switches [42]. In [43, 44] the space vector modulation for three-phase voltage source inverter is used in such a way that the time of application of the shoot-through state is totally subtracted from the zero vectors. If the same concept is applied in space vectors modulation for the five-phase voltage source inverter, it is noticed that the voltage waveform became more distorted and the THD% also increases, as will be shown in the simulation section. However, in the SVQ method for the five phase Z-source inverters, the application of the shoot-through state is divided among all the application time of all vectors, so that the time of application of the vectors shares equally the effect of the shoot-though period.

Figure 17.22 illustrates the difference between the switching period while subtracting the application time interval of the shoot-through state from zero state time interval (SVQ1 method) and the subtracting of the shoot-through state from all space vector time intervals (SVQ2).

17.2.4 Discontinuous Space Vector PWM for Five-Phase qZSI

To obtain sinusoidal output phase voltage, four active and two zero vectors are applied in each sector. The zero vectors are applied at the start of the switching period (00000) and the other one is applied at the end of the half switching period (11111 vector) [45]. If the active voltage pulse position moves within the half switching period, then this is called discontinuous modulation [45]. This is allowed to eliminate one zero output voltage pulse. This modulation strategy makes one or more inverter leg tie either to the positive or the negative dc bus. This method is particularly significant for high power applications, where switching losses are

(a)

| $\frac{T_0}{4}$ $\frac{T_{sh}}{4}$ | $\frac{T_{sh}}{10}$ | $\frac{T_{am}}{2}$ | $\frac{T_{sh}}{10}$ | $\frac{T_{bl}}{2}$ | $\frac{T_{sh}}{10}$ | $\frac{T_{al}}{2}$ | $\frac{T_{sh}}{10}$ | $\frac{T_{bm}}{2}$ | $\frac{T_{sh}}{10}$ | $\frac{T_0}{2}$ $\frac{T_{sh}}{2}$ | $\frac{T_{sh}}{10}$ | $\frac{T_{bm}}{2}$ | $\frac{T_{sh}}{10}$ | $\frac{T_{al}}{2}$ | $\frac{T_{sh}}{10}$ | $\frac{T_{bl}}{2}$ | $\frac{T_{sh}}{10}$ | $\frac{T_{am}}{2}$ | $\frac{T_{sh}}{10}$ | $\frac{T_0}{4}$ $\frac{T_{sh}}{4}$ |

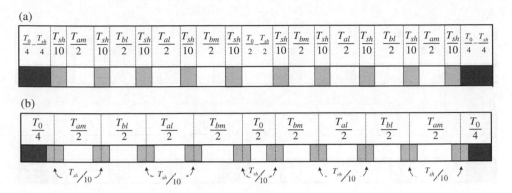

(b)

| $\frac{T_0}{4}$ | $\frac{T_{am}}{2}$ | $\frac{T_{bl}}{2}$ | $\frac{T_{al}}{2}$ | $\frac{T_{bm}}{2}$ | $\frac{T_0}{2}$ | $\frac{T_{bm}}{2}$ | $\frac{T_{al}}{2}$ | $\frac{T_{bl}}{2}$ | $\frac{T_{am}}{2}$ | $\frac{T_0}{4}$ |

$T_{sh}/10$ $T_{sh}/10$ $T_{sh}/10$ $T_{sh}/10$ $T_{sh}/10$

Figure 17.22 The application time intervals of space vectors for one switching cycle: (a) SVQ1, (b) SVQ2 (*Source*: Abduallah 2015 [42]. Reproduced with permission of IEEE).

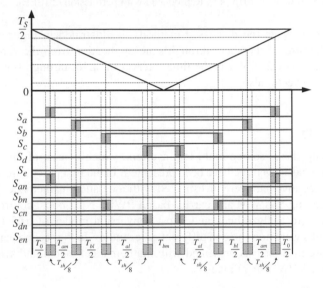

Figure 17.23 Discontinuous SVM for five-phase qZSI with large and medium vectors (DSVQ) (*Source*: Abduallah 2015 [42]. Reproduced with permission of IEEE).

significantly important [45]. Two discontinuous space vector PWM schemes are presented in this subsection to reduce the switching losses in the quasi Z-source inverter [45]. Nevertheless, both schemes rearrange the zero vector applications for the quasi ZSI. The first scheme (DSVQ1), uses the SVQ1 scheme time of application but it ties one of the legs to the negative dc bus as illustrated in Figure 17.23. Phase E is shown to be tied to negative dc rail and is not switching in the whole sampling period. The other legs will be switched off during different sectors. On average, each leg will not be switched equally in one fundamental cycle. The fundamental amplitude of the output voltage and the shape of phase voltage waveforms remain sinusoidal as that of continuous SVM with overall 20% reduction in the number of switching and subsequently corresponding switching losses [46].

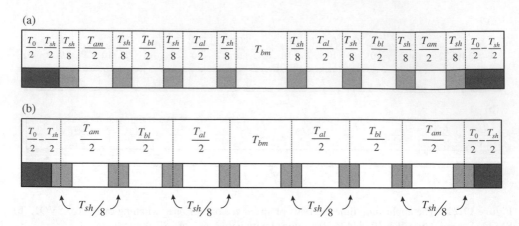

Figure 17.24 The application time intervals of the space vectors for one switching cycle: (a) DSVQ1, (b) DSVQ2 (*Source*: Abduallah 2015 [42]. Reproduced with permission of IEEE).

The difference between the discussed methods is illustrated in Figure 17.24, DSVQ1 subtracting all the time of the shoot-through state application from the zero vector time of application, while DSVQ2 subtracts the time interval of shoot-through state from all active vectors and the zero vector also.

17.3 Summary

In this chapter impedance source multi-leg inverters, which consist of three-phase four-leg inverters and five-phase inverters, have been presented. In the first part of this chapter, ZS/qZS three-phase four-leg inverters were investigated with a detailed mathematical model, switching schemes including SVPWM and carrier-based PWM. Next boost control methods were summarized. Then, to analyze buck and boost conversion modes of qZS four-leg inverter, experimental results were presented in different case studies. In the second part of this chapter, the qZS five-phase inverter was presented with mathematical model and switching schemes including various SVPWM techniques.

References

[1] V. Yaramasu, M. Rivera, M. Narimani, B. Wu, J. Rodriguez, "Model predictive approach for a simple and effective load voltage control of four-leg inverter with an output *RL* filter," *IEEE Trans. Ind. Electron.*, vol.61, no.10, pp.5259–5270, Oct. 2014.

[2] I. Vechiu, O. Curea, H. Camblong, "Transient operation of a four-leg inverter for autonomous applications with unbalanced load," *IEEE Trans. Power Electron.*, vol.25, no.2, pp.399–407, Feb. 2010.

[3] S. Ei-Barbari, W. Hofmann, "Digital control of a four leg inverter for standalone photovoltaic systems with unbalanced load," in *Proc. 26th Annual Conference of the IEEE Industrial Electronics Society (IECON)*, pp.729–734, 2000.

[4] A. Dolara, R. Faranda, S. Leva, G. C. Lazaroiu, "Power electronic converters for PV systems in extreme environmental conditions," in *Proc. IEEE Power and Energy Society General Meeting (PES)*, pp.1–5, 2013.

[5] Z. Wang, H. Li, "An integrated three-port bidirectional DC–DC converter for PV application on a DC distribution system," *IEEE Trans. Power Electron.*, vol.28, no.10, pp.4612–4624, Oct. 2013.

[6] B. Gu, J. Dominic, B. Chen, L. Zhang, J. Lai, "Hybrid transformer ZVS/ZCS DC–DC converter with optimized magnetics and improved power devices utilization for photovoltaic module applications," *IEEE Trans. Power Electron.*, vol.30, no.4, pp.2127–2136, April 2015.

[7] Y. Hu, W. Cao, S. J. Finney, W. Xiao, F. Zhang, S. F. McLoone, "New modular structure DC–DC converter without electrolytic capacitors for renewable energy applications," *IEEE Trans. Sustain. Energy*, vol.5, no.4, pp.1184–1192, Oct. 2014.

[8] D. Cao, W. Qian, F. Z. Peng, "A high voltage gain multilevel modular switched-capacitor DC-DC converter," in *Proc. 2014 IEEE Energy Conversion Congress and Exposition (ECCE)*, pp.5749–5756, 14–18 Sept. 2014.

[9] Y. Li, J. Anderson, F. Z. Peng, D. Liu, "Quasi-Z-source inverter for photovoltaic power generation systems," in *Proc. Twenty-Fourth Annual IEEE Applied Power Electronics Conference and Exposition (APEC)*, pp.918–924, 15–19 Feb. 2009.

[10] H. Abu-Rub, A. Iqbal, S. Moin Ahmed, F. Z. Peng, Y. Li, B. Ge, "Quasi-Z-source inverter-based photovoltaic generation system with maximum power tracking control using ANFIS," *IEEE Trans. Sustain. Energy*, vol.4, no.1, pp.11–20, Jan. 2013.

[11] M. Mosa, G. M. Dousoky, H. Abu-Rub, "A novel FPGA implementation of a model predictive controller for SiC-based quasi-Z-source inverters," in *Proc. Twenty-Ninth Annual IEEE Applied Power Electronics Conference and Exposition (APEC)*, pp.1293–1298, 16–20 March 2014.

[12] Y. Liu, B. Ge, H. Abu-Rub, F. Z. Peng, "Overview of space vector modulations for three-phase Z-source/quasi-Z-source inverters," *IEEE Trans. Power Electron.*, vol.29, no.4, pp.2098–2108, April 2014.

[13] M. Zhang, "Investigation of switching schemes for three-phase four-leg voltage source inverters," PhD Thesis, School of Electrical and Electronic Engineering, Newcastle University, June 2013.

[14] V. Yaramasu, M. Rivera, B. Wu, J. Rodriguez, "Model predictive current control of two-level four-leg inverters – Part I: Concept, algorithm, and simulation analysis," *IEEE Trans. Power Electron.*, vol.28, no.7, pp.3459–3468, July 2013.

[15] M. Rivera, V. Yaramasu, J. Rodriguez, B. Wu, "Model predictive current control of two-level four-leg inverters – Part II: experimental implementation and validation," *IEEE Trans. Power Electron.*, vol.28, no.7, pp.3469–3478, July 2013.

[16] I. Dzafic, H.-T. Neisius, M. Gilles, S. Henselmeyer, V. Landerberger, "Three-phase power flow in distribution networks using Fortescue transformation," *IEEE Trans. Power Syst.*, vol.28, no.2, pp.1027–1034, May 2013.

[17] M. Zhang, D. Atkinson, M. Armstrong, "A zero-sequence component injected PWM method with reduced switching losses and suppressed common-mode voltage for a three-phase four-leg voltage source inverter," in *Proc. 38th Annual Conference on IEEE Industrial Electronics Society (IECON)*, pp.5068–5073, 2012.

[18] R. Zhang, D. Boroyevich, V. H. Prasad, H. Mao, F. C. Lee, S. Dubovsky, "A three-phase inverter with a neutral leg with space vector modulation," in *Proc. Twelfth Annual IEEE Applied Power Electronics Conference and Exposition (APEC)*, pp.857–863 vol.2, 23–27 Feb 1997.

[19] D. Sreenivasarao, P. Agarwal, B. Das, "Neutral current compensation in three-phase, four-wire systems: A review," *Electric Power Systems Research*, vol.86, pp.170–180, May 2012.

[20] D. Fu, S. Wang, P. Kong, F. C. Lee, D. Huang, "Novel techniques to suppress the common-mode EMI noise caused by transformer parasitic capacitances in DC–DC converters," *IEEE Trans. Ind. Electron.*, vol.60, no.11, pp.4968–4977, Nov. 2013.

[21] J. Liang, T. C. Green, C. Feng, G. Weiss, "Increasing voltage utilization in split-link, four-wire inverters," *IEEE Trans. Power Electron.*, vol.24, no.6, pp.1562–1569, June 2009.

[22] K. Karanki, G. Geddada, M. K. Mishra, B. K. Kumar, "A modified three-phase four-wire UPQC topology with reduced DC-link voltage rating," *IEEE Trans. Ind. Electron.*, vol.60, no.9, pp.3555–3566, Sept. 2013.

[23] A. M. Hava, E. Un, "Performance analysis of reduced common-mode voltage PWM methods and comparison with standard PWM methods for three-phase voltage-source inverters," *IEEE Trans. Power Electron.*, vol.24, no.1, pp.241–252, Jan. 2009.

[24] B. Tamyurek, "A high-performance SPWM controller for three-phase UPS systems operating under highly non-linear loads," *IEEE Trans. Power Electron.*, vol.28, no.8, pp.3689–3701, Aug. 2013.

[25] H. Jou, J. Wu, K. Wu, W. Chiang, Y. Chen, "Analysis of zig-zag transformer applying in the three-phase four-wire distribution power system," *IEEE Trans. Power Del.*, vol.20, no.2, pp.1168–1173, April 2005.

[26] D. Suresh, S. Singh, "Reduced rating hybrid active power filter in a three-phase four-wire distribution system," in *Proc. International Conference on Computer Communication and Informatics (ICCCI)*, pp.1–5, 2014.

[27] I. Vechiu, H. Camblong, G. Tapia, B. Dakyo, O. Curea, "Control of four leg inverter for hybrid power system applications with unbalanced load," *Energy Conversion and Management*, vol.48, no.7, pp.2119–2128, July 2007.

[28] M. Rivera, V. Yaramasu, A. Llor, J. Rodriguez, B. Wu, M. Fadel, "Digital predictive current control of a three-phase four-leg inverter," *IEEE Trans. Ind. Electron.*, vol.60, no.11, pp.4903–4912, Nov. 2013.

[29] X. Li, Z. Deng, Z. Chen, Q. Fei, "Analysis and simplification of three-dimensional space vector PWM for three-phase four-leg inverters," *IEEE Trans. Ind. Electron.*, vol.58, no.2, pp.450–464, Feb. 2011.

[30] Y. Kumsuwan, S. Premrudeepreechacharn, V. Kinnares, "A carrier-based unbalanced PWM method for four-leg voltage source inverter fed unsymmetrical two-phase induction motor," *IEEE Trans. Ind. Electron.*, vol.60, no.5, pp.2031–2041, May 2013.

[31] F. Z. Peng, "Z-source inverter," *IEEE Trans. Ind. Appl.*, vol.39, no.2, pp.504–510, Mar/Apr 2003.

[32] Y. Liu, B. Ge, H. Abu-Rub, F. Z. Peng, "Phase-shifted pulse-width-amplitude modulation for quasi-Z-source cascade multilevel inverter-based photovoltaic power system," *IET Power Electron.*, vol.7, no.6, pp.1444–1456, June 2014.

[33] M. Mosa, H. Abu-Rub, J. Rodriguez, "High performance predictive control applied to three phase grid connected quasi-Z-source inverter," in *Proc. 39th Annual Conference of the IEEE Industrial Electronics Society (IECON)*, pp.5812–5817, 2013.

[34] D. Gan, O. Ojo, "Current regulation in four-leg voltage-source converters," *IEEE Trans. Ind. Electron.*, vol.54, no.4, pp.2095–2105, Aug. 2007.

[35] R. Zhang, V. H. Prasad, D. Boroyevich, F. C. Lee, "Three-dimensional space vector modulation for four-leg voltage-source converters," *IEEE Trans. Power Electron.*, vol.17, no.3, pp.314–326, May 2002.

[36] J. Kim, S. Sul, "Carrier-based PWM method for three-phase four-leg voltage source converters," *IEEE Trans. Power Electron.*, vol.19, no.1, pp.66,75, Jan. 2004.

[37] A. M. Hava, R. J. Kerkman, T. A. Lipo, "Simple analytical and graphical methods for carrier-based PWM-VSI drives," *IEEE Trans. Power Electron.*, vol.14, no.1, pp.49–61, Jan 1999.

[38] A. A. Abduallah, A. Iqbal, M. Meraj, L. Ben-Brahim, R. Alammari, H. Abu-Rub, "Discontinuous space vector pulse width modulation techniques for a five-phase quasi Z-source inverter," in *Proc. 41st Annual Conference of the IEEE Industrial Electronics Society (IECON)*, 2015.

[39] O. Ellabban, H. Abu-Rub, "Field oriented control of a five phase induction motor fed by a Z-source inverter," in *Proc. IEEE International Conference on Industrial Technology (ICIT)*, 2013, pp.1624–1629, 2013.

[40] A. Iqbal, R. Alammari, M. Mosa, H. Abu-Rub, "Finite set model predictive current control with reduced and constant common mode voltage for a five-phase voltage source inverter," in *Proc. IEEE 23rd International Symposium on Industrial Electronics (ISIE)*, pp.479–484, 2014.

[41] P. S. N. De Silva, J. E. Fletcher, B. W. Williams, "Development of space vector modulation strategies for five phase voltage source inverters," in *Proc. Second International Conference on Power Electronics, Machines and Drives (PEMD)*. vol.2, no., pp.650–655, 2004.

[42] A. A. Abduallah, A. Iqbal, M. Meraj, L. Ben-Brahim, R. Alammari, H. Abu-Rub, "Discontinuous space vector pulse width modulation techniques for a five-phase quasi Z-source inverter", to appear in the *IEEE Annual Conference of Industrial Electronics Society (IECON)*, Nov. 9–12, 2015.

[43] S. Thangaprakash, A. Krishnan, "Implementation and critical investigation on modulation schemes of three phase impedance source inverter," *International Journal of Electronics and Electrical Engineering (IJEEE)*, vol.6, no.2, pp.84–92, 2010.

[44] U. S. Ali, V. Kamaraj, "A novel space vector PWM for Z-source inverter," in *Proc. 1st International Conference on Electrical Energy Systems (ICEES)*, pp.82–85, 2011.

[45] M. A. Khan, S. K. M. Ahmed, A. Iqbal, H. Abu-Rub, S. K. Moinoddin, "Discontinuous space vector PWM strategies for a seven-phase voltage source inverter," in *Proc. 35th Annual Conference of IEEE Industrial Electronics (IECON)*, pp.397–402, 2009.

[46] A. R. Beig, S. Kanukollu, A. Dekka, "Space vector-based three-level discontinuous pulse-width modulation algorithm," *IET Power Electron.*, vol.6, no.8, pp.1475–1482, 2013.

[47] S. Bayhan, M. Trabelsi, H. Abu-Rub, "Model Predictive Control of Z-Source four-leg inverter for standalone Photovoltaic system with unbalanced load," in *Proc. Thirty-First Annual IEEE Applied Power Electronics Conference and Exposition (APEC)*, Long Beach, CA, USA, 2016, pp. 3663–3668.

[48] S. Bayhan, H. Abu-Rub, R. Balog, "Model Predictive Control of Quasi-Z Source Four-Leg Inverter," *IEEE Trans. Ind. Electron.*, vol.63, no.7, pp.4506–4516, Jul. 2016.

18

Model Predictive Control of Impedance Source Inverter

Sertac Bayhan[1,2], Mostafa Mosa[3,4], and Haitham Abu-Rub[2]

[1] Department of Electronics and Automation, Gazi University, Ankara, Turkey
[2] Electrical and Computer Engineering Program, Texas A&M University at Qatar, Qatar Foundation, Doha, Qatar
[3] Department of Electrical and Computer Engineering, Texas A&M University, College Station, TX, USA
[4] Department of Electrical Engineering, Aswan University, Aswan, Egypt

18.1 Introduction

The control strategy of the power converter plays a crucial role in ensuring reliable and efficient operation of renewable based distributed generation (DG) systems. In most applications, proportional-integral (PI) based cascaded control structures have been used to control current, voltage, etc. [1]. This control structure has two control loops consisting of an inner control loop (current controller) and an outer loop (voltage/power controller). Although this control technique is easy to implement, it has some drawbacks, the biggest of which is that the performance of the whole system substantially depends on the performance of the inner control loop. For this reason, the parameters of the inner controller must be chosen carefully to provide system stability at all operating points. Another drawback of this structure is that its linear nature does not consider the discrete operation of a voltage source converter (VSC) [2].

Another issue to be considered is that one additional switching state, called the shoot-through state, is required in order to control the impedance source inverter. To obtain this switching state, several modulation techniques have been proposed [3, 4]. These reconfigured modulation techniques lead to increased complexity of the control structure.

The model predictive control (MPC) method is an attractive alternative to classical control methods owing to its fast dynamic response, simple concept, and its ability to include different non-linearities and constraints. One advantage of predictive control is the possibility of

Impedance Source Power Electronic Converters, First Edition. Yushan Liu, Haitham Abu-Rub, Baoming Ge, Frede Blaabjerg, Omar Ellabban, and Poh Chiang Loh.

including non-linearities of the system in the predictive model, and hence calculating the behavior of the variables for different conduction states [5]. It is also worth mentioning that this scheme does not require internal current control loops and modulators, and thus it greatly reduces the complexity. Several studies have been presented under the name of the MPC for current control of traditional three-phase inverters [6], three-phase four-leg inverters [7], multilevel inverters [8, 9], quasi-Z-source three-leg inverters [10, 11], and several adjustable speed drives [12–14]. It is concluded that MPC is currently one of the most attractive control techniques for power converters and adjustable speed drives.

To eliminate the drawbacks of the cascaded control structure and difficulties of the modulation techniques, the MPC technique for impedance source – Z-source (ZS) and quasi-Z-source (qZS) – three-leg, four-leg, and multiphase inverters is examined in this chapter. It starts with an overview of the MPC technique. Then, ZS/qZS three-phase three-leg, four-leg, and multiphase (five-phase) inverter topologies are presented, along with a description of the topology, model, and switching schemes for control. After that, the MPC algorithm for ZS/qZS three-phase, three-leg, four-leg, and multiphase (five-phase) inverters are presented in detail. Finally, to prove the addressed MPC strategy, a performance investigation has been carried out with qZS four-leg inverter and simulation results are presented to demonstrate the capabilities of the MPC.

18.2 Overview of Model Predictive Control

Model predictive control has now seen approximately three decades of development and is considered one of the most important advances in process control. The MPC method for applications in power converters has received considerable attention in recent years [14]. The MPC techniques applied to power electronics have been classified into two main categories [14, 15]: continuous control set MPC (CCS-MPC) and finite control set MPC (FCS-MPC). In the first group, a modulator generates the switching states starting from the continuous output of the predictive controller. The FCS-MPC approach, by contrast, has the advantage of a limited number of switching states of the power converter for solving the optimization problem. A discrete model is used to predict the behavior of the system for every admissible actuation sequence up to the prediction horizon. The switching action that minimizes a predefined cost function is finally selected to be applied in the next sampling instant. The main advantage of the FCS-MPC lies in the direct application of the control action to the converter, without requiring a modulation stage [15].

The main characteristic of MPC is the use of a model of the system for predicting the future behavior of the controlled variables. This information is used by the controller to obtain optimal actuation of the system, according to a predefined optimization criterion. Using predictive control it is possible to avoid the cascaded structure, which is typically used in a linear control scheme, and thereby obtain a fast dynamic response [16].

The advantages of the MPC strategy can be summarized as

- concepts very intuitive and easy to understand
- can be applied to a great variety of systems
- the multivariable case can easily be considered
- dead times can be compensated
- easy inclusion of non-linearity in the model
- simple treatment of constraints

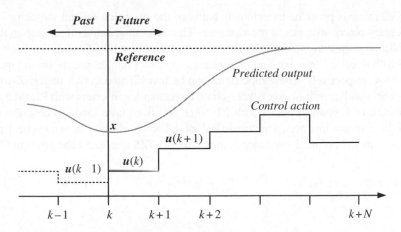

Figure 18.1   Working principle of model predictive control.

- resulting controller is easy to implement
- methodology suitable for the inclusion of modifications and extensions depending on specific applications.

However, there are some disadvantages, such as (a) a larger number of calculations compared to classic controllers, (b) the quality of the model has a direct influence on the quality of the resulting controller, and (c) if the parameters of the system change in time, some adaptation or estimation algorithm has to be considered [17].

The working principle of MPC is summarized in Figure 18.1. The future values of the states of the system are predicted until a predefined horizon in time $k+N$ using the system model and the available information (measurements) until time k. The sequence of the optimal actuations is calculated by minimizing a cost function and the first element of this sequence is applied. This whole process is repeated again for each sampling instant considering the new measured data [17].

In general, the MPC for power electronics converters can be designed using the following steps [18]:

- model the power converter, identifying all possible switching states and their relation to the input or output voltages or currents
- define a cost function that represents the desired behavior of the system
- obtain discrete-time models that allow prediction of the future behavior of the variables to be controlled.

18.3   Mathematical Model of the Z-Source Inverters

18.3.1   Overview of Topologies

Impedance source or Z-source inverters are special types of inverters that provide a voltage boost capability in conventional inverters. Conventional inverters work only buck as converter because the output voltage is always lower than the dc input voltage. Moreover, the upper and lower power switch cannot conduct simultaneously otherwise the dc source will be short-circuited.

Hence, a dead time is provided intentionally between the switching on and switching off of the complementary power switches of the same leg. This dead time causes distortion in the output voltage and also in the current. These shortcomings are eliminated in Z-source inverters [19, 20].

The ZS/qZS three-leg, four-leg, and multiphase inverter topologies are shown in Figures 18.2, 18.3, and 18.4, respectively. These topologies can be investigated in two stages: Z-source networks (Z-source and quasi Z-source), three-phase three/four leg inverters with RL filter and load, and multiphase (five-phase) inverter with RL filter and five-phase load. To describe the MPC scheme of the Z-source inverters, a mathematical model of the whole system must be determined. Mathematical models of the impedance source network (ZS and qZS) are given in Chapter 2.

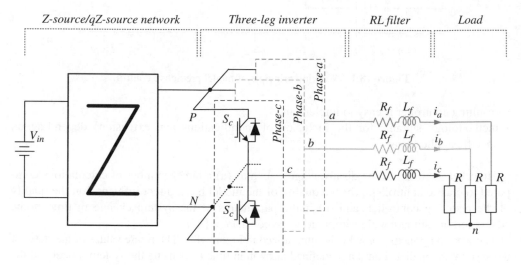

Figure 18.2   Impedance source three-phase three-leg inverter topology.

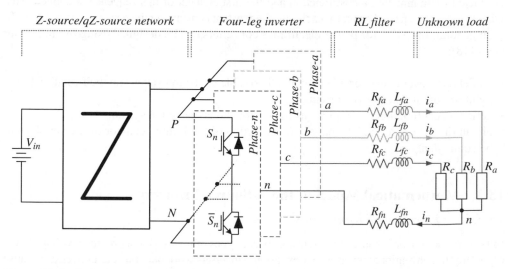

Figure 18.3   Impedance source three-phase four-leg inverter topology (*Source*: Sertac 2016 [40]. Reproduced with permission of IEEE).

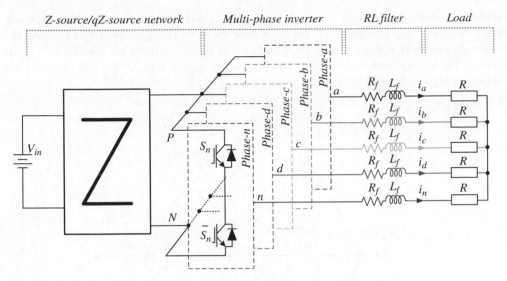

Figure 18.4 Impedance source *n*-phase inverter topology.

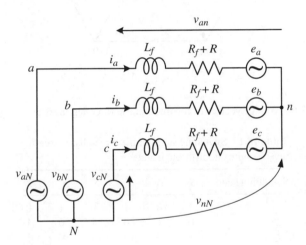

Figure 18.5 Equivalent circuit of the three-phase three-leg inverter in Figure 18.2.

In this section, the mathematical model of the three-phase inverters (three-leg and four-leg) and multiphase (five-phase) inverter models are described in detail. Note that the mathematical models of the Z-source network's inductor current and capacitor voltage are the same for both the ZS and qZS inverters. Therefore, the discussed MPC technique is suitable for both ZS and qZS inverters.

18.3.2 Three-Phase Three-Leg Inverter Model

The equivalent circuit of the three-phase three-leg inverter with an output *RL* filter is shown in Figure 18.5. The three control signals named S_a, S_b, and S_c form a total of eight (2^3) switching states (non-shoot-through states) of the inverter [21]. In addition to those switching states, one extra switching state is required in order to ensure the shoot-through state for controlling the Z-source's network voltage. Therefore, a total of nine switching states are used for this application. The voltage vectors and valid switching states for the ZS/qZS three-leg inverter are shown in Figure 18.6 and Table 18.1, respectively.

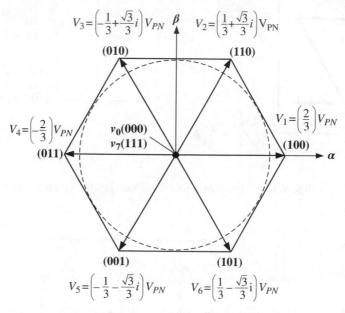

Figure 18.6   Voltage vectors of the three-phase three-leg inverter in Figure 18.2.

Table 18.1   Switching states of the three-phase three-leg inverter with shoot-through states

	Switching state	S_a	S'_a	S_b	S'_b	S_c	S'_c
Non-shoot-through states	0	1	0	1	0	1	0
	1	0	1	0	1	0	1
	2	1	0	0	1	0	1
	3	1	0	1	0	0	1
	4	0	1	1	0	0	1
	5	0	1	1	0	1	0
	6	0	1	0	1	1	0
	7	1	0	0	1	1	0
Shoot-through states	8	1	1	0	0	0	0
		0	0	1	1	0	0
		0	0	0	0	1	1
		1	1	1	1	0	0
		1	1	0	0	1	1
		0	0	1	1	1	1
		1	1	1	1	1	1

These switching signals define the value of the output voltages

$$v_{aN} = S_a v_{PN}$$
$$v_{bN} = S_b v_{PN}$$
$$v_{cN} = S_c v_{PN}$$

(18.1)

where v_{PN} is the dc-link voltage. The switching state can be written as a vector,

$$S = \frac{2}{3}\left(S_a + \mathbf{a}S_b + \mathbf{a}^2 S_c\right)$$

(18.2)

where $\mathbf{a} = e^{j2\pi/3}$ is a unitary vector, which represents the 120° phase displacement between the phases.

Taking into account the definitions of variables from the circuit shown in Figure 18.5, the equations for load current dynamics for each phase can be written as

$$
\begin{bmatrix} v_{aN} \\ v_{bN} \\ v_{cN} \end{bmatrix} = \begin{bmatrix} R+R_f & 0 & 0 \\ 0 & R+R_f & 0 \\ 0 & 0 & R+R_f \end{bmatrix} \begin{bmatrix} i_a \\ i_b \\ i_c \end{bmatrix} + \begin{bmatrix} L_f & 0 & 0 \\ 0 & L_f & 0 \\ 0 & 0 & L_f \end{bmatrix} \frac{d}{dt} \begin{bmatrix} i_a \\ i_b \\ i_c \end{bmatrix} + \begin{bmatrix} e_a \\ e_b \\ e_c \end{bmatrix} + \begin{bmatrix} 1 \\ 1 \\ 1 \end{bmatrix} v_{nN}
$$

(18.3)

where R is the load resistance, and R_f and L_f are the filter resistance and inductance, respectively. This equation can be written as

$$\mathbf{v} = \left(R_f + R\right)\mathbf{i} + L_f \frac{d\mathbf{i}}{dt} + \mathbf{e}$$

(18.4)

where $\mathbf{v}$ is the voltage vector generated by the inverter, $\mathbf{i}$ is the load current vector, and $\mathbf{e}$ the load back-emf vector. The derivative of the output current vector can be obtained from (18.4)

$$\frac{d\mathbf{i}}{dt} = \frac{1}{L_f}\left[\left(\mathbf{v}-\mathbf{e}\right)-\left(R_f+R\right)\mathbf{i}\right]$$

(18.5)

This equation can be used to obtain the discrete-time model of the load current.

18.3.3   Three-Phase Four-Leg Inverter Model

The equivalent circuit of the three-phase four-leg inverter with the output RL filter is shown in Figure 18.7, where L_{fj} is the filter inductance, R_{fj} is the filter resistance, and R_j is the load resistance ($j = a, b, c$). The four control signals named S_a, S_b, S_c, and S_n form a total of 16 (2^4) switching states (non-shoot-through states) of the converter [22]. As mentioned above, in addition to these switching states, for this application one extra switching state is required in order to ensure the shoot-through state. Therefore, a total of 17 switching states are used for this application. The voltage vectors and valid switching states for the ZS/qZS four-leg inverter are shown in Figure 18.8 and Table 18.2, respectively [23, 24]. As be seen, the voltage vectors

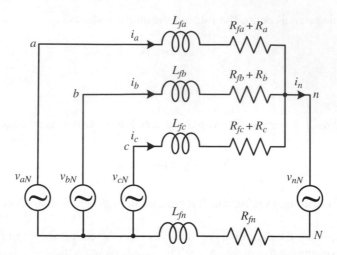

Figure 18.7 Equivalent circuit of the three-phase four-leg inverter in Figure 18.3 (*Source*: Sertac 2016 [40]. Reproduced with permission of IEEE).

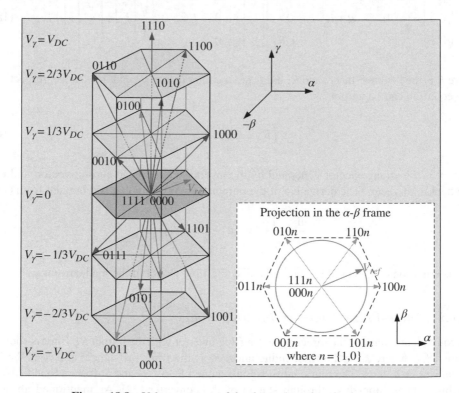

Figure 18.8 Voltage vectors of the three-phase four-leg inverter.

can be represented by a 3D representation with V_γ. More information about the 3D space vector for the four-leg inverter is given in Chapter 17.

The voltages in each leg of the inverter, measured from the negative point of dc-link N, can be expressed as

Table 18.2 Switching states of the three-phase four-leg inverter with shoot-through states

	Switching State	S_a	S'_a	S_b	S'_b	S_c	S'_c	S_n	S'_n
Non-shoot-through states	0	1	0	0	1	0	1	0	1
	1	1	0	1	0	0	1	0	1
	2	0	1	1	0	0	1	0	1
	3	0	1	1	0	1	0	0	1
	4	0	1	0	1	1	0	0	1
	5	1	0	0	1	1	0	0	1
	6	1	0	1	0	1	0	0	1
	7	0	1	0	1	0	1	0	1
	8	1	0	0	1	0	1	1	0
	9	1	0	1	0	0	1	1	0
	10	0	1	1	0	0	1	1	0
	11	0	1	1	0	1	0	1	0
	12	0	1	0	1	1	0	1	0
	13	1	0	0	1	1	0	1	0
	14	1	0	1	0	1	0	1	0
	15	0	1	0	1	0	1	1	0
Shoot-through states		1	1	0	0	0	0	0	0
		0	0	1	1	0	0	0	0
		0	0	0	0	1	1	0	0
		0	0	0	0	0	0	1	1
		1	1	1	1	0	0	0	0
		1	1	0	0	1	1	0	0
	16	1	1	0	0	0	0	1	1
		0	0	1	1	1	1	0	0
		0	0	1	1	0	0	1	1
		0	0	0	0	1	1	1	1
		1	1	1	1	1	1	0	0
		1	1	1	1	0	0	1	1
		1	1	0	0	1	1	1	1
		0	0	1	1	1	1	1	1
		1	1	1	1	1	1	1	1

$$\begin{bmatrix} v_{aN} \\ v_{bN} \\ v_{cN} \\ v_{nN} \end{bmatrix} = \begin{bmatrix} S_a \\ S_b \\ S_c \\ S_n \end{bmatrix} v_{PN} \qquad (18.6)$$

which can be written as

$$v_{jn} = S_j v_{PN}, \quad j = a,b,c,n. \qquad (18.7)$$

The voltage applied to the output RL filter, in terms of these voltages, is

$$\begin{bmatrix} v_{an} \\ v_{bn} \\ v_{cn} \end{bmatrix} = \begin{bmatrix} S_a - S_n \\ S_b - S_n \\ S_c - S_n \end{bmatrix} v_{PN} \qquad (18.8)$$

The above expression can be simplified to

$$v_{kn} = v_{kN} - v_{nN} = \left(S_k - S_n \right) v_{PN}, \quad k = a, b, c. \tag{18.9}$$

The inverter voltages are given as

$$
\begin{bmatrix} v_{aN} \\ v_{bN} \\ v_{cN} \\ v_{nN} \end{bmatrix} =
\begin{bmatrix} R_{fa} & 0 & 0 & 0 \\ 0 & R_{fb} & 0 & 0 \\ 0 & 0 & R_{fc} & 0 \\ 0 & 0 & 0 & R_{fn} \end{bmatrix}
\begin{bmatrix} i_a \\ i_b \\ i_c \\ i_n \end{bmatrix} +
\begin{bmatrix} L_{fa} & 0 & 0 & 0 \\ 0 & L_{fb} & 0 & 0 \\ 0 & 0 & L_{fc} & 0 \\ 0 & 0 & 0 & L_{fn} \end{bmatrix}
\frac{d}{dt}
\begin{bmatrix} i_a \\ i_b \\ i_c \\ i_n \end{bmatrix}
$$

$$
+ \begin{bmatrix} R_a & 0 & 0 & 0 \\ 0 & R_b & 0 & 0 \\ 0 & 0 & R_c & 0 \\ 0 & 0 & 0 & R_n \end{bmatrix}
\begin{bmatrix} i_a \\ i_b \\ i_c \\ i_n \end{bmatrix} +
\begin{bmatrix} 1 \\ 1 \\ 1 \\ 1 \end{bmatrix} v_{nN}
\tag{18.10}
$$

which can be written as

$$v_{jn} = \left(R_{fj} + R_j \right) i_j + L_{fj} \frac{di_j}{dt} + v_{nN}, \quad j = a, b, c, n. \tag{18.11}$$

The derivative of the output current vector can be obtained from (18.11) as

$$\frac{di_j}{dt} = \frac{1}{L_{fj}} \left[\left(v_{jn} - v_{nN} \right) - \left(R_{fj} + R_j \right) i_j \right], \quad j = a, b, c. \tag{18.12}$$

Based on (18.7) and (18.11), the load neutral voltage v_{nN} can be expressed as

$$v_{nN} = L_{eq} v_{PN} \sum_{j=a,b,c,n} \frac{S_j}{L_{fj}} - L_{eq} \sum_{j=a,b,c,n} \frac{R_{fj} + R_j}{L_{fj}} i_j \tag{18.13}$$

with

$$L_{eq} = \left(\frac{1}{L_{fa}} + \frac{1}{L_{fb}} + \frac{1}{L_{fc}} + \frac{1}{L_{fn}} \right)^{-1} \tag{18.14}$$

$$i_a + i_b + i_c + i_n = 0 \tag{18.15}$$

18.3.4   Multiphase Inverter Model

Multiphase drives are serious competitors to the traditional three-phase drives due to better fault tolerant properties and reduced per-phase converter rating that are especially suited to high power drives application. In the multiphase inverter, $2n$ switches are used, where n is the

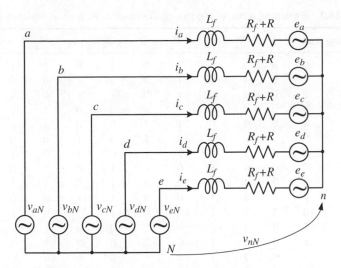

Figure 18.9   Equivalent circuit of the five-phase inverter.

number of phases. These switches are similar to the one used in the classical three-phase inverter, where two conditions have to be ensured: bidirectional current flow and unidirectional voltage blocking capability. Although there are numerous types of multiphase inverters, the five-phase inverter will be addressed in this subsection [25–27].

The equivalent circuit of the five-phase inverter with an output RL load is shown in Figure 18.9. Each of the two-level inverter legs has two switching states. Thus for the five-phase inverter, there are (2^5) 32 possible switching combinations. Of the 32 switching states, 30 states are active and two states are zero switching states (vectors). However, in addition to these switching states, for this application, one extra switching state is required in order to ensure the shoot-through state. Therefore, a total of 33 switching states are used for this application. The switching combinations of the five-phase inverter are shown in Table 18.3.

At any instant of time, the inverter can produce only one space vector. The space vector of phase voltages can be defined using the power variant transformations as

$$v_{\alpha\beta} = \frac{2}{5}\left(v_a + \mathbf{a}v_b + \mathbf{a}^2 v_c + \mathbf{a}^{*2} v_d + \mathbf{a}^* v_e\right) \tag{18.16}$$

where $\mathbf{a} = e^{\frac{j2\pi}{5}}$, $\mathbf{a}^2 = e^{\frac{j4\pi}{5}}$, $\mathbf{a}^* = e^{\frac{-j2\pi}{5}}$, $\mathbf{a}^{*2} = e^{\frac{-j4\pi}{5}}$, and * stands for a complex conjugate. The phase voltage space vectors thus obtained in α-β plane and shown in Figure 18.10. Since it is a five-phase system, the transformation is further done in order to obtain space vectors in x-y plane using equation (18.16) and the resulting space vectors are shown in Figure 18.11 [25].

$$v_{xy} = \frac{2}{5}\left(v_a + \mathbf{a}^2 v_b + \mathbf{a}^4 v_c + \mathbf{a}^6 v_d + \mathbf{a}^8 v_e\right) \tag{18.17}$$

The vectors are numbered by decimal values which, when converted to their equivalent binary, will give the switching state of each leg, where 1 corresponds to the upper switch being

Table 18.3 Switching states of the five-phase inverter with shoot-through states

Switching State	S_a	S'_a	S_b	S'_b	S_c	S'_c	S_d	S'_d	S_e	S'_e
0	0	1	0	1	0	1	0	1	1	0
1	0	1	0	1	0	1	1	0	0	1
2	0	1	0	1	0	1	1	0	1	0
3	0	1	0	1	1	0	0	1	0	1
4	0	1	0	1	1	0	0	1	1	0
5	0	1	0	1	1	0	1	0	0	1
6	0	1	0	1	1	0	1	0	1	0
7	0	1	1	0	0	1	0	1	0	1
8	0	1	1	0	0	1	0	1	1	0
9	0	1	1	0	0	1	1	0	0	1
10	0	1	1	0	0	1	1	0	1	0
11	0	1	1	0	1	0	0	1	0	1
12	0	1	1	0	1	0	0	1	1	0
13	0	1	1	0	1	0	1	0	0	1
14	0	1	1	0	1	0	1	0	1	0
15	1	0	0	1	0	1	0	1	0	1
16	1	0	0	1	0	1	0	1	1	0
17	1	0	0	1	0	1	1	0	0	1
18	1	0	0	1	0	1	1	0	1	0
19	1	0	0	1	1	0	0	1	0	1
20	1	0	0	1	1	0	0	1	1	0
21	1	0	0	1	1	0	1	0	0	1
22	1	0	0	1	1	0	1	0	1	0
23	1	0	1	0	0	1	0	1	0	1
24	1	0	1	0	0	1	0	1	1	0
25	1	0	1	0	0	1	1	0	0	1
26	1	0	1	0	0	1	1	0	1	0
27	1	0	1	0	1	0	0	1	0	1
28	1	0	1	0	1	0	0	1	1	0
29	1	0	1	0	1	0	1	0	0	1
30	1	0	1	0	1	0	1	0	1	0
31	0	1	0	1	0	1	0	1	0	1
	1	1	0	0	0	0	0	0	0	0
	0	0	1	1	0	0	0	0	0	0
	0	0	0	0	1	1	0	0	0	0
	0	0	0	0	0	0	1	1	0	0
	0	0	0	0	0	0	0	0	1	1
32	1	1	1	1	0	0	0	0	0	0
	1	1	0	0	1	1	0	0	0	0
	1	1	0	0	0	0	1	1	0	0
	1	1	0	0	0	0	0	0	1	1
	1	1	1	1	1	1	0	0	0	0

Non-shoot-through states (rows 0–31); Shoot-through states (row 32)

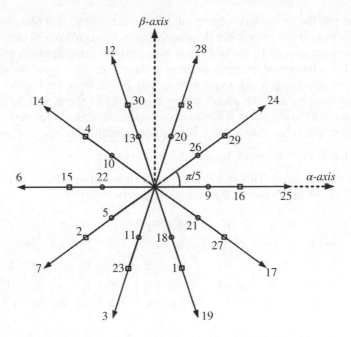

Figure 18.10   Phase voltage space vector in α-β axis for the five-phase inverter.

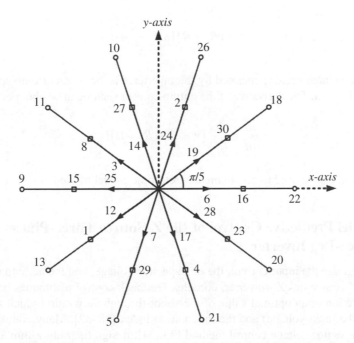

Figure 18.11   Phase voltage space vector in x-y axis for the five-phase inverter.

on and 0 represents the lower switch being on. Moreover the operation of each inverter leg is complementary in order to protect the dc source. It can be seen from Figure 18.10 that the outer decagon space vectors of the α-β plane map into the inner decagon of the x-y plane (Figure 18.11), the innermost decagon of the d-q plane forms the outer decagon of the x-y plane, and the middle decagon space vectors map into the same region. Further, it is observed from the above mapping that the phase sequence a,b,c,d,e of the α-β plane corresponds to a,c,e,b,d sequence of the x-y plane. It is also observed that there are three sets of vectors with three different lengths, referred to as large, medium, and small. The length of the large vectors is $\dfrac{1}{5}\cos\dfrac{\pi}{5}V_{dc}$, the medium vectors $\dfrac{2}{5}V_{dc}$, and the small vectors $\dfrac{1}{5}\cos\dfrac{2\pi}{5}V_{dc}$.

Taking into account the definitions of variables from the circuit shown in Figure 18.9, the equations for load current dynamics for each phase can be written as

$$
\begin{bmatrix} v_a \\ v_b \\ v_c \\ v_d \\ v_e \end{bmatrix} =
\begin{bmatrix}
R+R_f & 0 & 0 & 0 & 0 \\
0 & R+R_f & 0 & 0 & 0 \\
0 & 0 & R+R_f & 0 & 0 \\
0 & 0 & 0 & R+R_f & 0 \\
0 & 0 & 0 & 0 & R+R_f
\end{bmatrix}
\begin{bmatrix} i_a \\ i_b \\ i_c \\ i_d \\ i_e \end{bmatrix} +
\begin{bmatrix}
L_f & 0 & 0 & 0 & 0 \\
0 & L_f & 0 & 0 & 0 \\
0 & 0 & L_f & 0 & 0 \\
0 & 0 & 0 & L_f & 0 \\
0 & 0 & 0 & 0 & L_f
\end{bmatrix}
\frac{d}{dt}
\begin{bmatrix} i_a \\ i_b \\ i_c \\ i_d \\ i_e \end{bmatrix} +
\begin{bmatrix} e_a \\ e_b \\ e_c \\ e_d \\ e_e \end{bmatrix}
$$

$$(18.18)$$

where R is the load resistance, and R_f and L_f are the filter resistance and inductance, respectively. This equation can be written as

$$
\mathbf{v} = \left(R_f + R\right)\mathbf{i} + L_f\frac{d\mathbf{i}}{dt} + \mathbf{e}
\tag{18.19}
$$

where $\mathbf{v}$ is the voltage vector generated by the inverter, $\mathbf{i}$ is the load current vector, and $\mathbf{e}$ the load back-emf vector. The derivative of the output current vector can be obtained from (18.19)

$$
\frac{d\mathbf{i}}{dt} = \frac{1}{L_f}\left[\left(\mathbf{v}-\mathbf{e}\right) - \left(R_f + R\right)\mathbf{i}\right]
\tag{18.20}
$$

This equation can be used to obtain the discrete-time model of the load.

18.4 Model Predictive Control of the Z-Source Three-Phase Three-Leg Inverter

In order to regulate the input current, the dc capacitor voltage, and the ac output current in a single-stage Z-source or qZS inverter, complex feedback control techniques have been proposed by selection of an optimal value of the shoot-through duty ratio (which is responsible for boosting the input voltage) and the modulation index [28–32]. Many control techniques, such as the capacitor voltage control method [33, 34], design their algorithm according to a simplified small signal model, and the load is represented as a current source [35]. However, this model describes only the dynamic of the impedance network, but it fails to describe the dynamics of the load. To overcome this disadvantage, a third-order model is considered in the control design [36, 37]. Unfortunately, this method increases the complexity of the control, which requires much fine-tuning and system analysis.

To overcome these weaknesses of the traditional control of the ZS/qZS inverters, this section presents the MPC, which has emerged as a powerful control method for the control of power electronic converters. Predictive control is easy to implement and gives the ability to add different variables with constraints without major changes in the main control design. In addition, the switching frequency for the MPC is not constant, so the switching harmonic content for the three- or four-phase output currents is distributed across the frequency spectrum, reducing the individual harmonic component amplitudes. These advantages improve the overall performance and efficiency of power electronics inverters controlled by MPC for different applications.

Figure 18.12 shows the discussed MPC scheme of the ZS/qZS three-leg inverter. It consists of three layers, an extrapolation, a predictive model, and a cost function optimization. The extrapolation is used to extrapolate the reference values to a future state. Moreover, the discrete-time model of the system is used to predict the future behavior of the control variables. In addition, the cost function is used to minimize the error between the reference and the predicted control variables in the next sampling time. The MPC scheme has a number of steps: (1) measurements of signals, (2) determination of the references, (3) extrapolation, (4) build discrete time-models of the system, (5) cost function optimization, (6) prepare control algorithm to generate switching signals.

1. Measurements

As a first step in the implementation, the output currents, capacitor voltage, and inductor current are measured at the kth instant.

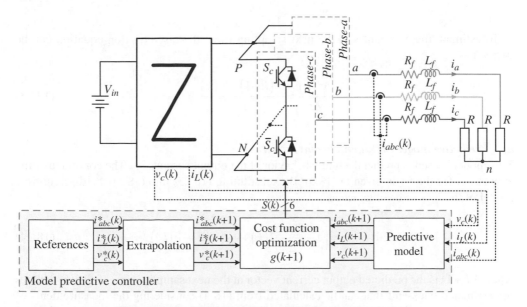

Figure 18.12 MPC scheme for a Z-source three-leg inverter.

2.   Determination of references

Depending upon the application, the references are generated. For grid-connected DG systems, the current references are obtained through maximum power point tracking and dc-link voltage control. The objective is to exhibit simple modeling and control of the three-leg inverter. For this reason, the references are chosen as being user defined. By changing the references, this control technique can be adapted to any specific application.

3.   Extrapolation

Once the references are obtained at the kth instant, they must be extrapolated to $k + 1$ instant for use in the cost function. When the sampling time T_s is sufficiently small ($<20\,\mu s$), no extrapolation is required. In this case, $x^*_{abc}(k+1)=x^*(k)$, where $x = i_{abc}, i_L, v_C$. When the sampling time T_s is greater than $20\,\mu s$, the following fourth-order Lagrange extrapolation [22] can be used:

$$x^*\left(k+1\right) = 4x^*\left(k\right)-6x^*\left(k-1\right)+4x^*\left(k-2\right)-x^*\left(k-3\right), x = i_{abc}, i_L, v_C. \quad (18.21)$$

4.   Predictive models

This control scheme is basically an optimization algorithm, and it can be digitally implemented in microprocessor-based hardware. Consequently, the analysis has to be developed using discrete mathematics in order to consider the additional restrictions such as delays, sampling time, and approximations. As a first consideration, it is important to obtain the discrete-time model from the continuous-time equations. To do that, the general structure of a forward-difference Euler equation (18.22) is used in order to compute the differential equations of the output currents, inductor current, and capacitor voltage.

$$\frac{df}{dt} \approx \frac{f\left(x_0+h\right)-f\left(x_0\right)}{h} \quad (18.22)$$

To estimate these variables in the next sampling time, the discretization equation can be defined as

$$\frac{\Delta f\left(k\right)}{\Delta t} \approx \frac{f\left(k+1\right)-f\left(k\right)}{T_s} \quad (18.23)$$

a)   Predictive model I (Output current model)

This model is used to predict the future behavior of the output currents i_{abc}. The continuous-time expression for the i_{abc} is given in (18.5). By substituting (18.23) into (18.5), the discrete-time model of the i_{abc} can be obtained as

$$\mathbf{i}\left(k+1\right) = \left(1-\frac{RT_s}{L}\right)\mathbf{i}\left(k\right)+\frac{T_s}{L}\left(\mathbf{v}\left(k\right)-\hat{\mathbf{e}}\left(k\right)\right) \quad (18.24)$$

where $\mathbf{i}(k+1)$ is the predicted output current vector at the next sampling time, and $\hat{\mathbf{e}}(k)$ denotes the estimated back-emf that can be calculated from (18.4) considering the measurements of the load voltage and current with the following expression:

$$\hat{\mathbf{e}}(k-1) = \mathbf{v}(k-1) - \frac{L}{T_s}\mathbf{i}(k) - \left(R - \frac{L}{T_s}\right)\mathbf{i}(k-1) \tag{18.25}$$

where $\hat{e}(k-1)$ is the estimated value of $e(k-1)$.

b) **Predictive model II (Inductor current model)**
This model is used to predict the future behavior of the inductor current (I_{L1}). The continuous-time model of the inductor voltage can be expressed as

$$v_{L1} = L_1 \frac{dI_{L1}}{dt} \tag{18.26}$$

where L_1 is the inductance of the inductor. Based on (18.26), the inductor current is derived as,

$$\frac{dI_{L1}}{dt} = \frac{1}{L_1} v_{L1} \tag{18.27}$$

By substituting (18.23) into (18.27), the discrete-time model of the I_{L1} can be obtained as

$$I_{L1}(k+1) = I_{L1}(k) + \frac{T_s}{L} v_{L1}(k) \tag{18.28}$$

where $I_{L1}(k+1)$ is the predicted inductor current at the next sampling time, and $v_{L1}(k)$ is the inductor voltage which depends on the states of the Z-source topology. According to operational principle of Z-source network that is explained in Chapter 2, for non-shoot-through and shoot-through states, the inductor voltage can be defined as follows:
During non-shoot-through state:

$$v_{L1} = V_{in} - V_{C1} \tag{18.29}$$

During shoot-through state:

$$v_{L1} = V_{in} + V_{C2} \tag{18.30}$$

c) **Predictive model III (Capacitor voltage)**
This model is used to predict the future behavior of the capacitor voltage (V_{C1}). The continuous-time model of the capacitor current can be expressed as

$$i_{C1} = C_1 \frac{d(V_{C1} - i_{C1}r_c)}{dt} \tag{18.31}$$

where C_1 and r_c are the capacitance and the equivalent series resistance (ESR) of the capacitor, respectively. Based on (18.31), the capacitor voltage is derived as,

$$\frac{dV_{C1}}{dt} = r_c \frac{di_{C1}}{dt} + \frac{1}{C_1} i_{C1} \tag{18.32}$$

By substituting (18.23) into (18.32), the discrete-time model of the V_{C1} can be obtained as

$$V_{C1}(k+1) = V_{C1}(k) + i_{C1}(k+1)r_c + i_{C1}(k)\left(\frac{T_s}{C} - r_c\right)$$ (18.33)

where $V_{C1}(k+1)$ is the predicted capacitor voltage at the next sampling time, and $i_{C1}(k)$ is the present capacitor current that depends on the states of the Z-source topology. According to the operational principle of the Z-source network which is explained in Chapter 2, for non-shoot-through and shoot-through states, the capacitor current can be defined as follows:
During non-shoot-through state:

$$i_{C1} = I_{L1} - \left(S_a i_a + S_b i_b + S_c i_c\right)$$ (18.34)

During shoot-through state:

$$i_{C1} = -I_{L1}$$ (18.35)

5. Cost function optimization

The selection of the cost function is a key part of the MPC scheme. The MPC has three cost functions, which are used to minimize output current, inductor current, and capacitor voltage errors in the next sampling time. The output current cost function is defined as

$$g_{i_o} = \left\| i_j^*(k+1) - i_j(k+1) \right\|^2$$
$$= \left[i_a^*(k+1) - i_a(k+1) \right]^2 + \left[i_b^*(k+1) - i_b(k+1) \right]^2 + \left[i_c^*(k+1) - i_c(k+1) \right]^2$$ (18.36)

where $i_j^*(k+1)$ is the reference output current vector and $i_j(k+1)$ is the predicted output current vector in the next step ($j = a, b, c$).
The cost function of the inductor current can also be defined as

$$g_{I_L} = \lambda_i \left| I_{L1}^*(k+1) - I_{L1}(k+1) \right|$$ (18.37)

where $I_{L1}^*(k+1)$ and $I_{L1}(k+1)$ are the reference and predicted inductor currents, respectively. λ_i is the weighting factor that has been determined by a trial and error method.
The cost function of the capacitor voltage can also be defined as

$$g_{v_C} = \lambda_v \left| v_{C1}^*(k+1) - v_{C1}(k+1) \right|$$ (18.38)

where $v_{C1}^*(k+1)$ and $v_{C1}(k+1)$ are the reference and predicted capacitor voltages, respectively, λ_v is the weighting factor that has been determined by trial and error method. The final cost function can be defined as

$$g(k+1) = g_{i_o}(k+1) + g_{I_L}(k+1) + g_{v_C}(k+1)$$ (18.39)

To illustrate how the cost function works, sketch maps of the reference and predicted variables are given in Figure 18.13. The output load current has been converted from the

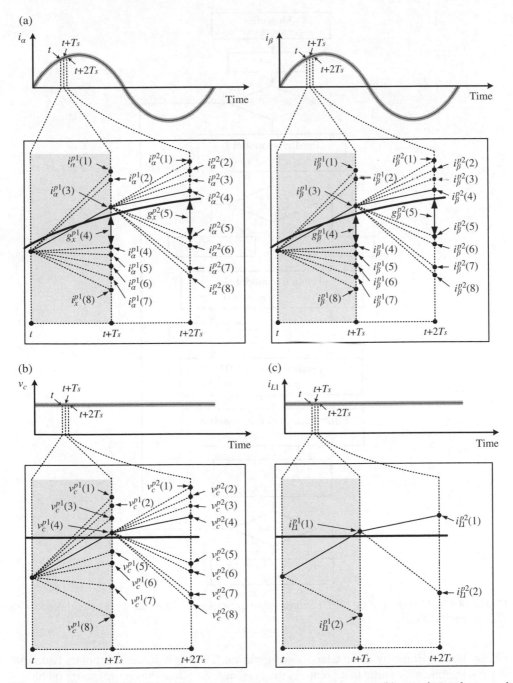

Figure 18.13 Schematic illustration of the prediction (a) output current, (b) capacitor voltage, and (c) inductor current [10, 11]. (*Source*: Mosa 2013 [10]. Reproduced with permission of IEEE.)

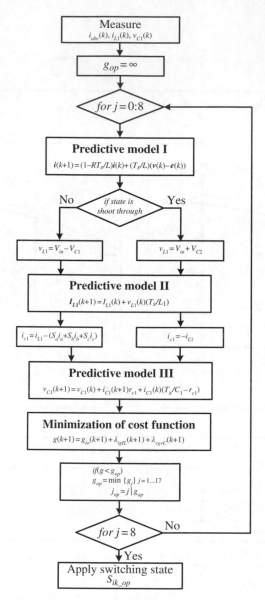

Figure 18.14 Flowchart of the MPC for three-phase qZSI (*Source*: Mosa 2014 [11]. Reproduced with permission of IEEE).

abc to the α-β frame by using Clark transformation. Then two successive points from this wave were taken. Assume that point t is the current time when the control starts its observation. The optimizer chooses the closest state to the future reference. Figure 18.13(a) shows the prediction schematic for i_α and i_β. From t to $t + T_s$ there are eight possible predictive states for the real and complex parts of the output current, which can be calculated from (18.24). Moreover, Figure 18.13(b) shows the prediction for the capacitor voltage, which also has eight states. However, Figure 18.13(c) shows the inductor current that has two prediction states because the behavior of it is not dependent on the SVM, which is shown in Figure 18.6. The distance between any state and the reference is known as the cost function with unity weighting factor which can be calculated from (18.36)–(18.38). The selector chooses the nearest state for the reference, which can optimize the cost function for the minimum value. As shown from Figure 18.13(a), the nearest future state current to the reference is $i_\alpha^{p1}(3)$, and so on for the rest of figures. Depending on this selection, the proper switching state will be chosen. Similarly, the controller will repeat the sequence again from $t + T_s$ to $t + 2T_s$ to calculate the future behavior of the constraints and to choose the optimal switching state. In other words, the corresponding voltage vector at each sample time is determined by the cost function, according to the minimum error or distance between reference and predicted values.

6. Control algorithm

The control algorithm is illustrated in Figure 18.14 as a flowchart. As shown in the flowchart, the minimization of the cost function is implemented as a repeat loop to predict each voltage vector, evaluate the cost function, and store the minimum value and the index value of the corresponding switching state.

The control algorithm can be summarized in the following steps:

- Sample the output currents (i_{abc}) then convert them by using Clark transformation matrix to $(i_{\alpha\beta})$, inductor current (i_{L1}), and capacitor voltage (v_{CI}).
- The values are used to predict output currents, inductor current, and capacitor voltage using the predictive model I (18.24), II (18.28), and III (18.33), respectively.
- All predictions are evaluated using the cost function (18.39).
- The optimal switching state that corresponds to the optimal voltage vector that minimizes the cost function is selected to be applied at the next sampling time.

18.5   Model Predictive Control of the Z-Source Three-Phase Four-Leg Inverter

The MPC scheme of the ZS/qZS four-leg inverter is shown in Figure 18.15 [38–40]. The structure is similar to the three-phase three-leg inverter MPC structure, which is given in the previous section. However, the model of the four-leg inverter (predictive model I) is different from the three-leg inverter. For that reason, the discrete-time model of the four-leg inverter and the flowchart of the prepared control algorithm will be described. Other steps are the same as for the three-leg inverter's MPC scheme, which are described in the previous section.

18.5.1   Discrete-Time Model of the Output Current for Four-Leg Inverter

This model is used to predict the future behavior of the output currents i_{abc}. The continuous-time expression for i_{abc} is given in (18.12). By substituting (18.23) into (18.12), the discrete-time model of the i_{abc} can be obtained as

$$i_j(k+1) = A_v . v_j(k+1) + A_i i_j(k), \quad j = a,b,c. \tag{18.40}$$

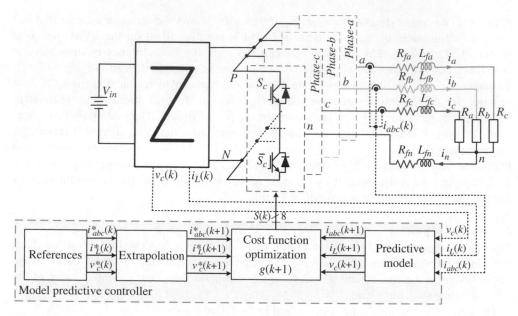

Figure 18.15 MPC scheme for a Z-source four-leg inverter (*Source*: Bayhan 2016 [38]. Reproduced with permission of IEEE).

where $i_j(k+1)$ is the predicted output current vector at the next sampling time, A_v and A_i are constant and defined as

$$A_v = \frac{T_s}{L_{fj} + \left(R_j + R_{fj}\right)T_s}$$

$$A_i = \frac{L_{fj}}{L_{fj} + \left(R_j + R_{fj}\right)T_s}$$

(18.41)

18.5.2   Control Algorithm

The control algorithm is illustrated in Figure 18.16 as a flowchart. It is similar to Figure 18.14, but the mathematical model of the output current is different from the three-phase inverter model. As shown in the flowchart, the minimization of the cost function is implemented as a repeat loop to predict each voltage vector, evaluate the cost function, and store the minimum value and the index value of the corresponding switching state.

18.6   Model Predictive Control of the Z-Source Five-Phase Inverter

The MPC scheme of the ZS/qZS five-phase inverter is shown in Figure 18.17. The structure is similar to three-phase three-leg and four-leg inverter's MPC structure, which is given in the previous sections. However, the switching schemes and the load model of

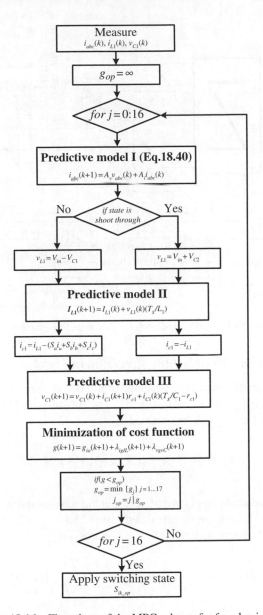

Figure 18.16 Flowchart of the MPC scheme for four-leg inverter.

the five-phase inverter (predictive model I) are different from other inverters. For that reason, the discrete-time model of the load, cost function of the load current, and flowchart of the prepared control algorithm will be given. Other steps should be the same as the three-phase three-leg and four-leg inverter's MPC schemes, as described in the previous sections.

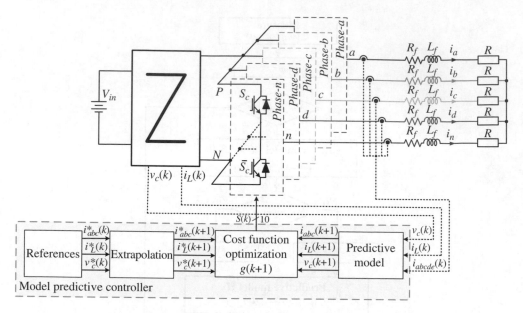

Figure 18.17　MPC scheme for a Z-source five-phase inverter.

18.6.1　Discrete-Time Model of the Five-Phase Load

This model is used to predict the future behavior of the output currents i_{abcde}. The continuous-time expression for the i_{abcde} is given in (18.20). By substituting (18.23) into (18.20), the discrete-time model of the i_{abcde} can be obtained as

$$\mathbf{i}(k+1) = \left(1 - \frac{RT_s}{L}\right)\mathbf{i}(k) + \frac{T_s}{L}\left(\mathbf{v}(k) - \hat{\mathbf{e}}(k)\right) \tag{18.42}$$

where $\mathbf{i}(k+1)$ is the predicted output current vector at the next sampling time, and $\hat{\mathbf{e}}(k)$ denotes the estimated back-emf, which can be calculated from (18.19), considering the measurements of the load voltage and current with the following expression;

$$\hat{\mathbf{e}}(k-1) = \mathbf{v}(k-1) - \frac{L}{T_s}\mathbf{i}(k) - \left(R - \frac{L}{T_s}\right)\mathbf{i}(k-1) \tag{18.43}$$

where $\hat{\mathbf{e}}(k-1)$ is the estimated value of $\mathbf{e}(k-1)$.

The discrete-time model of the load is used to pre-calculate the behavior of the current in the next sampling interval. The pre-calculated current sample is then fed to the optimizer along with the reference current (obtained from the external as user defined). The optimizer calculates the cost function for all the possible switching combinations of the inverter. Thus it generates the optimal switching pattern corresponding to the global minimum cost function in each sampling interval and passes it on to the gate drive of the inverter. Owing to the fact that the five-phase inverter generates a large number of space vectors (30 active and two zero vectors), many possible solutions will exist to be implemented to the model predictive control.

18.6.2   Cost Function for the Load Current

The selection of the cost function is the most important aspect in the MPC. A judicious choice leads to the optimal solution of the control objective. Thus the cost function should include all parameters to be optimized within the imposed constraints. In the current control the most important variable is the current tracking error. Thus the most simple and straightforward choice is the absolute value of the current error. The other choices could be such as the square of the current error, the integral of the current error, or the rate of change of the current error. Specifically in a five-phase drive system, there are two orthogonal subspaces namely α-β and x-y. Thus in case of a five-phase drive system the current errors in both planes have to be considered for deriving a cost function. In general, for the square of the current error, the cost function is given as

$$\hat{g}_{i\_\alpha\beta} = \left|i_{\alpha}^{*}(k) - \hat{i}_{\alpha}(k+1)\right| + \left|i_{\beta}^{*}(k) - \hat{i}_{\beta}(k+1)\right|$$
$$\hat{g}_{i\_xy} = \left|i_{x}^{*}(k) - \hat{i}_{x}(k+1)\right| + \left|i_{y}^{*}(k) - \hat{i}_{y}(k+1)\right| \tag{18.44}$$

The final cost function of the load current can be expressed as

$$g_{io} = \left\|\hat{g}_{\alpha\beta}\right\|^{2} + \left\|\gamma\,\hat{g}_{xy}\right\|^{2} \tag{18.45}$$

where γ is a tuning parameter that offers a degree of freedom in order to put emphasis on α-β or x-y subspaces.

18.6.3   Control Algorithm

The control algorithm is illustrated in Figure 18.18 as a flowchart. It is similar to Figure 18.14 and Figure 18.16. However, the output current model is different from the three-phase three-leg and four-leg inverter models. Although the Z-source network current and voltage cost functions are the same as the three-phase three-leg and four-leg inverters, the cost function of the load current is different from the others, as is the switching scheme of this inverter. It has 30 active states, two zero states, and one shoot-through state (Table 18.3). As shown in the flowchart, the minimization of the cost function is implemented as a for loop to predict each voltage vector, evaluate the cost function, and store the minimum value and the index value of the corresponding switching state.

18.7   Performance Investigation

In order to verify the MPC scheme for qZS four-leg inverter, a simulation model has been developed using the MATLAB/Simulink software. The parameters used in the simulations are given in Table 18.4 and the block diagram of the implemented system is given in Figure 18.15.

Steady-state results of the discussed MPC scheme under balanced reference output currents and balanced loads are shown in Figure 18.19. During the test, the reference currents (i_a^*, i_b^*, i_c^*) and reference capacitor voltage (V_{C1}^*) are set to 10A and 150V, respectively. It can be observed

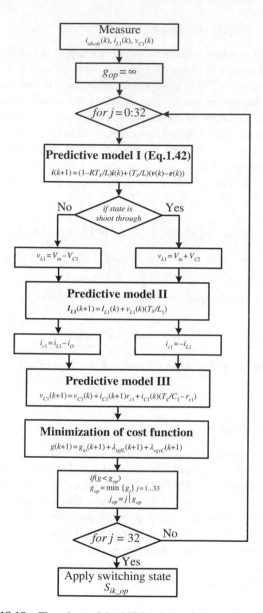

Figure 18.18   Flowchart of the MPC scheme for five-phase inverter.

from Figure 18.19 that the output current tracks its reference value very well while the capacitor and dc-link voltages are kept constant. Furthermore, the neutral current is zero, since the reference output currents are balanced as shown in Figure 18.19(b).

The MPC scheme is also tested to verify the robustness of the technique against unbalanced reference currents ($i_a^* = 10$A, $i_b^* = i_c^* = 5$A) and unbalanced loads ($R_a = 10\,\Omega$, $R_b = 7.5\,\Omega$, $R_c = 5\,\Omega$). This is a typical application for three-phase four-wire systems, where the load demand

Table 18.4 qZS four-leg inverter and load parameters.

Parameter	Value
Input dc voltage (V_{in})	100 V
qZS network inductances (L_1, L_2)	1.5 mH
qZS network capacitances (C_1, C_2)	1000 μF
Load resistance (R)	5–10 Ω
Filter inductance (L_f)	10 mH
Filter resistance (R_f)	0.05 Ω
Nominal frequency (v_{ln})	50 Hz
Sampling time (T_s)	40 μs

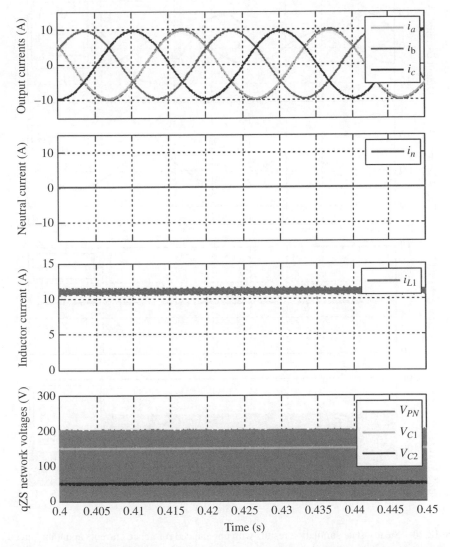

Figure 18.19 Steady-state simulation results with balanced reference currents and balanced loads for four-leg inverter: (a) three-phase output currents (i_a, i_b, i_c), (b) neutral current (i_n), (c) inductor current (I_L), and (d) qZS network voltages (V_{C1}, V_{C2}, v_{PN}).

on each phase is different. Steady-state results of the discussed MPC scheme under unbalanced reference output currents and unbalanced loads are shown in Figure 18.20. The results show that the controller is able to control each phase current independently while the capacitor and dc-link voltages are kept constant. The neutral current, which is the sum of the three-phase currents, flows through the fourth leg because of the unbalanced reference currents.

The transient-state performance of the discussed MPC scheme for the reference output currents step from 10 to 5 A are shown in Figure 18.21. For this test the reference currents and

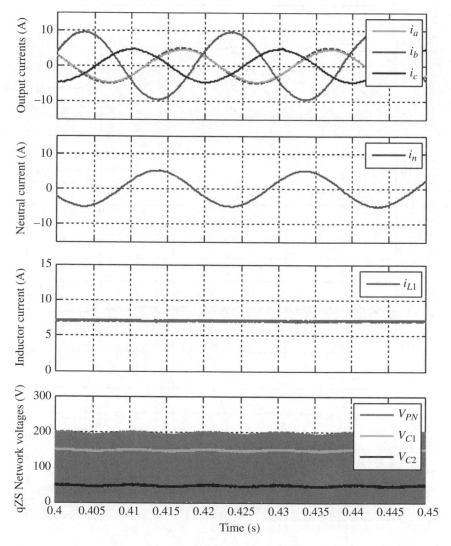

Figure 18.20 Steady-state simulation results with unbalanced reference currents and unbalanced loads for four-leg inverter: (a) three-phase output currents (i_a, i_b, i_c), (b) neutral current (i_n), (c) inductor current (I_L), and (d) qZS network voltages (V_{C1}, V_{C2}, v_{PN}).

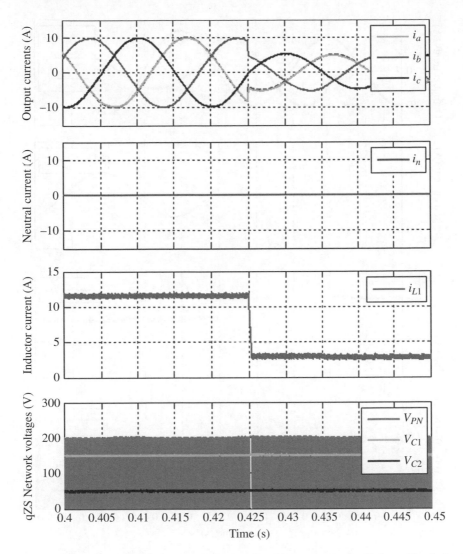

Figure 18.21 Transient simulation results with balanced reference currents and balanced loads for four-leg inverter: (a) three-phase output currents (i_a, i_b, i_c), (b) neutral current (i_n), (c) inductor current (I_L), and (d) qZS network voltages (V_{C1}, V_{C2}, v_{PN}).

loads are balanced. These results show that the transient time is very short, and the output current tracks its reference, while the capacitor and dc-link voltages are kept constant. The neutral current is zero because of the balanced reference load currents.

In Figure 18.22, the results are presented with an unbalanced reference current step change and unbalanced loads ($R_a = 5\,\Omega$, $R_b = R_c = 10\,\Omega$). For this test, all reference output currents are set to 10 A at the beginning. Then, the reference currents are set to $i_a^* = 8\,A$, $i_b^* = 5\,A$, and $i_c^* = 12\,A$. The results of this test show that the controller handles each phase current independently, and the output currents (i_{abc}), inductor current (I_{L1}), and the capacitor voltage (V_{C1})

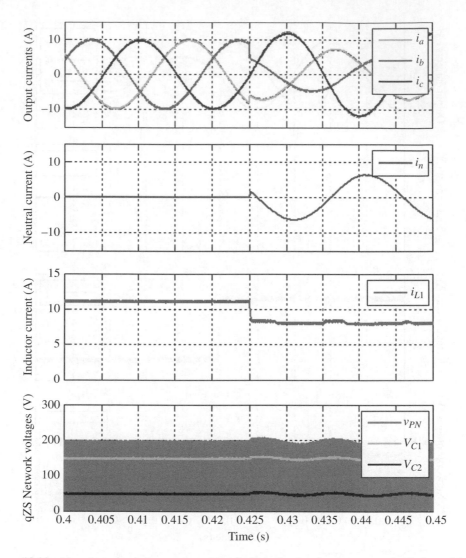

Figure 18.22 Transient simulation results with unbalanced reference currents and unbalanced loads for four-leg inverter: (a) three-phase output currents (i_a, i_b, i_c), (b) neutral current (i_n), (c) inductor current (I_L), and (d) qZS network voltages (V_{C1}, V_{C2}, v_{PN}).

track their references with high accuracy while the dc-link voltage (V_{PN}) is kept constant. However, the double-line frequency ($2*f_o = 100\,\text{Hz}$) ripples exist on the Z-source network due to the unbalanced current.

The zoomed version of the inductor current and the dc-link voltage are shown in Figure 18.23 to illustrate the impact of the shoot-through states. It can be observed from this figure, during the shoot-through states, the inductor current is increasing, while dc-link voltage is zero because of short-circuit of the dc-link. Then, during active states (non shoot-through states), the inductor current is decreasing, while dc-link voltage is kept constant (200 V).

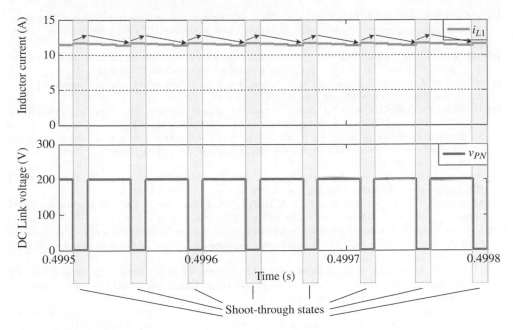

Figure 18.23   Simulation result of shoot-through states in boost conversion mode.

18.8   Summary

The advances in modern control theory and in microprocessors have made it possible to apply MPC in power electronics and drives in a natural and simple way. MPC offers a high flexibility to control different circuit topologies and to manage several control objectives, without adding significant complexity. This chapter presents a model predictive control (MPC) scheme for Z-source/qZ-source three-phase three-leg, four-leg inverters and Z-source/qZ-source multiphase inverters. The mathematical models of these inverters are presented with switching scheme and voltage vectors. To control these inverters three different MPC schemes have been presented with details. The presented MPC schemes have three cost functions to control output current, inductor current, and capacitor voltage. In order to verify the dynamic and steady-state performances of the discussed MPC scheme, simulation studies have been carried out with qZS four-leg inverter. Simulation results show that the discussed MPC scheme provides simplicity and excellent control performance under different reference current and load conditions. The MPC scheme consists of only a one step ahead prediction horizon. In order to improve the MPC scheme more steps ahead can be used in future research.

References

[1] L. Wang, S. Chai, D. Yoo, L. Gan, K. Ng, *PID and Predictive Control of Electrical Drives and Power Converters using Matlab/Simulink*, Wiley & IEEE Press, 2015.

[2] S. Bayhan, H. Abu-Rub, "Model predictive sensorless control of standalone doubly fed induction generator," in *Proc. 40th Annual Conference of the IEEE Industrial Electronics Society (IECON)*, pp.2166–2172, 2014.

[3] Y. Liu, B. Ge, H. Abu-Rub, Fang Zheng Peng, "Phase-shifted pulse-width-amplitude modulation for quasi-Z-source cascade multilevel inverter-based photovoltaic power system," *IET Power Electron.*, vol.7, no.6, pp.1444–1456, June 2014.

[4] Y. Liu, B. Ge, H. Abu-Rub, F. Z. Peng, "Overview of space vector modulations for three-phase Z-source/quasi-Z-source inverters," *IEEE Trans. Power Electron.*, vol.29, no.4, pp.2098–2108, 2014.

[5] H. Abu-Rub, J. Guzinski, Z. Krzeminski, H. A. Toliyat, "Predictive current control of voltage-source inverters," *IEEE Trans. Ind. Electron.*, vol.51, no.3, pp.585–593, June 2004.

[6] J. Rodriguez, J. Pontt, C. A. Silva, P. Correa, P. Lezana, P. Cortes, U. Ammann, "Predictive current control of a voltage source inverter," *IEEE Trans. Ind. Electron.*, vol.54, no.1, pp.495–503, Feb. 2007.

[7] V. Yaramasu, M. Rivera, M. Narimani, B. Wu, J. Rodriguez, "Model predictive approach for a simple and effective load voltage control of four-leg inverter with an output *RL* filter," *IEEE Trans. Ind. Electron.*, vol.61, no.10, pp.5259–5270, 2014.

[8] P. Cortes, A. Wilson, S. Kouro, J. Rodriguez, H. Abu-Rub, "Model predictive control of multilevel cascaded H-bridge inverters," *IEEE Trans. Ind. Electron.*, vol.57, no.8, pp.2691–2699, 2010.

[9] M. A. Perez, J. Rodriguez, E. J. Fuentes, F. Kammerer, "Predictive control of AC–AC modular multilevel converters," *IEEE Trans. Ind. Electron.*, vol.59, no.7, pp.2832–2839, July 2012.

[10] M. Mosa, H. Abu-Rub, J. Rodriguez, "High performance predictive control applied to three phase grid connected quasi-Z-source inverter," in *Proc. 39th Annual Conference of the IEEE Industrial Electronics Society (IECON)*, pp.5812–5817, 2013.

[11] M. Mosa, G. M. Dousoky, H. Abu-Rub, "A novel FPGA implementation of a model predictive controller for SiC-based quasi-Z-source inverters," in *Proc. Twenty-Ninth Annual IEEE Applied Power Electronics Conference and Exposition (APEC)*, pp.1293–1298, 2014.

[12] J. Guzinski, H. Abu-Rub, "Predictive current control implementation in the sensorless induction motor drive," in *Proc. 2011 IEEE International Symposium on Industrial Electronics (ISIE)*, pp.691–696, 2011.

[13] J. Guzinski, H. Abu-Rub, "Speed sensorless induction motor drive with predictive current controller," *IEEE Trans. Ind. Electron.*, vol.60, no.2, pp.699–709, Feb. 2013.

[14] J. Rodriguez, M. P. Kazmierkowski, J. R. Espinoza, P. Zanchetta, H. Abu-Rub, H. A. Young, C. A. Rojas, "State of the art of finite control set model predictive control in power electronics," *IEEE Trans. Ind. Informat.*, vol.9, no.2, pp.1003–1016, 2013.

[15] H. A. Young, M. A. Perez, J. Rodriguez, H. Abu-Rub, "Assessing finite-control-set model predictive control: A comparison with a linear current controller in two-level voltage source inverters," *IEEE Ind. Electron. Mag.*, vol.8, no.1, pp.44–52, 2014.

[16] P. Cortes, M. P. Kazmierkowski, R. M. Kennel, D. E. Quevedo, J. Rodriguez, "Predictive control in power electronics and drives," *IEEE Trans. Ind. Electron.*, vol.55, no.12, pp.4312–4324, 2008.

[17] J. Rodriguez, P. Cortes, *Predictive Control of Power Converters and Electrical Drives*, Wiley & IEEE Press, 2012.

[18] M. B. Shadmand, R. S. Balog, H. Abu-Rub, "Model predictive control of PV sources in a smart DC distribution system: maximum power point tracking and droop control," *IEEE Trans. Energy Convers.*, vol.29, no.4, pp.913–921, Dec. 2014

[19] F. Z. Peng, "Z-source inverter," *IEEE Trans. Ind. Appl.*, vol.39, no.2, pp.504–510, 2003.

[20] Y. Li, J. Anderson, F. Z. Peng, L. Dichen, "Quasi-Z-source inverter for photovoltaic power generation systems," in *Proc. Twenty-Fourth Annual IEEE Applied Power Electronics Conference and Exposition (APEC)*, pp.918–924, 2009.

[21] M. Mosa, O. Ellabban, A. Kouzou, H. Abu-Rub, J. Rodriguez, "Model predictive control applied for quasi-Z-source inverter," in *Proc. 2013 Twenty-Eighth Annual IEEE Applied Power Electronics Conference and Exposition (APEC)*, pp.165–169, 2013.

[22] V. Yaramasu, M. Rivera, B. Wu, J. Rodriguez, "Model predictive current control of two-level four-leg inverters – Part I: concept, algorithm, and simulation analysis," *IEEE Trans. Power Electron.*, vol.28, no.7, pp.3459–3468, 2013.

[23] M. Rivera, V. Yaramasu, J. Rodriguez, B. Wu, "Model predictive current control of two-level four-leg inverters – Part II: experimental implementation and validation," *IEEE Trans. Power Electron.*, vol.28, no.7, pp.3469–3478, July 2013.

[24] X. Li, Z. Deng, Z. Chen, Q. Fei, "Analysis and simplification of three-dimensional space vector PWM for three-phase four-leg inverters," *IEEE Trans. Ind. Electron.*, vol.58, no.2, pp.450–464, 2011.

[25] A. Iqbal, R. Alammari, M. Mosa, H. Abu-Rub, "Finite set model predictive current control with reduced and constant common mode voltage for a five-phase voltage source inverter," in *Proc. 2014 IEEE 23rd International Symposium on Industrial Electronics (ISIE)*, pp.479–484, 2014.

[26] S. Moinoddin, H. Abu-Rub, A. Iqbal, A. Moin, O. Dordevic, E. Levi, "Space vector pulse-width modulation technique for an eleven-phase voltage source inverter with sinusoidal output voltage generation," *IET Power Electron.*, vol.8, no.6, pp.1000–1008., 2015.

[27] S. Moinoddin, H. Abu-Rub, A. Iqbal, "Modelling and implementation of SVPWM technique for a thirteen-phase voltage source inverter-sinusoidal output waveform," in *Proc. 2014 Twenty-Ninth Annual IEEE Applied Power Electronics Conference and Exposition (APEC)*, pp.268–273, 2014.

[28] Y. Liu, B. Ge, H. Abu-Rub, F. Z. Peng, "Control system design of battery-assisted quasi-Z-source inverter for grid-tie photovoltaic power generation," *IEEE Trans. Sustain. Energy*, vol.4, no.4, pp.994–1001, 2013.

[29] Y. Li, F. Z. Peng, J. G. Cintron-Rivera, S. Jiang, "Controller design for quasi-Z-source inverter in photovoltaic systems," in *Proc. 2011 IEEE Energy Conversion Congress and Exposition (ECCE)*, 2010, pp.3187–3194.

[30] H. Abu-Rub, A. Iqbal, M. Ahmed, F. Z. Peng, Y. Li, B. Ge, "Quasi-Z-source inverter-based photovoltaic generation system with maximum power tracking control using ANFIS," *IEEE Trans. Sustain. Energy*, vol.4, no.1, pp.11–20, Jan. 2013.

[31] J. H. Park, H. G. Kim, E. C. Nho, T. W. Chun. "Power conditioning system for a grid connected PV power generation using a quasi Z-source inverter," *Journal of Power Electronics*, vol.10, pp.79–84, 2010.

[32] M. S. Shen, J. Wang, A. Joseph, F. Z. Peng, "Maximum constant boost control of the Z-source inverter," in *Proc. 39th IEEE Industry Applications Conference*, pp.142–147, 2004.

[33] P. Jong-Hyoung, K. Heung-Geun, N. Eui-Cheol, C. Tae-Won, "Capacitor voltage control for MPPT range expansion and efficiency improvement of grid-connected quasi Z-source inverter," in *Proc. IEEE International Power Electronics Conference (IPEC)*, 2010, pp.927–931.

[34] X. Ding, Z. Qian, S. Yang, B. Cui, F. Z. Peng, "A PID control strategy for dc-link boost voltage in Z-source inverter," in *Proc. Annual IEEE Applied Power Electronics Conference and Exposition (APEC)*, 2007, pp.1145–1148.

[35] P. C. Loh, D. M. Vilathgamuwa, C. J. Gajanayake, L. Yih Rong, T. Chern Wern, "Transient modeling and analysis of pulse-width modulated Z-source inverter," *IEEE Trans. Power Electron.*, vol.22, pp.498–507, 2007.

[36] O. Ellabban, J. V. Mierlo, P. Lataire, "A DSP-based dual-loop peak dc-link voltage control strategy of the Z-source inverter," *IEEE Trans. Power Electron.*, vol.27, pp.4088–4097, Sep.2012.

[37] M. Shen, Q. Tang, F. Z. Peng, "Modeling and controller design of the Z-source inverter with inductive load," in *Proc. IEEE Power Electron. Spec. Conf. (PESC)*, pp.1804–1809, Jun. 2007.

[38] S. Bayhan, H. Abu-Rub, M. Trabelsi, "Model predictive control of Z-source four-leg inverter for standalone photovoltaic system with unbalanced load," to appear in the *Thirty-first Annual IEEE Applied Power Electronics Conference and Exposition (APEC)*, 19–26 March 2016.

[39] S. Bayhan, H. Abu-Rub, "Model predictive control of quasi-Z source three-phase four-leg inverter," to appear in the *IEEE Annual Conference of Industrial Electronics Society (IECON)*, Nov. 9–12, 2015.

[40] S. Bayhan, H. Abu-Rub, R. Balog, "Model predictive control of quasi-Z source four-leg inverter," *IEEE Trans. Ind. Electron.*, vol.63, no.7, pp.4506–4516, Jul. 2016.

19

Grid Integration of Quasi-Z Source Based PV Multilevel Inverter

Mohamed Trabelsi and Haitham Abu-Rub
Electrical and Computer Engineering Program, Texas A&M University at Qatar,
Qatar Foundation, Doha, Qatar

In this chapter, a quasi Z-source (qZS) based cascaded H-bridge (CHB) multilevel inverter is presented for the grid integration of PV systems. The mixed topology is characterized by high-quality staircase output voltage with low harmonic distortions, independent dc-link voltage compensation with special voltage step-up/down function in a single-stage power conversion, and independent control of power delivery with high reliability. Moreover, a model predictive control (MPC) technique is used to transfer the power to the grid with unity power factor, low THD, low voltage ride-through (LVRT) capability, and anti-islanding protection.

19.1   Introduction

Distributed generators (DGs) are essential in producing energy with low or zero greenhouse gases (GHGs) [1]. Also, they add the much-needed flexibility to the energy resources by decreasing the dependency on fossil fuels. Thus, many DG alternatives are being explored for different industrial applications [2–5]. Among these DGs, PV systems are becoming the most clean and attractive options for isolated and remote locations due to their abundance and good performance compared to other renewable sources. Furthermore, grid integration of PV power generation systems has the advantage of immediate and efficient utilization of generated power [6]. However, the efficiency of the grid integration system depends essentially on the performance of the power conditioner (full exploitation of the PV power to meet the increasing demand of load for extracting the maximum power) and the capability of the adopted control strategies in achieving high performance. This is not only during normal operating conditions of the grid (achieving grid current injection with low THD and unity power factor), but also under fault conditions (essentially voltage sags and islanding). Hence, there are many power converter

Impedance Source Power Electronic Converters, First Edition. Yushan Liu, Haitham Abu-Rub, Baoming Ge,
Frede Blaabjerg, Omar Ellabban, and Poh Chiang Loh.
© 2016 John Wiley & Sons, Ltd. Published 2016 by John Wiley & Sons, Ltd.

topologies employed in interconnecting the PV systems to the grid, which are classified into two-stage or single-stage and two-level or multilevel inverters (MLI). MLIs are characterized by the use of advanced medium-power semiconductor technology compared with the conventional two-level converters, having higher power quality at the AC side, reducing the filter size significantly, operating at higher voltage levels, and reducing switching losses [7–10].

In this chapter, a single-stage qZS-CHB MLI for grid-tied PV systems is presented [11–14]. This mixed topology is characterized by high-quality staircase output voltage with low harmonic distortion, independent dc-link voltage compensation with special voltage step-up/down function in a single-stage power conversion, and independent control of power delivery with high reliability [15]. Moreover, an MPC technique is used to transfer the power to the grid with unity power factor, low THD, LVRT capability, and anti-islanding protection.

19.2 Topology and Modeling

The discussed single-phase qZS based power conditioning system (PCS) is illustrated in Figure 19.1. The system consists of two-cell grid-tied CHB inverters, where each cell is connected to a qZS network. This network consists of two inductors, two capacitors, and one diode. This topology is used to interface the PV output voltage with the inverter input while having a boost capability. The output voltage of the qZS-CHB inverter is a multilevel type (five-level output voltage) resulting from the sum of the two module output voltages. This number of levels L is given by

$$L = 2n + 1 \tag{19.1}$$

with n the number of cells.

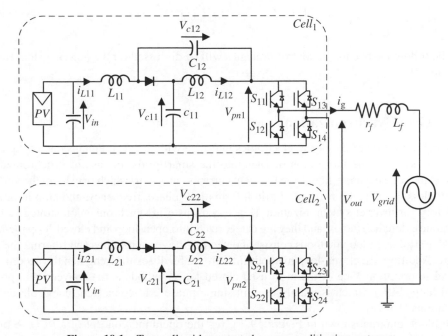

Figure 19.1 Two-cell grid-connected power conditioning system.

As discussed in the previous chapters, the qZS converter's operation modes can be divided into two states: non-shoot-through and shoot-through states.

During the non-shoot-through state, the inverter is controlled as a conventional H-bridge inverter and the derivations of the state variables i_{Li1} and V_{Ci1} for each cell i ($i=1, 2$) are obtained from

$$C\frac{dV_{Ci1}}{dt} = i_{Li1}(t) - i_g(t) \tag{19.2}$$

$$L\frac{di_{Li1}}{dt} = V_{in} - V_{Ci1}(t) \tag{19.3}$$

In the shoot-through state, the two switches in the same leg are turned on simultaneously. The system equations for this state are given by

$$C\frac{dV_{Ci1}}{dt} = -i_{Li1}(t) \tag{19.4}$$

$$L\frac{di_{Li1}}{dt} = V_{Ci1}(t) \tag{19.5}$$

Also, according to Figure 19.1, the output voltage V_{out} in terms of its current and filter parameters is given by

$$V_{out}(t) = L_f\frac{di_g(t)}{dt} + r_f i_g(t) + V_{grid}(t) \tag{19.6}$$

In the following, two major control strategies will be discussed for the power system shown in Figure 19.1.

19.3   Grid Synchronization

The synchronization of the power electronics interface with the grid remains one of the most important issues in the integration of DGs into the smart grids. In case of synchronization failure at the time of connection, large transient currents can occur, which may harm the system. Nevertheless, accurate knowledge of grid information (phase, frequency, and amplitude) are mandatory for inverter synchronization. Hence, various grid synchronization strategies have undergone extensive studies, and they are categorized into open-loop and closed-loop methods [16]. Classical open-loop methods consist of detecting the zero crossings and filtering the grid voltage. Additionally, closed-loop methods introduce a feedback to strengthen the accuracy of the grid information. One of the most used closed-loop methods is the conventional phase-locked loop (PLL), which is mainly used in single-phase grid-connected renewable energy applications [16–18].

The block diagram shown in Figure 19.2 gives the operating concept of a PLL. A phase error detector (PD) produces a voltage proportional to the phase difference between the input

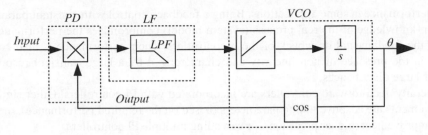

Figure 19.2 Block diagram of conventional PLL using phase detector, loop filter, and voltage controlled oscillator.

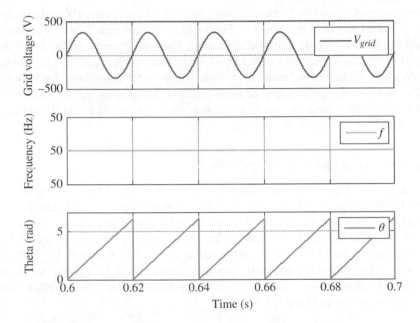

Figure 19.3 Tracking of the grid frequency using the conventional PLL.

signal and the reproduced output signal. This voltage upon filtering through the loop filter (LF) is used as the control signal for the voltage controlled oscillator (VCO) – which is mainly based on a PI controller – to generate the frequency of the output signal. If the error between the frequency of the output signal is locked with the input frequency, then the phase of the output signal is locked to the input signal. Figure 19.3 shows an example of the grid phase ($V_{grid\_rms} = 240\,\text{V}, f_{grid} = 50\,\text{Hz}$) calculated by the conventional PLL.

19.4 Power Flow Control

Proportional integral (PI) controllers are the most applied controllers in industrial applications. However, while PI controllers are applicable to many fields and often perform suitably, they are in general considered as non-optimal control strategies and can show poor performance in some applications. Being a feedback control with constant parameters, the non-knowledge of the real process (system model) compromises the performance, and this is the fundamental drawback of these controllers. Their performance is also compromised in the presence of non-linearity and characterized by a large transient response in case of large disturbances.

Generally, feed-forward controllers are incorporated with PI controllers. Other significant improvements are adaptive tuning parameters (based on the required performance), measurement improvement (higher precision), or cascading multiple PI controllers.

19.4.1   Proportional Integral Controller

This section presents a suitable PI based control strategy for integrating PV systems to the grid through the qZS-CHB inverter shown in Figure 19.1. In the control structure, unity power factor and low THD current is delivered to the grid. Also, the amount of total power delivered to the grid is regulated through a dual-loop dc-link peak voltage control [19] employed in every qZS-CHB module to balance the dc-link voltages (by controlling the shoot-through ratio). The complete design is described in this section. The overall control scheme shown in Figure 19.4 is to fulfill these purposes. Simulation results based on a 1.2 kW two-cell prototype are carried out to validate the methods.

19.4.1.1   DC-Link Peak Voltage Regulation

The dc-link peak voltage is regulated through the control of the shoot-through duty ratio for each qZS-CHB module. As shown in Figure 19.4, a dual loop concept is employed. In the inductor current loop, a proportional (P) controller is used to improve the dynamic response. Also, a PI controller is used in the voltage loop to ensure the tracking of the dc-link peak voltage reference. From [20], for each qZS module, the transfer functions $G_{vpn}(s)$ (giving the dc-link peak voltage from the shoot-through duty ratio) and $G_{iL}(s)$ (giving the L_1 inductor current from the shoot-through duty ratio) can be calculated. Thus, the modified dual loop block diagram is shown in Figure 19.5.

19.4.1.2   Output Current Regulation

The output current is controlled to ensure power injection to the grid with unity power factor and low harmonic distortion. Figure 19.6 shows the block diagram of the output current controller.

At the output of the qZS-CHB inverter, the grid-tied system has the following dynamics:

$$V_{out}(t) = V_{grid}(t) + L_f \frac{di_g(t)}{dt} + r_f i_g(t) \tag{19.7}$$

where V_{out} is the CHB output voltage, V_{grid} is the grid voltage, i_g is the output current, L_f is the filter inductance, and r_f is its parasitic resistance. Thus, the output current transfer function $G_f(s)$ can be given by

$$G_f(s) = \frac{i_g(s)}{V_{out}(s) - V_{grid}(s)} = \frac{1}{L_f s + r_f} \tag{19.8}$$

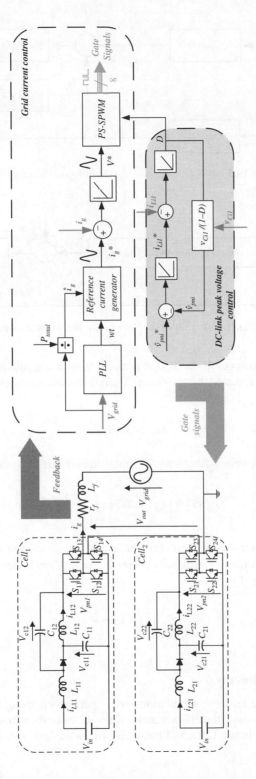

Figure 19.4 Dual loop block diagram for the control of the injected grid current and dc-link voltage.

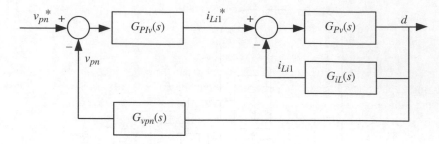

Figure 19.5 Block diagram of the dc-link peak voltage control.

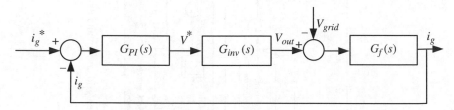

Figure 19.6 Block diagram of the grid current loop.

At the qZS network output, all the dc-link peak voltages are regulated around the same reference value. Thus, it is assumed that the two qZS-CHB inverter modules have the same transfer function given by

$$G_{inv}\left(s\right) = \hat{v}_{pn} \tag{19.9}$$

Then, the combined inverter–filter transfer function is obtained from

$$G_{sys}\left(s\right) = G_{inv}\left(s\right).G_f\left(s\right) = \frac{\hat{v}_{pn}}{L_f s + r_f} \tag{19.10}$$

Also, a PI current controller is applied on the actual output current to guarantee a suitable tracking of the desired reference with unity power factor and low harmonic distortions. The PI transfer function is given by

$$G_{PI}\left(s\right) = k_p + \frac{k_i}{s} \tag{19.11}$$

19.4.1.3 Simulation Results

In order to validate the concept, simulations were performed using Matlab/Simulink, and results are presented to show the effectiveness of the discussed method in achieving grid-tie current injection, low current THD, and balancing of the dc-link voltage for both qZS-CHB inverter modules. The system parameters are reported in Table 19.1. In the studied case, the peak value of the qZS-CHB reference grid current is set to 7 A and the dc-link voltage reference $v_{pn}*$ is fixed at 240 V. Thus, for the boost conversion mode, the qZS network is boosting the input voltage V_{in} (150 V) to the required reference value $v_{pn}*$ ($B = 1.6$) by varying the shoot-through value D. Figures 19.7 and 19.8 show the simulation results of the regulated dc-link voltage and capacitor voltages. It can be noted that the dc-link voltage is properly regulated around the reference value (240 V). In Figure 19.9, the injected grid current is depicted with respect to its reference with very suitable tracking quality and the current THD is given by Figure 19.10. The five-level output voltage is shown in Figure 19.11. Figure 19.12 shows the transient performance of the PI controller during a 50% step-down change of the active power reference. It can be noted that the PI controller scheme reacts with an acceptable response time in tracking the new grid and inductor current references. Finally, Figure 19.13 shows the grid synchronization and the unity power factor.

Table 19.1 Simulation parameters for the power flow control of the two-cell qZS-CHB inverter

Parameter	Value
Total output power (P_{total})	1.2 kW
AC grid RMS voltage (V_{grid})	240 V
qZS inductances (L_1, L_2)	2.5 mH
qZS capacitances (C_1, C_2)	4.7 mF
Filtering inductance (L)	1 mH
PV array voltage for qZS-HBI module (V_{in})	150 V
Grid frequency (f)	50 Hz

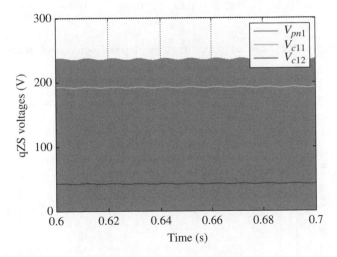

Figure 19.7 Regulated dc-link voltage using the dual loop control strategy.

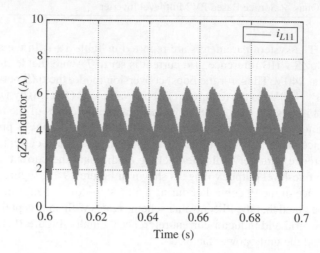

Figure 19.8 qZS inductor current using the dual loop control strategy.

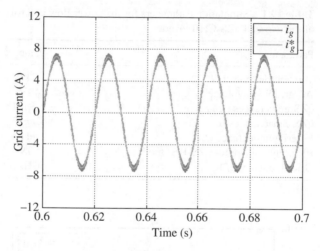

Figure 19.9 Injected grid current using the dual loop control strategy.

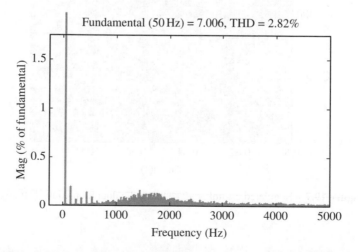

Figure 19.10 Grid current THD using the dual loop control strategy.

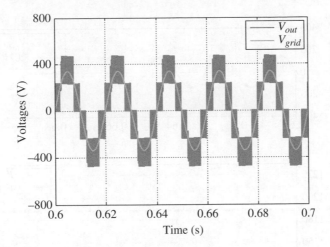

Figure 19.11 Five-level output voltage with the grid voltage using the dual loop control strategy.

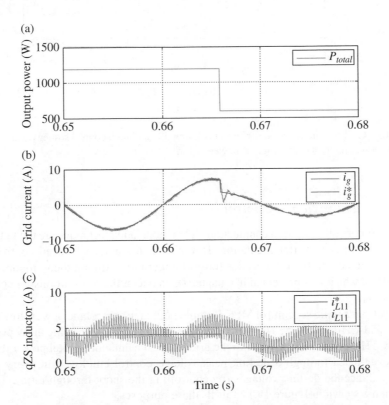

Figure 19.12 Dynamic test using the dual loop control strategy: (a) Total output power, (b) qZS-CHB output current, (c) qZS inductor current.

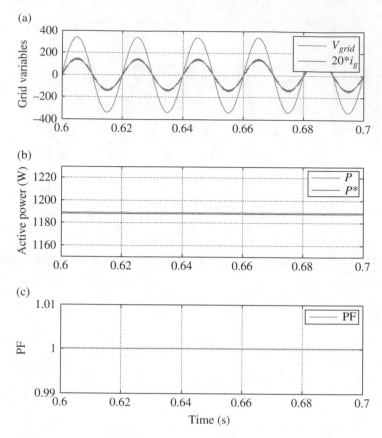

Figure 19.13 Grid current injection performance using the dual loop control strategy: (a) grid synchronization, (b) power reference tracking, (c) power factor.

19.4.2 Model Predictive Control

Usually, the PI controllers used in conjunction with a modulation technique are the main solution for power electronic converter regulation. However, many other advanced control methods have been developed in the past decades for grid integration. Among them, model predictive control (MPC), which was presented in Chapter 18, is one of the most promising control techniques [21–27].

In this section, a finite set control MPC is implemented to control the power flow of the grid connected qZS-CHB topology. The MPC structure is capable of generating unity power factor while a low THD current is delivered to the grid. Also, the amount of total power delivered to the grid is regulated through a dc-link peak voltage control employed in every qZS-CHB module to balance the dc-link voltages (by controlling the shoot-through ratio). The overall control scheme shown in Figure 19.14 fulfills these purposes.

The scheme shown consists of three main stages: (1) generation of state variable references, (2) model prediction, (3) cost function optimization.

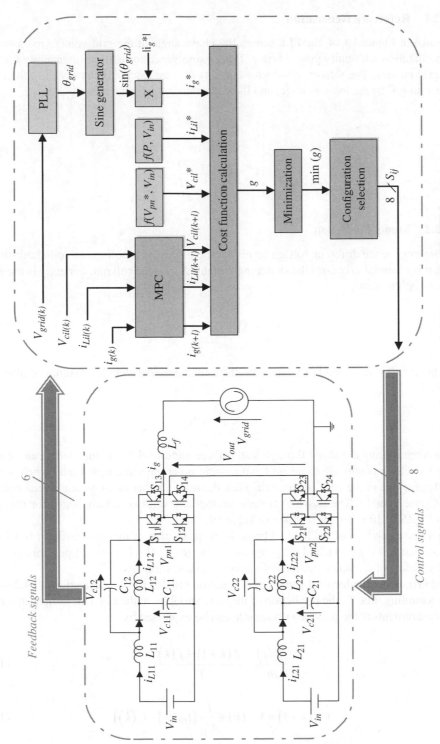

Figure 19.14 MPC block diagram for the control of the two-cell qZS-CHB inverter.

19.4.2.1   Reference Generation

As shown in Figure 19.14, the PLL senses the phase angle of the grid voltage to ensure grid synchronization with unity power factor. Then a sine generator is used to generate the reference grid current. The reference inductor currents i_{L1}* are calculated by dividing the system power rating P by the input voltage, and the capacitor voltage references V_{c1}* are

$$
\begin{cases}
i_{L1}{}^* = \dfrac{P}{2V_{in}} \\[2mm]
V_{c1}{}^* = V_{in}\dfrac{1-D}{1-2D}
\end{cases}
\tag{19.12}
$$

19.4.2.2   Model Prediction

In order to generate different voltage levels for the discussed topology, in non-shoot-through states, it is essential to control the switching variable S_{ij} (i is the cell number and j is the switch number) in line with

$$
V_{out} = V_{pni}\sum_{i=1}^{n}\left(S_{i1} - S_{i3}\right)
\tag{19.13}
$$

Including all voltage level redundancies, the total number of switching signal combinations is given by

$$
C_S = 2^{2n} = 16
\tag{19.14}
$$

Moreover, during the shoot-through states, three additional switching states can be noted. These switching combinations result from turning on at the same time both switches on the same leg for one of the cells or for both. Then the total number of switching signal combinations C_s becomes 19. The possible voltage vectors and valid switching states for the shown two-cell qZS-CHB inverter are given in Table 19.2.

Then, the main idea of the MPC scheme is the prediction of the grid current $i_g(k+1)$, qZS inductor currents $i_{Li1}(k+1)$, and qZS capacitor voltages $V_{ci1}(k+1)$ for each possible switching state by means of discrete equations of the system state variables.

To do so, using the forward Euler approximation (19.15) for the derivations (19.2)–(19.6) with a sampling time T_s, the prediction of the state variables at the $(k+1)$th sample in terms of the measurements at the previous (k) sample can be expressed by

$$
\frac{df(t)}{dt} = \frac{f(k+1)-f(k)}{T_s}
\tag{19.15}
$$

$$
V_{ci1}(k+1) = V_{ci1}(k) + \frac{T_s}{C_{i1}}\left(i_{Li1}(k) - i_g(k)\right)
\tag{19.16}
$$

Table 19.2 Possible switching states of the two-cell qZS-CHB inverter

	Switching state	V_{out}	S_{11}	S_{12}	S_{13}	S_{14}	S_{21}	S_{22}	S_{23}	S_{24}
Non-shoot-through states	0		0	1	0	1	0	1	0	1
	1		0	1	0	1	1	0	1	0
	2	0	0	1	1	0	1	0	0	1
	3		1	0	0	1	0	1	1	0
	4		1	0	1	0	0	1	0	1
	5		1	0	1	0	1	0	1	0
	6		0	1	0	1	0	1	1	0
	7		0	1	1	0	0	1	0	1
	8	$-V_{in}$	0	1	1	0	1	0	1	0
	9		1	0	1	0	0	1	1	0
	10		0	1	0	1	1	0	0	1
	11		1	0	0	1	0	1	0	1
	12	V_{in}	1	0	0	1	1	0	1	0
	13		1	0	1	0	1	0	0	1
	14	$-2V_{in}$	0	1	1	0	0	1	1	0
	15	$2V_{in}$	1	0	0	1	1	0	0	1
Shoot-through states		V_{in}	1	1	0	1	1	0	0	1
		V_{in}	1	0	0	1	1	1	0	1
	16	0	1	1	0	1	1	1	0	1

$$i_{Li1}(k+1) = i_{Li1}(k) + \frac{T_s}{L_{i1}}\left(V_{out}(k) - V_{ci1}(k)\right) \qquad (19.17)$$

$$i_g(k+1) = \left(1 - \frac{r_f}{L_f}T_s\right)i_g(k) + \frac{T_s}{L_f}\left(V_{out}(k) - V_{grid}(k)\right) \qquad (19.18)$$

19.4.2.3 Cost Function

The cost function has two objectives: minimize the error between the predicted grid current $i_g(k+1)$, inductor currents $i_{Li1}(k+1)$, and capacitor voltages $V_{ci1}(k+1)$ and their references, and also balance the dc-link capacitor voltages. These control objectives are represented as

$$g = \lambda_{ig}\left|i_g{}^* - i_g(k+1)\right| + \sum_{i=1}^{2}\lambda_{Vc}\left|V_{c1}{}^* - V_{ci1}(k+1)\right| + \lambda_{iL}\left|i_{L1}{}^* - i_{Li1}(k+1)\right| \qquad (19.19)$$

where λ_{ig}, λ_{Vc}, and λ_{iL} are weighting factors, which can be adjusted according to the desired performance. The switching state giving the minimum of the cost function is selected and then applied at the next sampling instant.

19.4.2.4 Simulation Results

In the studied case, the simulation parameters are the same as shown in Table 19.1. Figure 19.15 shows the simulation results of the regulated dc-link voltage and capacitor voltages. Note that the dc-link voltage is properly regulated around the reference value (240 V), V_{c1} and V_{c2} are controlled around the reference values given by (19.12) ($V_{c1}{}^* = 195$ V, $V_{c2}{}^* = 45$ V). Figure 19.16 shows that the qZS inductor current is maintained in the continuous mode allowing reducing the input stress. In addition, Figure 19.17 denotes a zoom of Figure 19.15 and Figure 19.16 presents the evolution of the qZS variables with the shoot-through state. The lower part of the figure shows that the dc-link voltage is properly regulated around the reference value (240 V) while the upper part shows that the qZS inductor current is increasing during the shoot-through, decreasing during the non-shoot-through state, and always maintained in the continuous mode. The five-level output voltage is shown in Figure 19.18. In Figure 19.19, the grid current is plotted against its reference showing an excellent tracking quality. The current THD is given in Figure 19.20. Figure 19.21 shows the grid synchronization and power factor

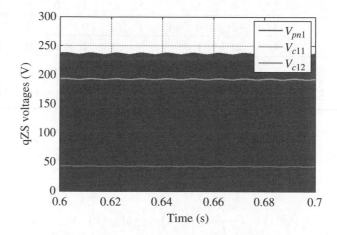

Figure 19.15 Regulated dc-link voltage using model predictive control.

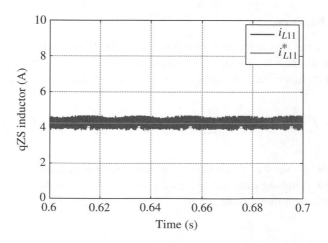

Figure 19.16 qZS inductor current using model predictive control.

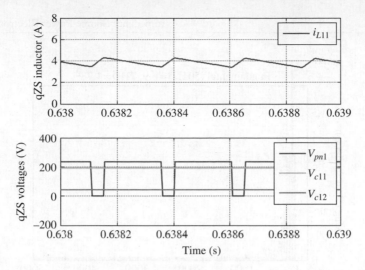

Figure 19.17 Zoomed snapshot of the qZS inductor current and dc-link voltage using model predictive control.

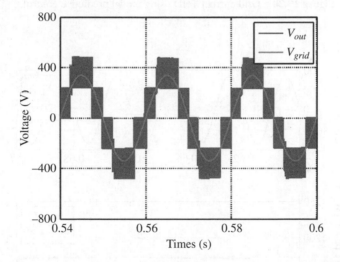

Figure 19.18 Five-level output voltage with the grid voltage using model predictive control.

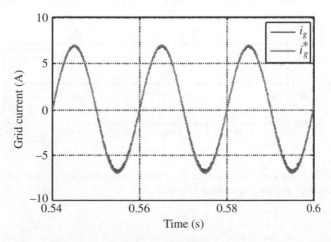

Figure 19.19 Injected grid current using model predictive control.

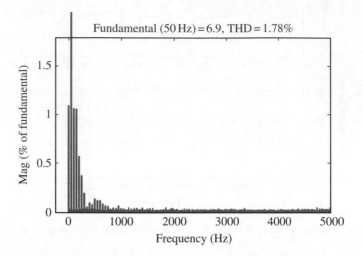

Figure 19.20   Grid current THD using model predictive control.

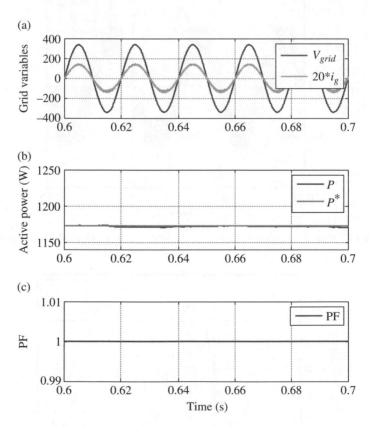

Figure 19.21   Grid current injection performance using model predictive control: (a) grid synchronization, (b) power reference tracking, (c) power factor.

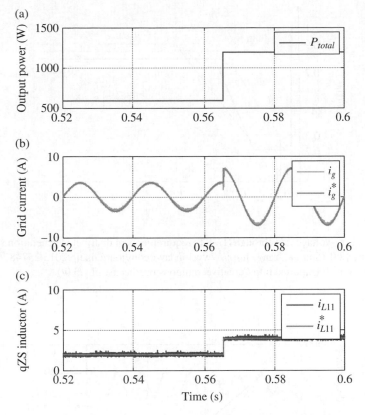

Figure 19.22 Dynamic test (MPC): (a) total output power, (b) qZS-CHB output current, (c) qZS inductor current.

measurement (unity power factor). Finally, Figure 19.22 shows the dynamic performance of the MPC scheme during a 50% step-up change of the active power reference. It can be noted that the system achieves a very fast response time (much faster than the PI controller) in tracking the grid and inductor current references.

19.5 Low Voltage Ride-Through Capability

In international grid codes, DGs should offer a low-voltage ride-through (LVRT) capability. That is to say, the DGs should stay connected to the grid for a short time (half a cycle to a few seconds) and inject some reactive power to support the grid when a voltage fault occurs. Different LVRT curves with defined stay-connected time are shown in Figure 19.23.

Voltage sag is the most probable common voltage fault, which is defined as a temporary voltage drop that lasts between half a cycle and a few seconds (30 cycles as per many standards), and has a typical magnitude of 10% to 90% per unit range. Also, voltage sag is often accompanied by voltage distortion, current overshoot, transients, and phase shifts. Therefore, inverters' control techniques should achieve high performance not only during normal

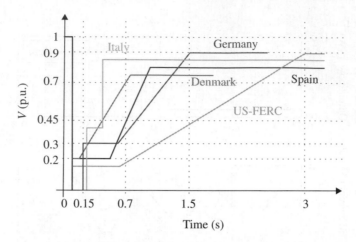

Figure 19.23 Low-voltage ride-through (LVRT) requirements of distributed generation systems in different countries [28] (*Source*: Yang, http://www.hindawi.com/journals/ijp/2013/257487/. Used under CC BY Attribution 3.0 Unported http://creativecommons.org/licenses/by/3.0/).

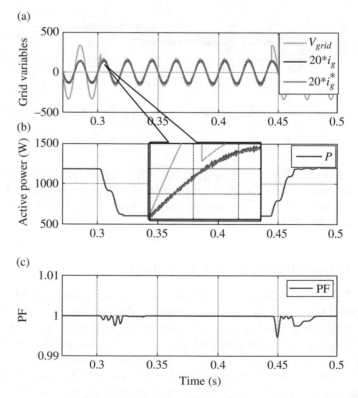

Figure 19.24 Simulations results under voltage sag using model predictive control: (a) grid voltage and current, (b) active power, (c) Power factor.

operating conditions of the grid, but also under fault conditions. Several control strategies such as vector current controller with feedforward (VCCF) of negative sequence grid voltage [29] and dual vector current controllers (DVCC) [30, 31] have been proposed to deal with this issue. These methods can produce the required reactive current to meet the LVRT requirement, but the interfacing inverter is often stressed by a higher peak current during the fault time (risk of overcurrent protection tripping). To avoid this problem, positive and negative sequence reactive current compensation control was proposed in [32].

In this section, the LVRT capability of the MPC strategy is tested. During the simulations, a grid voltage sag (50% of the nominal voltage) is introduced at the time $t=0.3$ s and the recovery time was set to $t=0.44$ s (Figure 19.24). As shown in this figure, soon after the fault time, the system starts to inject reactive current and limit the active power output to prevent the inverter from overcurrent (the MPC technique achieves smooth response, faster re-tracking of the reference, and negligible current overshoot). After the fault is cleared (voltage rises to 100% of its nominal value), the system goes back to its normal operation.

19.6 Islanding Protection

Islanding protection is an important safety requirement to be addressed in grid connected power electronic systems. Islanding occurs when a portion of the utility system remains energized while isolated from the utility system [33]. This phenomenon can cause safety issues to utility service personnel or related equipment. Hence, various islanding detection methods (IDMs) [34–37] have undergone extensive studies. These techniques can be classified into three main types: passive, active, and remote methods [38]. Under islanding conditions, the magnitude and frequency of the voltage at the point of common coupling (PCC) tend to drift from the rated values. Passive IDMs rely on the detection of the disturbance in the voltage at the PCC. They are effective in preventing islanding in systems with large power imbalances. Conversely, islanding tends to occur for systems with small power imbalances. In this case, the passive IDMs can fail to detect the islanding (these schemes have a relatively large non-detection zone [39]). Active IDMs (low voltage applications) use a variety of methods in an attempt to cause an abnormal condition (disturbance) in the PCC voltage's magnitude and frequency.

In this section, three main active IDMs are applied for the protection of the grid connected qZS-CHB inverter shown in Figure 19.25, where the power flow is illustrated [40]. Moreover, the performance assessment and the anti-islanding protection performance is evaluated for the described active IDMs.

Thus, the qZS-CHB inverter is controlled to detect islanding when the grid is disconnected. In this case, it stops supplying the load and the grid. This technique is based on the detection of the V_{PCC} frequency f_{PCC} for each zero crossing. An under/overvoltage and under/over frequency protection block is used to compare the rms and the frequency values of the voltage at the PCC to the threshold values corresponding to the IEEE standards (IEEE std 929-2000). As a result, as shown in Figure 19.26, this block generates a fault signal when the f_{PCC} or the rms values exceed the fixed limits. However, the islanding phenomenon can occur while the mentioned parameters are within the fixed limits. In this case, the inverter fails to detect islanding, which is the major drawback of passive IDMs. In this chapter, three active IDMs are presented to decrease the non-detection zone (NDZ), increase the islanding detection

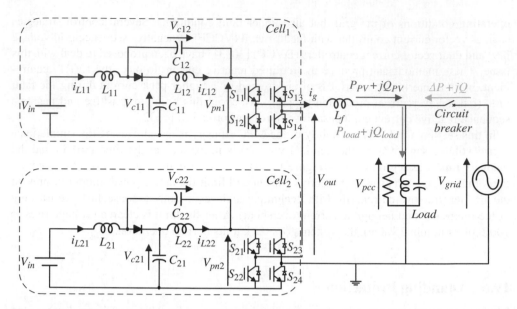

Figure 19.25 Grid-connected system configuration showing the power flow (*Source*: Trabelsi 2015 [40]. Reproduced with permission of IEEE).

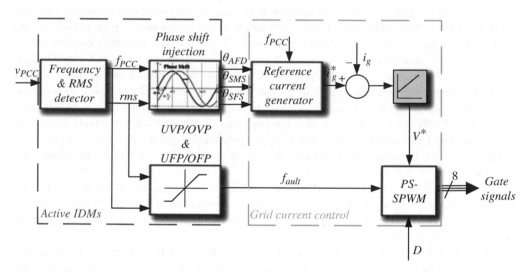

Figure 19.26 Combined controller for anti-islanding protection (*Source*: Trabelsi 2015 [40]. Reproduced with permission of IEEE).

possibilities, and to take into consideration the load parameters. These techniques are named active frequency drift (AFD), slip-mode frequency shift (SMS), and Sandia frequency shift (SFS), and are based on the detection of the voltage frequency at the PCC and the injection of a phase shift in the output inverter current in order to cause an abnormality at the PCC. Simulation results are given for the combined PI-IDMs [40] and MPC-IDMs controllers.

19.6.1 Active Frequency Drift (AFD)

This method is also known as the frequency bias technique. It consists of placing a slight distortion onto the reference output current for a drift-up operation. The AFD is implemented by forcing the current frequency to be higher than the voltage frequency in the previous cycle by a frequency drift. Then, the reference output current can be expressed by

$$I_{ref}(t) = I_{max} \sin\left(2\pi\left(f + \delta f\right)t\right) \tag{19.20}$$

where f is the frequency of the PCC voltage in the previous cycle and δf is the frequency drift. However, an AFD with large frequency drift degrades the output power quality by injecting reactive power while a small drift could fails to detect islanding.

19.6.2 Sandia Frequency Shift (SFS)

This method is also called active frequency drift with positive feedback (AFDPF). It can be considered as an extension of the AFD method, which uses the f_{PCC} as a positive feedback to prevent islanding. Then, the reference output current can be expressed by

$$I_{ref}(t) = I_{max} \sin\left(2\pi f t + \theta_{SFS}\right) \tag{19.21}$$

The phase angle θ_{SFS} is given by

$$\theta_{SFS} = \frac{\pi}{2}\left[cf_0 + k_{SFS}\left(f - f_g\right)\right] \tag{19.22}$$

where cf_0 is the chopping factor at zero frequency error, k_{SFS} is an accelerating gain, and f_g is the grid frequency. Basically, SFS is an improved version of the AFD technique with good islanding detection effectiveness.

19.6.3 Slip-Mode Frequency Shift (SMS)

Using this method, the phase angle of the reference current is calculated as a function of the deviation of the f_{PCC} of the previous cycle from the nominal operating frequency of the grid. Then, the reference output current can be expressed by

$$I_{ref}(t) = I_{max} \sin\left(2\pi f t + \theta_{SMS}\right) \tag{19.23}$$

The phase angle θ_{SMS} is given by

$$\theta_{SMS} = \frac{2\pi}{360}\theta_m \sin\left(\frac{\pi}{2}\frac{f - f_g}{f_m - f_g}\right) \tag{19.24}$$

where θ_m is the maximum phase shift in degrees, f_m is the frequency at which the maximum phase shift occurs, and f_g is the grid frequency. This technique is considered as one of the best active IDMs since it provides a good compromise between islanding detection effectiveness and output power quality.

19.6.4   Simulation Results

Figure 19.27 shows the anti-islanding protection test results using the SMS method [40]. It is worth noting that the three tested IDMs are effective in detecting islanding. The upper part shows the qZS-CHB five-level output voltage, the PCC voltage as well as the grid voltage.

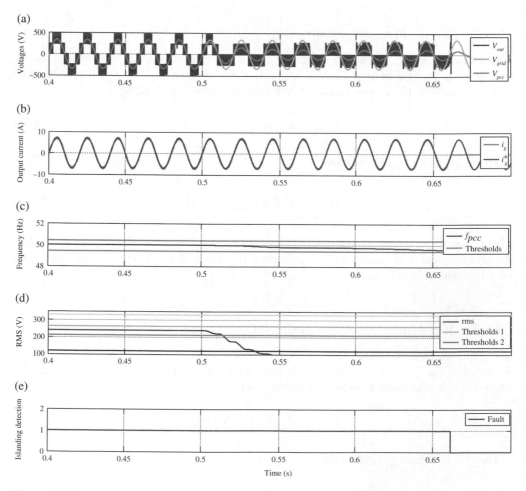

Figure 19.27 Anti-islanding protection test results using the SMS method: (a) voltages, (b) injected grid current, (c) frequency at the common coupling point, (d) rms value of the voltage at the common coupling point, (e) fault signal (*Source*: Trabelsi 2015 [40]. Reproduced with permission of IEEE).

An islanding (grid disconnection) event was set at time 0.5 s. Beyond this point, the PCC voltage will undergo frequency and rms variations. According to the IEEE std 929-2000, if the variation of the frequency and rms values exceeds a fixed threshold for a certain number of cycles, the inverter should trip and stop energizing the utility system (distributed resource and load). Thus, as shown in the lowest part of Figure 19.27, a fault signal was generated and the inverter stopped energizing the utility system. Moreover, Figures 19.28, 19.30, and 19.32 show the output current THD using three IDMs combined with the PI controller while Figures 19.29, 19.31, and 19.33 show the same results using the combined MPC-IDM controller. As expected, the SMS method combined with the MPC controller offers the lower THD.

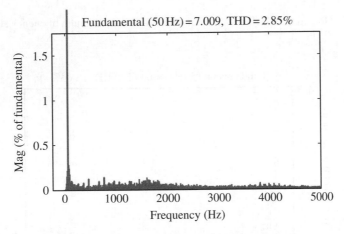

Figure 19.28 Grid current THD with the combined dual loop – slip mode frequency shift controller.

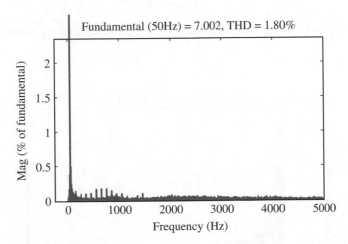

Figure 19.29 Grid current THD with the combined model predictive control – slip mode frequency shift.

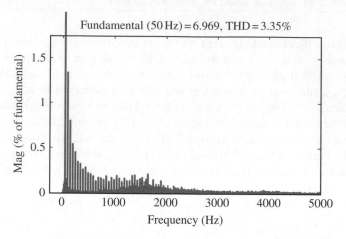

Figure 19.30 Grid current THD with the combined dual loop – Sandia frequency shift controller.

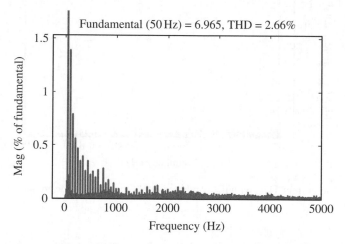

Figure 19.31 Grid current THD with the combined model predictive control – Sandia frequency shift controller.

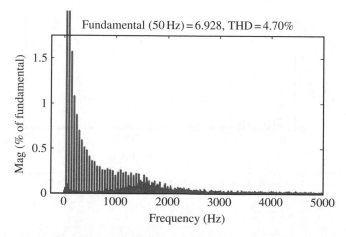

Figure 19.32 Grid current THD with the combined dual loop – Active Frequency Drift controller.

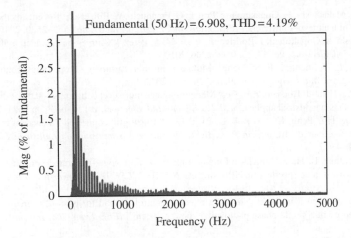

Figure 19.33   Grid current THD with the combined Model Predictive Control – Active Frequency Drift controller.

19.7   Conclusion

Distributed generators such as fuel cells, solar, and wind play a crucial role in producing energy with low or zero CO_2 emissions and are explored for various standalone and grid connected applications. Since inverters are essential for connecting DGs to the grid, and with the continuous increase of DG power, the multilevel inverters are becoming the standard topologies for grid connections.

In this chapter, by introducing an appropriate model predictive controller structure, a quasi-Z-source based cascaded H-bridge multilevel inverter is used for connecting a PV system to a single-phase grid. In the control structure, unity power factor and low THD current is delivered to the grid during normal operating conditions. Moreover, under grid faults, the controller is characterized by low voltage ride-through capability and anti-islanding protection. The anti-islanding protection performance was evaluated for three phase-shift based active islanding detection methods: active frequency drift, slip-mode phase shift, and Sandia frequency shift.

References

[1] H. Abu-Rub, M. Malinowski, K. Al-Haddad, *Power Electronics for Renewable Energy Systems, Transportation and Industrial Applications*, John Wiley & Sons, 2014.

[2] H. Doukas, K. D. Patlitzianas, A. G. Kagiannas, J. Psarras, "Renewable energy sources and rationale use of energy development in the countries of GCC: myth or reality?" *Renewable Energy*, vol.31, pp.755–770, 2006.

[3] A. H. Marafia, "Feasibility study of photovoltaic technology in Qatar," *Renewable Energy*, vol.24, pp.565–567, 2001.

[4] K. D. Patlitzianas, H. Doukas, J. Psarras, "Enhancing renewable energy in the Arab States of the Gulf: Constraints & efforts," *Energy Policy*, vol.34, pp.3719–3726, 2006.

[5] M. Trabelsi and L. Ben-Brahim, "Experimental photovoltaic power supply based on flying capacitors multilevel inverter," in *Proc. International Conference on Clean Electrical Power (ICCEP)*, 2011, pp.578–583.

[6] M. Trabelsi and L. Ben-Brahim, "Development of a grid connected photovoltaic power conditioning system based on flying capacitors inverter," in *Proc. 2011 8th International Multi-Conference on Systems, Signals and Devices (SSD)*, 2011, pp.1–6.

[7] S. Kouro, M. Malinowski, K. Gopakumar, J. Pou, L. G. Franquelo, W. Bin, et al., "Recent advances and industrial applications of multilevel converters," *IEEE Trans. Ind. Electron.*, vol.57, pp.2553–2580, 2010.

[8] M. Malinowski, K. Gopakumar, J. Rodriguez, X. Pe, M. A. Perez, "A survey on cascaded multilevel inverters," *IEEE Trans. Ind. Electron.*, vol.57, pp.2197–2206, 2010.

[9] J. Rodriguez, L. Jih-Sheng, P. Fang Zheng, "Multilevel inverters: a survey of topologies, controls, and applications," *IEEE Trans. Ind. Electron.*, vol.49, pp.724–738, 2002.

[10] H. Abu-Rub, J. Holtz, J. Rodriguez, B. Ge, "Medium-voltage multilevel converters – state of the art, challenges, and requirements in industrial applications," *IEEE Trans. Ind. Electron.*, vol.57, no.8, pp.2581–2596, Aug. 2010.

[11] D. Sun, B. Ge, F. Z. Peng, H. Abu-Rub, D. Bi, Y. Liu, "A new grid-connected PV system based on cascaded H-bridge quasi-Z source inverter," in *Proc. IEEE International Symposium on Industrial Electronics (ISIE)*, 2012, pp.951–956.

[12] Z. Yan, L. Liming, L. Hui, "A high-performance photovoltaic module-integrated converter (MIC) based on cascaded quasi-Z-source inverters (qZSI) Using eGaN FETs," *IEEE Trans. Power Electron.*,, vol.28, pp.2727–2738, 2013.

[13] Y. Liu, B. Ge, H. Abu-Rub, F. Z. Peng, "An effective control method for quasi-Z-source cascade multilevel inverter-based grid-tie single-phase photovoltaic power system," *IEEE Trans. Ind. Informat.*, vol.10, pp.399–407, 2014.

[14] Y. Liu, B. Ge, F. Z. Peng, H. Abu-Rub, A. T. de Almeida, F. J. T. E. Ferreira, "Quasi-Z-source inverter based PMSG wind power generation system," in *Proc. IEEE Energy Conversion Congress and Exposition (ECCE)*, 2011, pp.291–297.

[15] Y. Li, S. Jiang, J. G. Cintron-Rivera, F. Z. Peng, "Modeling and control of quasi-Z-source inverter for distributed generation applications," *IEEE Trans. Ind. Electron.*, vol.60, pp.1532–1541, 2013.

[16] Q. Zhong, T. Hornik, *Control of Power Inverters in Renewable Energy and Smart Grid Integration*, John Wiley & Sons, 2013.

[17] M. Kesler, E. Ozdemir, "Synchronous-reference-frame-based control method for UPQC under unbalanced and distorted load conditions," *IEEE Trans. Ind. Electron.*, vol.58, no.9, pp.3967–3975, Sept. 2011.

[18] R. Teodorescu, F. Blaabjerg, "Flexible control of small wind turbines with grid failure detection operating in standalone and grid-connected mode," *IEEE Trans. Power Electron.*, vol.19, no.5, pp.1323–1332, Sept. 2004.

[19] O. Ellabban, J. Van Mierlo, P. Lataire, "A DSP based dual-loop peak dc-link voltage control of the Z-source inverter," *IEEE Trans. Power Electron.*, vol.27, no.9, pp.4088–4097, Sept. 2012.

[20] Y. Liu, B. Ge, F. Z. Peng, H. Abu-Rub, A. T. de Almeida, F. J. T. E. Ferreira, "Quasi-Z-source inverter based PMSG wind power generation system," in *Proc. IEEE Energy Conversion Congress and Exposition (ECCE)*, 2011, pp.291–297.

[21] J. Lee, "Model predictive control: Review of the three decades of development," *International Journal of Control, Automation and Systems*, vol.9, pp.415–424, 2011.

[22] J. Rodriguez, M. P. Kazmierkowski, J. R. Espinoza, P. Zanchetta, H. Abu-Rub, H. A. Young, et al., "State of the art of finite control set model predictive control in power electronics," *IEEE Trans. Ind. Informat.*, vol.9, pp.1003–1016, 2013.

[23] J. Guzinski and H. Abu-Rub, "Speed sensorless induction motor drive with predictive current controller," *IEEE Trans. Ind. Electron.*, vol.60, pp.699–709, 2013.

[24] J. Rodriguez, H. Abu-Rub, M. A. Perez, S. Kouro, "Application of predictive control in power electronics: an AC-DC-AC converter system," in *Advanced and Intelligent Control in Power Electronics and Drives*, Springer, 2014.

[25] M. Trabelsi, K. A. Ghazi, N. Al-Emadi, L. Ben-Brahim, "A weighted real-time predictive controller for a grid connected flying capacitors inverter," *International Journal of Electrical Power & Energy Systems*, vol.49, pp.322–332, 2013.

[26] A. Sanchez, A. De Castro, J. Garrido, "A comparison of simulation and hardware-in-the-loop alternatives for digital control of power converters," *IEEE Trans. Ind. Informat.*, vol.8, pp.491–500, 2012.

[27] M. P. Kazmierkowski, M. Jasinski, G. Wrona, "DSP-based control of grid-connected power converters operating under grid distortions," *IEEE Trans. Ind. Informat.*, vol.7, pp.204–211, 2011.

[28] Y. Yang, F. Blaabjerg, "Low-voltage ride-through capability of a single-stage single-phase photovoltaic system connected to the low-voltage grid," *International Journal of Photo Energy*, vol.2013.

[29] S. Alepuz, S. Busquets-Monge, J. Bordonau, J. A. Martinez-Velasco, C. A. Silva, J. Pontt, J. Rodriguez, "Control strategies based on symmetrical components for grid-connected converters under voltage dips," *IEEE Trans. Ind. Electron.*, vol.56, pp.2162–2173, 2009.

[30] F. A. Magueed, A. Sannino, J. Svensson, "Transient performance of voltage source converter under unbalanced voltage dips," in *Proc. IEEE Power Electronics Specialists Conference (PESC)*, pp.1163–1168, 2004.

[31] M. Bongiorno, J. Svensson, A. Sannino, "Dynamic performance of vector current controllers for grid-connected VSC under voltage dips," in *Proc. Industry Applications Conference*, pp.904–909, 2005.

[32] C. W. Hsu, C. T. Lee, P. T. Cheng, "A low voltage ride-through technique for grid-connected converters of distributed energy resources," in *Proc. IEEE Energy Conversion Congress and Exposition (ECCE)*, pp.3388–3395, 2010.

[33] "IEEE Recommended Practice for Utility Interface of Photovoltaic (PV) Systems," IEEE Std 929–2000, 2000.

[34] V. Menon, M. H. Nehrir, "A hybrid islanding detection technique using voltage unbalance and frequency set point," *IEEE Trans. Power Syst.*, vol.22, no.1, pp.442–448, Feb. 2007.

[35] B. Wen, D. Boroyevich, R. Burgos, Z. Shen, P. Mattavelli, "Impedance-based analysis of active frequency drift islanding detection method for grid-tied inverter system," in *Proc. International Power Electronics Conference (IPEC-Hiroshima ECCE-ASIA)*, pp.3850–3856, 2014.

[36] S. S. Ahmed, N. C. Sarker, A. B. Khairuddin, M. R. B. A. Ghani, H. Ahmad, "A scheme for controlled islanding to prevent subsequent blackout," *IEEE Trans. Power Syst.*, vol.18, no.1, pp.136–143, Feb 2003.

[37] S. Lee, J. Park, "Improvement on stability and islanding detection performances by advanced inverter control of DG," *IEEE Trans. Power Syst.*, vol.28, no.4, pp.3954–3963, Nov. 2013.

[38] H. Vahedi, R. Noroozian, A. Jalilvand, G. B. Gharehpetian, "A new method for islanding detection of inverter-based distributed generation using DC-link voltage control," *IEEE Trans. Power Del.*, vol.26, pp.1176–1186, 2011.

[39] M. E. Ropp, M. Begoviv, A. Rohatgi, "Prevention of islanding in grid-connected photovoltaic systems," *Progress in Photovoltaics: Research and Applications*, vol.7, no.1, pp.39–59, 1999.

[40] M. Trabelsi, H. Abu-Rub, "A unique active anti-islanding protection for a quasi-Z-source based power conditioning system," in *Proc. 30th Applied Power Electronics Conference and Exposition (APEC)*, pp.2237–2243, 2015.

20

Future Trends

20.1 General Expectation

20.1.1 Volume and Size Reduction by Wide Band-Gap Devices

The increasing energy demands, investments in solar energy, and semiconductor manufacturing technology have stimulated new studies on developing highly efficient photovoltaic (PV) power conversion based on wide band gap (WBG) semiconductor materials [1–4], for instance, silicon carbide (SiC), gallium nitride (GaN), and enhanced GaN (eGaN).

WBG semiconductors provide very interesting characteristics compared to the traditional silicon (Si) semiconductor, such as higher blocking voltage, reduced switching energy loss, and especially against temperature [5]. All of these make them appropriate for higher switching frequency without significant prejudice to the efficiency [3]. As a result, higher efficiency and higher power density can be brought to power electronic systems. GaN theoretically enables a tremendous reduction of the specific on-resistance, which has an even lower theoretical limit than SiC [6].

In three-phase dc-dc or ac-ac impedance source inverters/converters, where the switching frequency is dominant for Z-source/quasi-Z-source (ZS/qZS) inductance and capacitance, the high switching frequency SiC devices are expected to help to reduce the inductor volume and size, and the low loss of SiC devices improves the efficiency, especially in high power systems [6–8]. A comparison result from the authors' publication of [7] demonstrates that for a 3 kVA/380 V qZS indirect matrix converter (IMC), the inductance is 4.58 mH using 10 kHz Si-based devices, whereas, it will be 1.47 mH using 30 kHz SiC-based devices. A detailed comparison will be illustrated in the following section.

Impedance Source Power Electronic Converters, First Edition. Yushan Liu, Haitham Abu-Rub, Baoming Ge, Frede Blaabjerg, Omar Ellabban, and Poh Chiang Loh.
© 2016 John Wiley & Sons, Ltd. Published 2016 by John Wiley & Sons, Ltd.

20.1.2   Parameter Minimization for Single-Phase qZS Inverter

The single-phase qZS inverter can operate as a module to form cascaded inverter systems [9–11] or as independent inverter systems [12–18]. However, the second-order harmonic (2ω) pulsating power appears in the single-phase qZSI power module's dc-link and transfers to qZS capacitors and qZS inductors, which introduces low-order harmonics into the ac outputs. To limit the 2ω voltages and currents within tolerance ranges, large qZS capacitance and inductance are applied first when using the traditional carrier pulse-width modulation (PWM) of the single-phase qZSI [12, 13]; a complete set of models is established to disclose the relationship between passive component values with the 2ω ripple on qZS inductor current, qZS capacitor voltage, and PV-panel voltage and current in PV applications [14], which is also presented in Section 14.3 of this book. Recently, possibilities are emerging to optimize the qZS inductance and capacitance values in the following ways.

A hybrid pulse-width modulation combining the traditional PWM and pulse-amplitude modulation (PAM) with its grid-tie control method in PV systems, proposed by the authors in [15] and [16], significantly reduces the impedance values and power loss, without any requirement to limit the dc-link 2ω ripple. For instance, in a 500 W power rating and 108 V peak output voltage single-phase qZSI system, the conventional PWM-based one requires the qZS inductance 1000 μH (or 500 μH coupled inductor) and qZS capacitance 4400 μF to buffer the 2ω dc-link voltage ripple within 5%, and inductor current ripple within 20%. Whereas, at the same power and voltage ratings, the hybrid modulation reduces the qZS inductance to 500 μH (or 250 μH coupled inductor) and qZS capacitance to 10 μF, owing to no longer restraining the dc-link 2ω ripple while maintaining ac output quality. It also benefits the efficiency improvement. Shortcomings are that a large capacitance is demanded at the PV panel terminal to prevent the 2ω ripple from propagating to the PV panel, and the high 2ω ripple on the qZS capacitance will degrade the performance of the energy storage battery when a battery is paralleled for power balance.

The 2ω power flow of single-phase qZSI is investigated, and its current ripple damping control algorithm is proposed in the authors' publication of [17]. In that method, the qZS inductance is proposed to only restrict the switching frequency ripple, while the qZS capacitance handles the 2ω pulsating power, because the designed control strategy damps the 2ω ripple of inductor current. A similar concept can be found in [18], but those control methods are supposed to store the 2ω power in the qZS network in an optimal way, without totally solving it.

Recalling the conventional single-phase H-bridge inverter, there is no way to reduce its dc-link capacitance as small as one of three-phase systems; that is, eliminating 2ω ripple power, if without extra circuits. Further, an active filter integrated single-phase qZSI to eliminate the 2ω ripple from the dc side is proposed by the authors [19]. Through paralleling a third leg with an inductor–capacitor branch, the active filter integrated qZSI directly transfers the 2ω power to the ac capacitor of the LC branch, so that both the qZS impedance and the active filter capacitance could be small in value and size. In addition, because the 2ω ripple no longer flows into the dc side of qZSI, there will be no ripple in the battery current when an energy storage battery parallels one of qZS capacitors, benefitting battery performance and lifetime [19]. The analysis, modeling, parameter design method of the qZS network and the active filter, and control strategy of that topology are investigated.

In summary, parameter optimization for single-phase qZSI is still an open topic owing to its superior advantages and widespread applications. Future research is expected to combine the energy storage battery into the optimized single-phase qZSI for use either as an independent power systems or as a power module in CMI, taking consideration of the above-mentioned techniques. In addition, the active filter integrated single-phase qZSI presents prospects in high-power applications and development.

20.1.3 Novel Control Methods

The control strategy is crucial to ensuring reliable and effective operation of power converters. In Chapter 5, the closed-loop control of the shoot-through duty cycle of impedance source inverters/converters were overviewed, which is the basis for the voltage boosting of all impedance network formed converters. In later chapters, the system control methods are presented for impedance source matrix converters, multilevel inverters, with energy storage battery, PV applications, wind applications, and electric motor drives, etc. These are based on the conventional proportional-integral (PI) controllers.

In recent years, owing to its fast dynamic response, simple concept, and also its ability to include different non-linearities and constraints, model predictive control (MPC) has been applied to the impedance source three-phase inverter, multilevel inverter, four-leg inverter, and five-phase inverter, etc., as discussed in the previous chapters of the book and in [20–24]. As an emerging technique, the current contributions focus on the basic operation of impedance source converters, such as fast response speed, low current harmonics, and load balancing in grid-connected applications. It is known that the cost function is essential to the system performance in MPC, and different constraints can be added to improve the entire system operation. Therefore, efforts are expected to incorporate the advantageous MPC with the performance improvement of the impedance source converters, such as the fixed switching frequency, energy storage battery integration, and maximum power point tracking of renewable energy power systems [25], etc.

20.1.4 Future Applications

The ever-increasing installed capacity of large-scale wind and PV power plants, ranging from a few hundred kilowatts (kWs) up to several hundred megawatts (MWs), has driven the research and development of utility wind and PV converter topologies moving toward multilevel structure, which is capable of better harmonic spectra, lower weight of filtering components, and higher power by low-voltage low-power devices. The modular multilevel converter (MMC), formed by a cascade connection of multiple identical converter modules, has been recognized as the most appropriate topology in medium-voltage (MV) and high-voltage (HV) applications; it is praised for its modular structure, scalable voltage and power, ease of installation and maintenance, low expense for redundancy and fault tolerant operation, etc. [26–27].

Considering PV's natural dc characteristics, the series connection of PV converters at the distribution side is expected to minimize the power conversion stages, extend the dc voltage level, and reduce power loss for the dc transmission of modern high-power PV systems. However, owing to the PV panels' insulation demands, i.e. between 1 kV and 600 V for current EU and US standards, front-end isolation is inevitable in the dc-dc HB cascaded converter when the dc-bus voltage is higher than 1 kV [28]. Whereas, computation burden and control complexity will become significant when hundreds of such SMs are used in series to achieve the MV/HV dc because both the isolation stage and output-stage converters require a control algorithm to fulfill the PV MPPT, the dc-link voltage balance between modules, and entire power integration.

Owing to its single-stage topology dealing with wide dc voltage variations, high-power applicability, and short-circuit immunity of phase legs, the qZS network is expected to improve the two-stage dc-dc converter module for the use of higher voltage dc integration and transmission of high-power PV systems [29]. Further work is needed on the system control of dc power integration, evaluation, and validation of the concept.

20.2 Illustration of Using Wide Band Gap Devices

Figure 20.1 shows the characteristics of SiC material with respect to Si material. The SiC material is characterized by electrical field strength almost nine times that of normal Si, allowing the design of semiconductor devices with very thin drift layers and as a consequence low on-state resistance and reduced switching losses. In other words, this characteristic can be translated into the possibility of operating at higher blocking voltages with reduced losses. Increased robustness, especially against temperature and cosmic radiation-induced failure [3] are additional highlights of this new technology.

Within the power electronics community, the use of WBG semiconductor devices in next generation inverters is considered favorable due to the high thermal conductivity and reduced switching losses associated with WBG. Nevertheless, the cost of WBG semiconductors is still viewed as prohibitively high for mainstream commercial applications, even though WBG devices can operate at larger current density than the legacy silicon devices. Therefore, the right selection of topology and peripheral devices, such as switches, passive devices, drivers and cooling systems, is important in order to maximize the benefits of using the WBG devices in power conversion systems [1].

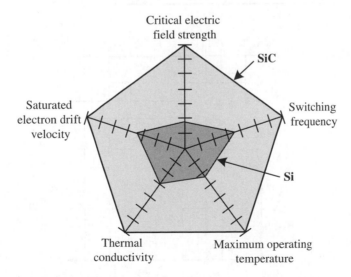

Figure 20.1 Characteristics of SiC material with respect to Si material (*Source*: Biela 2011 [3]. Reproduced with permission of IEEE).

20.2.1 Impact on Z-Source Network

The qZS diode is an important power device in the qZSI, where as soon as there is shoot-through action in the inverter phase legs, the qZS diode will turn off. Figure 20.2(b) shows an example of the qZS diode current in the single-phase qZSI of Figure 20.2(a), when using SPWM [11]. In the figure, S_{11}, S_{12}, S_{13}, and S_{14} are the gating signals of the four switches. It can be seen that the qZS diode presents two-time shoot-through (ST) switching events even though each H-bridge only switches one time for the ST. Meanwhile, the conduction current is two-time qZS inductor current i_L in the traditional zero states, and is $2i_L - i_{ac}$ in the active states (ACT). Hence, the power loss in the qZS diode may be significant, and using SiC power devices makes it possible to reduce qZS inductance and power dissipation on H-bridge power devices and the qZS diode.

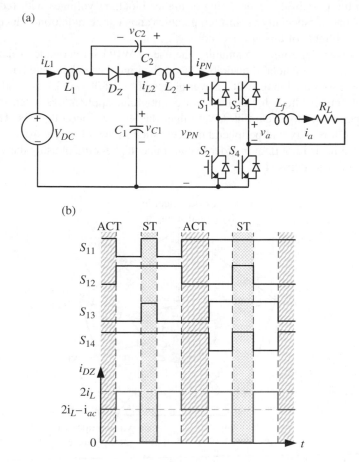

Figure 20.2 qZS-diode current of single-phase qZSI in one control cycle when using SPWM: (a) topology, (b) qZS-diode current (*Source*: Liu 2014 [30]. Reproduced with permission of IEEE).

20.2.2   Analysis and Evaluation of SiC Device Based qZSI

As an example, the power loss and efficiency of a fuel cell fed ZSI for an induction motor (IM) system, shown in Figure 20.3, are compared when using conventional Si IGBT modules, Si IGBTs/SiC Schottky diode hybrid modules, and also full SiC modules [8]. The system specifications are shown in Table 20.1.

Based on the resulting maximum voltage and current, the Si switches chosen for the ZSI are 600 V/300A 6MBP300RA060IGBT-IPMs, from Fuji Electric. Cree has developed 600 V/75A SiC Schottky diodes. Since the ZSI used in this paper has 600 V/300A, four 75 A SiC Schottky diodes are employed to replace one 300 A Si diode of 6MBP300RA060. In the SiC-based system, the IGBTs are replaced with SiC MOSFETs and the Si diodes are replaced with SiC

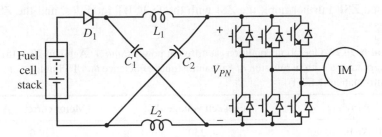

Figure 20.3   Fuel cell fed ZSI for induction motor system.

Table 20.1   System specifications (*Source*: Aghdam 2011 [8]. Reproduced with permission of Korean Institute of Science and Technology Information (KISTI))

System parameters	Values
System capacity	55 kVA
Fuel cell voltage at maximum power	250 V
Maximum fuel cell output voltage	420 V
IM power factor (PF) at maximum power	0.9
Switching frequency	10 kHz

IM parameters	Values
Peak torque	240 Nm
Maximum RMS current	250 A
Peak power at 312 V	78 kW
Peak efficiency	93%
Continuous torque	55 Nm
Continuous power	34 kW
Nominal speed	2500 rpm
Stator and rotor leakage inductance	0.2 mH
Magnetizing inductance	3.64 mH
Stator resistance	11.7 mΩ
Rotor resistance	82 mΩ
Corresponding resistance to the core loss	150 Ω

Schottky diodes. In the SiC-based drive system, a 1.2 kV 4H-SiC DMOSFET is used as the main power switch [8].

A comparison in terms of losses and efficiency is conducted based on the specifications of the inverter and the induction motor as introduced above. The operation conditions at different power levels are listed in Table 20.2. The semiconductor power losses are calculated using methods introduced in Chapter 13.

Figure 20.4 shows the total component losses as a function of the power in the three investigated scenarios. It shows that the ZSI with the SiC-based system features less losses and therefore superior efficiency, compared to the other modules in the three investigated scenarios. Figure 20.5(a) shows the loss distribution of the considered ZSI at 50 kW with a constant carrier frequency of 10 kHz. It is noticeable that the ZSI losses of the Si-based IGBT modules and hybrid modules increase by 238%, and 230%, compared to the SiC module-based ZSI. Furthermore, the ZSI with the Si IGBT modules, and the ZSI with the

Table 20.2 Operation conditions at different power (*Source*: Aghdam 2011 [8]. Reproduced with permission of Korean Institute of Science and Technology Information (KISTI))

Power rating	Fuel cell voltage (V)	Motor current (A)
50 kW PF=0.90	250	139.6
40 kW PF=0.85	280	113.0
30 kW PF=0.80	305	86.9
20 kW PF=0.74	325	60.9
10 kW PF=0.70	340	31.5

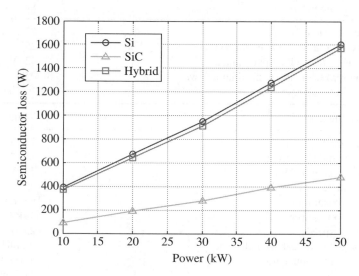

Figure 20.4 ZSI power loss versus different power when using three different power modules (*Source*: Aghdam 2011 [8]. Reproduced with permission of Korean Institute of Science and Technology Information (KISTI)).

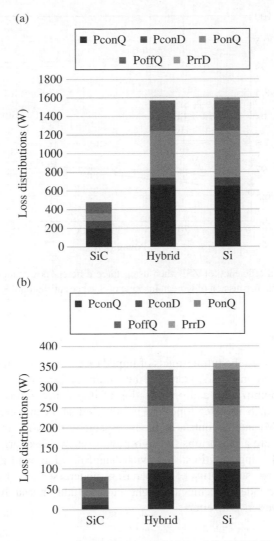

Figure 20.5 ZSI semiconductor loss distribution at (a) 50 kW and (b) 10 kW (*Source*: Aghdam 2011 [8]. Reproduced with permission of Korean Institute of Science and Technology Information (KISTI)).

hybrid modules generate 359% and 332% more semiconductor losses, compared to the ZSI with the SiC modules, if a power of 10 kW is applied, seen from Figure 20.5(b). Note that the parameters of the semiconductor devices are considered at junction temperature $T_j = 125\,°C$.

Figure 20.6 shows the calculated efficiencies of the ZSI in the three cases. It can be seen that the ZSI with the SiC materials provides the highest efficiency that reduces both the thermal dissipation requirements and volume. Moreover, the much lower power loss and short turn on/off time of SiC devices at the 10 kHz switching frequency provide potential of higher switching frequency, which will also benefit to reduce the qZS inductance and capacitance values, as a result of further reduction of power loss, volume, and costs on passive components.

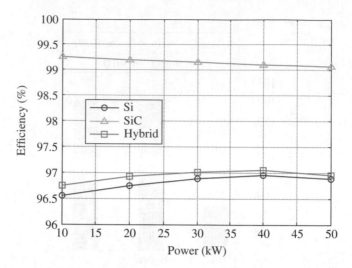

Figure 20.6 Calculated efficiency of ZSI when using three different power modules (*Source*: Aghdam 2011 [8]. Reproduced with permission of Korean Institute of Science and Technology Information (KISTI)).

20.3 Conclusion

This chapter has presented the future trends of impedance source inverters/converters. A general expectation was demonstrated in terms of the volume and size reduction by applying wide band gap devices; minimization and optimization of impedance source network parameters for the single-phase qZS inverter topology being used as an independent power system and submodule in CMI; novel control methods, especially the MPC to improve the system performance; and future applications for high-power renewable energy power conversion. A comparison of ZSI using traditional Si devices and SiC wide band gap devices is further illustrated, verifying the superiority of cooperating modern semiconductor technique. This chapter has mainly provided a technical insight into the future research and development of impedance source inverters/converters.

References

[1] J. Holtz, M. Holtgen, J. O. Krah, "A space vector modulator for the high-switching frequency control of three-level SiC inverters," *IEEE Trans. Power Electron.*, vol.29, no.5, pp.2618–2626, May 2014.

[2] C. N.-M. Ho, H. Breuninger, S. Pettersson, G. Escobar, F. Canales, "A comparative performance study of an interleaved boost converter using commercial Si and SiC diodes for PV applications," *IEEE Trans. Power Electron.*, vol.28, no.1, pp.289–299, Jan. 2013.

[3] J. Biela, M. Schweizer, S. Waffler, J. W. Kolar, "SiC versus Si – evaluation of potentials for performance improvement of inverter and DC-DC converter systems by SiC power semiconductors," *IEEE Trans. Power Electron.*, vol.58, no.7, pp.2872–2882, July 2011.

[4] J. Millan, P. Godignon, X. Perpinya, A. Perez-Tomas, J. Rebollo, "A survey of wide band gap power semiconductor devices," *IEEE Trans. Power Electron.*, vol.29, no.5, pp.2155–2163, May 2014.

[5] M. A. Briere, "GaN based power conversion: A new era in power electronics," in *Proc. PCIM Europe Conference*, 2009.

[6] Y. Liu, H. Abu-Rub, B. Ge, "Z-Source/quasi-Z-source inverters – derived networks, modulations, controls, and emerging applications to photovoltaic conversion," *IEEE Ind. Electron. Mag.*, vol.8, no.4, pp.32–44, Dec. 2014.

[7] M. Li, Y. Liu, H. Abu-Rub, B. Ge, Z. Salam, "SiC power devices and applications in quasi-Z-source converters/inverters," to appear in the *2015 IEEE Conference on Energy Conversion (CENCON)*, Johor Bahru, Malaysia, 19–20 October 2015.

[8] M. G. H. Aghdam, "Z-source inverter with SiC power semiconductor devices for fuel cell vehicle applications," *Journal of Power Electron.*, vol.11, no.4, pp.606–611, 2011.

[9] Y. Xue, B. Ge, F. Z. Peng, "Reliability, efficiency, and cost comparisons of MW-scale photovoltaic inverters," in *Proc. IEEE Energy Conversion Congress and Exposition (ECCE)*, pp.1627–1634, 2012.

[10] Y. Zhou, L. Liu, H. Li, "A high-performance photovoltaic module-integrated converter (MIC) based on cascaded quasi-Z-source inverters (qZSI) using eGaN FETs," *IEEE Trans. Power Electron.*, vol.28, no.6, pp.2727–2738, June 2013.

[11] D. Sun, B. Ge, X. Yan, D. Bi, H. Zhang, Y. Liu, H Abu-Rub, L. Ben-Brahim, F. Peng, "Modeling, impedance design, and efficiency analysis of quasi-Z source module in cascade multilevel photovoltaic power system," *IEEE Trans. Ind. Electron.*, vol.61, no.11, pp.6108–6117, 2014.

[12] Y. Yu, Q. Zhang, B. Liang, S. Cui, "Single-phase Z-source inverter: analysis and low-frequency harmonics elimination pulse width modulation," in *Proc. IEEE Energy Conversion Congress and Exposition (ECCE)*, pp.2260–2267, 2011.

[13] D. Sun, B. Ge, X. Yan, D. Bi, H. Abu-Rub, F. Z. Peng, "Impedance design of quasi-Z source network to limit double fundamental frequency voltage and current ripples in single-phase quasi-Z source inverter," in *Proc. IEEE Energy Conversion Congress and Exposition (ECCE)*, pp.2745–2750, 2013.

[14] Y. Liu, B. Ge, H. Abu-Rub, D. Sun, "Comprehensive modeling of single-phase quasi-Z-source photovoltaic inverter to investigate low-frequency voltage and current ripple," *IEEE Trans. Ind. Electron.*, vol.62, no.7, pp.4194–4202, 2015.

[15] H. Zhang, B. Ge, Y. Liu, D. Sun, H. Abu-Rub, F. Z. Peng, "A hybrid modulation method for single-phase quasi-Z source inverter," in *Proc. IEEE Energy Conversion Congress and Exposition (ECCE)*, pp.4444–4449, 2014.

[16] Y. Liu, H. Abu-Rub, B. Ge, "Hybrid pulse-width modulated single-phase quasi-Z-source grid-tie photovoltaic power system," in *Proc. Thirtieth Annual IEEE Applied Power Electronics Conference and Exposition (APEC)*, vol., pp.1763–1767, 2015.

[17] B. Ge, H. Abu-Rub, Y. Liu, R. Balog, "Minimized quasi-Z source network for single-phase inverter," in *Proc. Thirtieth Annual IEEE Applied Power Electronics Conference and Exposition (APEC)*, pp.806–811, 2015.

[18] Y. Zhou, H. Li, H. Li, X. Lin, "A capacitance minimization control strategy for single-phase PV quasi-Z-source inverter," in *Proc. Thirtieth Annual IEEE Applied Power Electronics Conference and Exposition (APEC)*, pp.1730–1735, 15–19 March 2015.

[19] B. Ge, H. Abu-Rub, Y. Liu, R. Balog, "An active filter method to eliminate DC-side's low-frequency power ripples for single-phase quasi-Z-source inverter," in *Proc. Thirtieth Annual IEEE Applied Power Electronics Conference and Exposition (APEC)*, pp.827–832, 15–19 March 2015.

[20] W. Mo, P. C. Loh, F. Blaabjerg, "Model predictive control for Z-source power converter," in *Proc. IEEE 8th International Conference on Power Electronics and ECCE Asia (ICPE & ECCE)*, pp.3022–3028, 2011.

[21] M. Mosa, H. Abu-Rub, J. Rodriguez, "High performance predictive control applied to three phase grid connected quasi-Z-source inverter," in *Proc. 39th Annual Conference of the IEEE Industrial Electronics Society (IECON)*, pp.5812–5817, 2013.

[22] O. Ellabban, M. Mosa, H. Abu-Rub, J. Rodriguez, "Model predictive control of a grid connected quasi-Z-source inverter," in *Proc. IEEE International Conference on Industrial Technology*, pp.1591–1596, 2013.

[23] S. Bayhan, H. Abu-Rub, "Model predictive control of quasi-Z source three-phase four-leg inverter," to appear in the *IEEE Annual Conference of Industrial Electronics Society (IECON)*, Nov. 9–12, 2015.

[24] S. Bayhan, H. Abu-Rub, M. Trabelsi, "Model predictive control of Z-source four-leg inverter for standalone photovoltaic system with unbalanced load," to appear in the *Thirty-first Annual IEEE Applied Power Electronics Conference and Exposition (APEC)*, 19–26 March 2016.

[25] M. M. Metry, M. B. Shadmand, Y. Liu, R. Balog, H. Abu-Rub, "Maximum power point tracking of photovoltaic systems using sensorless current-based model predictive control," in *Proc. IEEE Energy Conversion Congress and Exposition (ECCE)*, pp.6635–6641, 20–24 Sept. 2015.

[26] A. Nami, J. Q. Liang, F. Dijkhuizen, G. D. Demetriades, "Modular multilevel converters for HVDC applications: review on converter cells and functionalities," *IEEE Trans. Power Electron.*, vol.30, pp.18–36, Jan 2015.

[27] H. Liu, K. Ma, Z. Qin, P. C. Loh, F. Blaabjerg, "Lifetime estimation of MMC for offshore wind power HVDC application," to appear in the *IEEE Journal of Emerging and Selected Topics in Power Electronics.*

[28] J. Echeverria, S. Kouro, M. Perez, H. Abu-Rub, "Multi-modular cascaded DC-DC converter for HVDC grid connection of large-scale photovoltaic power systems," in *Proc. 39th Annual Conference of the IEEE Industrial Electronics Society (IECON)*, pp.6999–7005, 10–13 Nov. 2013.

[29] Y. Liu, H. Abu-Rub, B. Ge, "Front-end isolated quasi-Z-source DC-DC converter modules in series for photovoltaic high voltage DC applications," to appear in the *Thirty-first Annual IEEE Applied Power Electronics Conference and Exposition (APEC)*, 19–26 March 2016.

[30] Y. Liu, B. Ge, H. Abu-Rub, F. Z. Peng, "Phase-shifted pulse-width-amplitude modulation for quasi-Z-source cascade multilevel inverter based photovoltaic power system," *IET Power Electron.*, vol.7, no.6, pp.1444–1456, June. 2014.

Index

active filter, 391, 392
active state, 10, 20, 40–43, 46–49, 62, 124, 141, 143, 211, 233, 235, 258, 353, 358, 394
adjustable speed drive (ASD), 15, 52, 175, 176, 266, 267, 269, 276, 277, 283, 284, 290, 300, 319, 330
alternative phase opposition disposition (APOD), 200, 201, 203, 206
anti-islanding protection, 362, 363, 381, 382, 384, 387

battery energy management, 179, 187, 192
bidirectional power flow, 35, 41, 51, 52, 79, 86, 87, 158, 277
boost factor, 22, 41, 55, 64, 88, 90, 92, 93, 98, 103, 104, 140, 155, 163, 167, 171, 182, 198, 238, 305, 307, 316–318

closed-loop control, 27, 31, 47, 74, 129, 153, 164, 182, 183, 217, 253, 270, 392
common-mode voltage (CMV), 105, 151, 266, 268, 302
compensation, 132–135, 153, 183–185, 362, 363, 381
conduction loss, 233, 236
continuous conduction mode, 180, 208
continuous input current, 14, 20, 21, 33, 39, 41, 82, 84–86, 90, 91, 94–97, 99, 101, 104, 126, 127, 153, 168, 269
cost function, 77, 283, 284, 330, 341, 343, 344, 346, 348–354, 359, 372, 373, 375, 392
current-fed Z-source inverter/quasi-Z-source inverter (CF-ZSI/qZSI), 13–15, 35–41, 44, 45, 47, 48, 51, 52, 79, 113

current ripple, 48, 56, 65, 71, 151, 153, 156, 164, 206, 226, 232, 233, 241, 242, 247–249, 258, 301, 391
current source inverter (CSI), 35, 36, 38–41, 43–47
current source rectifier (CSR), 128, 158, 277

dc-link peak voltage, 32, 64, 71, 76, 134, 182, 192, 206, 213, 219, 221, 222, 224, 238, 248, 249, 259, 261, 262, 366, 368, 372
dead time, 13, 23, 175, 179, 244, 266, 330, 332
direct torque control (DTC), 269, 272, 276, 279, 280, 282, 283
discrete-time model, 322, 331, 335, 342–346, 349, 351, 352
distributed Z-network, 13, 14, 85
double-line-frequency ripple, 226, 245
dual-loop control, 75, 369–372
dynamic model, 20, 25, 33, 119, 129, 130, 132, 137, 179

efficiency, 1, 5, 22, 35, 63, 78, 128, 206, 239, 240, 243, 390, 391, 395–397
electromagnetic interference (EMI), 6, 12, 13, 23, 85, 128, 299, 302, 308
embedded source, 85
energy storage, 4, 5, 80, 179, 180, 192, 268, 391, 392
energy stored Z-source/quasi-Z source inverters, 179
equivalent circuit
  of current-fed qZSI, 40–42, 49
  of five-phase inverter, 319, 339
  of four-leg inverter, 308

Impedance Source Power Electronic Converters, First Edition. Yushan Liu, Haitham Abu-Rub, Baoming Ge, Frede Blaabjerg, Omar Ellabban, and Poh Chiang Loh.
© 2016 John Wiley & Sons, Ltd. Published 2016 by John Wiley & Sons, Ltd.